Urban Sustainability

The Urban Sustainability Book Series is a valuable resource for sustainability and urban-related education and research. It offers an inter-disciplinary platform covering all four areas of practice, policy, education, research, and their nexus. The publications in this series are related to critical areas of sustainability, urban studies, planning, and urban geography.

This book series aims to put together cutting-edge research findings linked to the overarching field of urban sustainability. The scope and nature of the topic are broad and interdisciplinary and bring together various associated disciplines from sustainable development, environmental sciences, urbanism, etc. With many advanced research findings in the field, there is a need to put together various discussions and contributions on specific sustainability fields, covering a good range of topics on sustainable development, sustainable urbanism, and urban sustainability. Despite the broad range of issues, we note the importance of practical and policy-oriented directions, extending the literature and directions and pathways towards achieving urban sustainability.

The series will appeal to urbanists, geographers, planners, engineers, architects, governmental authorities, policymakers, researchers of all levels, and to all of those interested in a wide-ranging overview of urban sustainability and its associated fields. The series includes monographs and edited volumes, covering a range of topics under the urban sustainability topic, which can also be used for teaching materials.

Sushobhan Majumdar · Moharana Choudhury ·
Ali Cheshmehzangi
Editors

Sustainable Urban Future

Embracing Circular Economy in Waste Management

Springer

Editors
Sushobhan Majumdar
Department of Geography
Pandit Raghunath Murmu Smriti Mahavidyalaya
Bankura, West Bengal, India

Moharana Choudhury
Environmental Research and Management Division
Voice of Environment
Guwahati, Assam, India

Ali Cheshmehzangi
School of Architecture, Design and Planning
The University of Queensland
Brisbane, QLD, Australia

ISSN 2731-6483 ISSN 2731-6491 (electronic)
Urban Sustainability
ISBN 978-981-95-2975-9 ISBN 978-981-95-2976-6 (eBook)
https://doi.org/10.1007/978-981-95-2976-6

The original submitted manuscript has been translated into English. The translation was done using artificial intelligence. A subsequent revision was performed by the author(s) to further refine the work and to ensure that the translation is appropriate concerning content and scientific correctness. It may, however, read stylistically different from a conventional translation.

This Springer imprint is published by the registered company Springer Nature Singapore Pte Ltd.
The registered company address is: 152 Beach Road, #21-01/04 Gateway East, Singapore 189721, Singapore

Contents

About the Editors

Sushobhan Majumdar is an Assistant Professor in the Department of Geography at Pandit Raghunath Murmu Smriti Mahavidyalaya (Affiliated to Bankura University, India). He has over 12 years of research and professional experience in various environmental planning and research projects. Urban geography, social geography, regional development and planning, remote sensing and GIS, and geography of transport are among his main areas of study. Previously, he worked on various development projects for the Socio-Economic Planning Unit of the Kolkata Metropolitan Development Authority (KMDA), Government of West Bengal. He has authored more than 90 research papers and a book chapter, published in a variety of reputable journals and conference proceedings, and has edited more than 15 books. He is currently serving as a Reviewing Editor in Springer. He is now editor and reviewer of various renowned publishing houses like Springer, Elsevier, Royal Society of Chemistry, CRC Press, AAP Press, Wiley Scrivener, Cambridge Scholars Publishing etc. In excess of 35 national and international seminars and conferences, both in India and overseas, he has presented papers. He is presently the Co-Principal Investigator of a major research project supported by the Indian Council of Social Science Research (ICSSR). He gave guest or special talks as part of numerous national and regional programs. He holds memberships in the Geographical Society of India, the Regional Science Association of India (RSAI), the Institute of Indian Geographers (IIG), and the National Association of Geographers of India (NAGI).

Moharana Choudhury is an Indian researcher and environmentalist. He is a Research Coordinator in the Environmental Research and Management Division of "Voice of Environment" (VoE). Choudhury has over 14 years of experience in environmental research and services, spanning both research and professional work. His research interests encompass environmental research, waste management, wastewater treatment, environmental risk assessment, pollution control, plastic pollution, and climate change mitigation. He has authored or edited more than 15 books with internationally renowned publishing houses and published more than 40 research articles, review articles, and published more than 50 book chapters in internationally recognized journals and publishers. He has been working as a reviewer and editor of various prestigious international publishing houses worldwide. He contributes to various environmental research, management, social media mass awareness, and research studies. He is also well known in India for his efforts to implement environmental education and safeguard the privileges of environmental degree holders and Environmental professionals in India. He has completed his education as an MSc, PGDGIS, PGDEM, and PGCGARD from Gurukula Kangri Deemed to be University, Haridwar, Uttarakhand; North-Eastern Hill University (NEHU), Shillong, Meghalaya; Tezpur University, Assam; and National Institute of Rural Development and Panchayati Raj (NIRDPR), Hyderabad, India, respectively.

Ali Cheshmehzangi is a Professor and Head of the School of Architecture, Design and Planning (ADP) at the University of Queensland, where he leads the school and researches across the fields of architecture, design, and urbanism. He has been in the world's top 2% field leader since 2022, recognized by Stanford University. He is among the top 25 global scholars in the urban sustainability research area. Ali has a wealth of experience, expertise, and a strong commitment to innovation and excellence in the fields of architecture, design, and planning, the three core disciplines of his School at UQ. With a career spanning over two decades, he has made significant contributions to the academic and professional communities, with a focus on sustainable and environmentally conscious design. Prior to joining UQ, Ali held several strategic leadership and senior managerial roles, such as Vice President for the International Engagement and Global Partnership, Founding Director and Head of the Center for Innovation in Education and Research, Head of the Department of Architecture and Built Environment, Founding Director of the Urban Innovation Lab, Director of a university-wide teaching and learning platform, Director of the International Research Network for Rural and Urban Development, Founder and Head of Research Group for Sustainable Built Environment, co-director of university-wide research priority areas, Director of the Center for Sustainable Energy Technologies, and Interim Director of the Digital Design Lab. So far, Ali has published over 500 journal papers, articles, conference papers, book chapters, and reports. This is his 42nd academic book, alongside his many poetry collections that he has written in Parsi/Persian and English. Some of his previous academic books have received awards at the national, provincial, and municipal levels. He has also received international awards and recognition for his research on urban resilience studies and sustainability research, as well as a 2018 Vice Chancellor's award for his impactful contribution to higher education.

The Circular Economy Paradigm: Core Concepts and Value

Moharana Choudhury, Ali Cheshmehzangi, and Sushobhan Majumdar

Abstract This chapter explores the foundational concepts, value propositions, and critical enablers of the circular economy (CE) paradigm as an alternative to the traditional linear economic model. The circular economy emphasizes designing out waste, keeping products and materials in use, and regenerating natural systems. The chapter begins by outlining the core principles of CE and the environmental, economic, and social benefits it offers. It then addresses key challenges, such as technological and financial barriers, cultural resistance, the risk of circular washing, and the complexities of measuring circularity and impact. Recognizing these limitations, the chapter discusses enabling tools and frameworks essential for advancing circularity, including digital technologies (IoT, blockchain, AI), life cycle assessment, material flow analysis, design thinking, and sustainability certifications. Furthermore, it emphasizes the importance of cross-sector collaboration, innovation, and supportive policy in scaling circular models across regions and industries. Looking ahead, the chapter envisions a global circular economy supported by harmonized standards, robust measurement systems, and inclusive governance. By integrating systemic thinking with practical tools and strategies, the circular economy holds the potential to reshape value creation and consumption for long-term sustainability.

Keywords Circular economy · Sustainability · Life cycle assessment · Digital technologies · Systems thinking

M. Choudhury
Environmental Research and Management Division, Voice of Environment (VoE), Guwahati, Assam, India

A. Cheshmehzangi
School of Architecture, Design and Planning (ADP), University of Queensland, St Lucia, QLD, Australia

S. Majumdar (✉)
Department of Geography, Pandit Raghunath Murmu Smriti Mahavidyalaya, Bankura, West Bengal, India

S. Majumdar et al. (eds.), *Sustainable Urban Future*, Urban Sustainability,
https://doi.org/10.1007/978-981-95-2976-6_1

1 Introduction

1.1 Definition of the Circular Economy (CE)

A circular economy (CE) is a regenerative economic framework designed to reduce waste and pollution by maximizing the longevity of materials and products through methods such as reuse, recycling, and remanufacturing. It diverges from the conventional linear "take-make-dispose" model by advocating for resource efficiency, minimizing environmental impact, and fostering a more sustainable and resilient economy. The Circular Economy prioritizes the optimal use of resources, aiming to reduce waste production throughout the entire product life cycle, from design to disposal. A circular economy seeks to establish closed-loop processes in which materials are continually recycled, reused, or repurposed, rather than being discarded. A fundamental tenet of the Circular Economy is to eliminate waste and pollution from the outset, rather than addressing it subsequently [1]. Diverse tactics are utilized, encompassing sharing, leasing, repairing, refurbishing, remanufacturing, and recycling. The Circular Economy aims to rehabilitate and rejuvenate natural systems by minimizing resource extraction and promoting actions that enhance ecological well-being. In addition to environmental advantages, the circular economy can foster economic expansion, generate employment, and to improve resource security. The linear economy model, defined by resource extraction, production, consumption, and disposal, stands in stark contrast to the circular economy's emphasis on resource sustainability and regeneration [2].

1.2 Trends and Insights in Circular Economy Research

A comprehensive bibliometric analysis was conducted using the Scopus database to investigate the scholarly evolution of circular economy research, with a focus on core concepts and value creation mechanisms. The search strategy—TITLE-ABS-KEY (Circular economy OR core concepts AND value) AND PUBYEAR > 2014 AND PUBYEAR < 2025—targeted peer-reviewed publications from 2015 to 2024, a period characterised by intensifying global efforts toward sustainable development and systemic economic transformation. The analysis explores multiple dimensions, including publication output by country, citation impact, and thematic trends.

As shown in Fig. 1, the geographic distribution of research reveals that India leads in publication volume, followed closely by the United Kingdom and Italy, reflecting the growing commitment to circular practices of both emerging and established economies. Citation impact (Fig. 2) underscores Europe's dominance, with Italy, the United Kingdom, and the Netherlands leading in scholarly influence, while countries such as India and Brazil also demonstrate rising impact. Furthermore, Fig. 3 illustrates the thematic evolution of the field. Dominant and persistent themes include '*circular economy*,' '*sustainable development*,' and '*sustainability*,' while emerging focal areas such as '*closed-loop*,' '*green economy*,' and '*design/methodology/approach*' suggest a recent pivot toward systems thinking and integrated design strategies. Foundational concepts like '*recycling*,' '*industrial ecology*,' and '*cradle to cradle*' remain relevant, highlighting the field's interdisciplinary.

Collectively, these insights offer a comprehensive understanding of how academic discourse on circular economy principles has evolved—geographically, conceptually, and methodologically—over the past decade in response to escalating environmental and economic challenges.

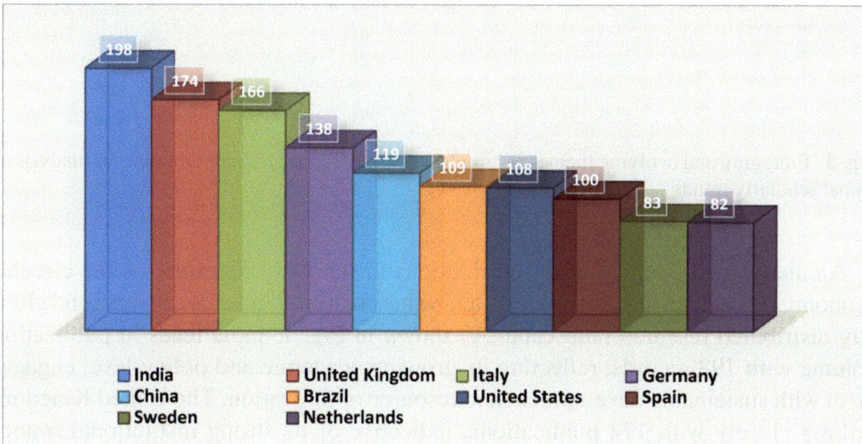

Fig. 1 Publication Country wise

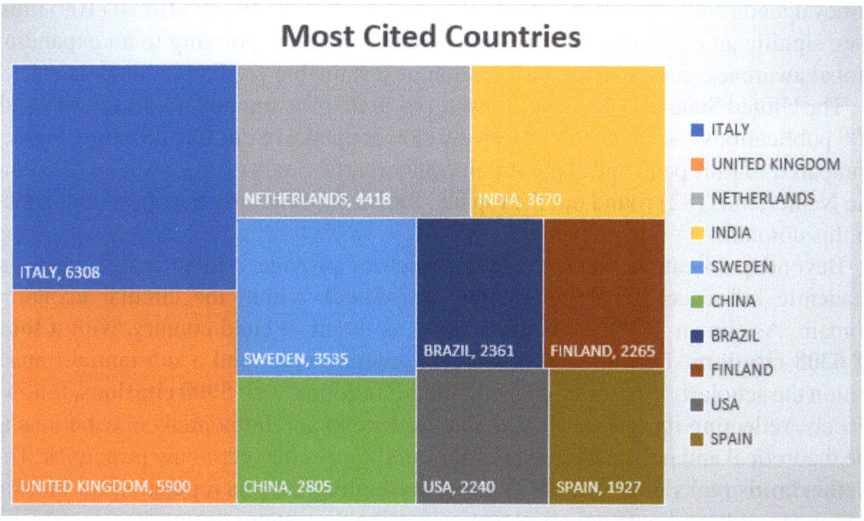

Fig. 2 Citation impact by country

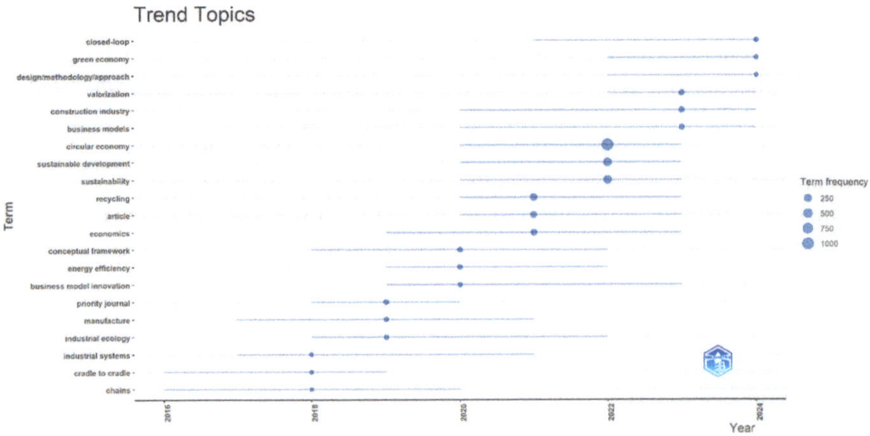

Fig. 3 Emerging and evolving themes in circular economy research using bibliometric analysis of global scholarly trends

An analysis of country-wise contributions to scholarly literature on the circular economy focused on its core concepts and value creation—reveals a diverse and globally distributed research landscape. As shown in Fig. 1, India leads in publication volume with 198 records, reflecting its growing academic and policy-level engagement with sustainable development and resource optimization. The United Kingdom follows closely with 174 publications, indicative of its strong institutional frameworks and policy-driven emphasis on circular economy transitions. Italy (166) and Germany (138) also feature prominently, underscoring Europe's overall leadership in circular economy research, likely influenced by the European Union's strategic policy agenda and circular economy action plans. China (119) and Brazil (109) illustrate significant contributions from emerging economies, pointing to an expanding global awareness and localized innovation in sustainable production models.

The United States, despite its extensive research infrastructure, ranks seventh with 108 publications—suggesting a relatively slower uptake of circular economy themes compared to European and some Asian counterparts. Spain (100), Sweden (83), and the Netherlands (82) round out the top ten, further highlighting Europe's dominance in this domain.

Beyond publication volume, citation metrics provide critical insight into the academic influence and dissemination of research within the circular economy domain. As shown in Fig. 2, **Italy** emerges as the most cited country, with **a total of 6308 citations**, indicating both a high research output and a substantial impact within the scholarly community. The **United Kingdom**, with **5900 citations**, follows closely, reflecting its sustained academic leadership and influential contributions to the theoretical and policy frameworks surrounding circular economy principles. The **Netherlands** ranks third with **4418 citations**, highlighting its reputation for innovation and interdisciplinary collaboration in sustainable systems design and closed-loop models. **India** (3670) and **Sweden** (3535) also exhibit substantial citation counts, indicating an increasing international visibility of their research, despite differing levels of infrastructure and funding.

China (2805), **Brazil** (2361), and **Finland** (2265) demonstrate strong regional academic engagement, with a growing citation footprint that reflects both localised innovation and broader global relevance. Notably, the **USA**, with **2240 citations**, appears relatively lower in this ranking, reinforcing observations from the publication analysis that circular economy research has not yet been a dominant thematic priority in its academic agenda. Finally, **Spain**, with **1927 citations**, rounds out the top ten, making a significant contribution to the European research landscape. These findings underscore that while Europe continues to dominate both in volume and impact, emerging economies such as India, China, and Brazil are increasingly shaping the global research narrative on circular value creation and resource circularity.

The thematic evolution of circular economy literature over the past decade reveals an increasingly diverse and interdisciplinary research landscape. As illustrated in Fig. 3, the trend topic analysis highlights key terms based on their frequency and temporal emergence, offering insight into both foundational and emerging areas of scholarly focus. Terms such as '*circular economy*,' '*sustainable development*,' and '*sustainability*' dominate the discourse, with their presence peaking around 2021–2023 and showing high term frequencies (approaching 1000 mentions), suggesting that these continue to serve as central pillars of the academic conversation. The prominence of '*business models*' *and* '*valorisation*' reflects a concurrent interest in practical and economic mechanisms for achieving circularity.

Notably, newer or resurging terms such as '*closed-loop*,' '*green economy*,' and '*design/methodology/approach*' have gained traction in more recent years (2023–2024), indicating a shift toward systems-thinking and integrated design strategies. Meanwhile, foundational themes like '*recycling*,' '*energy efficiency*,' *and* '*business model innovation*' were more prevalent in earlier years (2016–2020), but continue to maintain relevance. Interdisciplinary concepts such as '*industrial ecology*,' '*cradle to cradle*,' and '*industrial systems*' have experienced more focused bursts of attention, primarily concentrated in the 2017–2019 period, suggesting a maturation of these areas or their integration into broader research themes. The emergence of methodological terms like '*conceptual framework*,' '*design/methodology/approach*,' and '*article*' highlights a growing emphasis on theoretical structuring and academic rigour in the field. Additionally, the frequency and persistence of terms like '*construction industry*' and '*chains*' reflect application-oriented research, particularly in the context of supply chain transformation and sector-specific circular strategies.

1.2.1 Historical Background and Evolution from the Linear Model

The "linear model" across disciplines such as communication, innovation, and statistics denotes a sequential, unidirectional transmission of information or a process. Despite its simplicity, it has been extensively utilized as a core framework; however, it frequently faces criticism for its inadequacies in encapsulating complexity. The progression from this fundamental linear model entails acknowledging feedback loops, non-linear interactions, and the interdependence of various components within a system.

The linear model of communication, also known as the transmission model, conceptualizes communication as a unidirectional process in which a sender conveys a message across a channel to a receiver who subsequently decodes it [3]. It highlights the sender's responsibility and the message's clarity, frequently neglecting the intricacies of interpretation and response. Critiques encompass its inability to consider noise, context, and the proactive involvement of the receiver in meaning construction. The linear model of innovation, commonly utilized in science and technology policy, posits a sequential process that commences with fundamental scientific study, progresses to applied research and development, and ultimately culminates in production and dissemination. This concept posits that scientific knowledge is the principal catalyst for innovation and economic expansion. Nonetheless, it has faced criticism for its reductionist approach to the innovation process, overlooking the interconnections among various stages, the significance of user requirements, and the intricate dynamics within the innovation ecosystem. Some contend that the linear model never existed as a broadly acknowledged framework in practice, but rather as a construct employed for critique and discourse—Linear Model in Statistics. In statistics, the linear model refers to a comprehensive category of models that assume a linear relationship between a dependent variable and one or more independent variables. This encompasses models such as linear regression, ANOVA, and generalized linear models. The evolution of linear models in statistics is based on the notion of algebraic invariants, which originated in the nineteenth century. The theory of algebraic invariants focuses on identifying quantities that remain constant during linear transformations, thereby establishing a foundation for understanding linear relationships in data analysis. The progression of linear models in statistics has encompassed the enhancement of the fundamental linear regression model to accommodate more intricate situations, such as numerous variables, non-normally distributed data, and hierarchical frameworks. Generalised linear models (GLMs) extend linear models to accommodate response variables with non-normal distributions, utilising link functions to relate the linear predictor to the predicted value of the response variable.

Software such as SAS has been crucial in the advancement and application of linear model techniques, with procedures like GLM, REG, VARCOMP, NLIN, MIXED, and GENMOD emerging to tackle increasingly complex statistical challenges.

The progression from linear models entails transitioning to increasingly intricate, dynamic, and interactive models that accurately reflect the complexities of real-world systems. This encompasses the identification of feedback loops, non-linear interactions, and the interdependence of various components within a system. In communication, interactive models highlight the active participation of the receiver and the dynamic interchange of information. The linear model of innovation has been superseded by more complex models that acknowledge the iterative nature of innovation, the importance of user feedback, and the value of networks and collaborations. In statistics, the progression has transitioned from basic linear models to the integration of non-linear relationships, random effects, and additional complexities, utilising methodologies such as generalised additive models, mixed-effects models, and machine learning techniques [4].

1.2.2 Relevance in the Context of Global Sustainability Challenges

The linear economic model, characterised by its "take-make-dispose" methodology, is increasingly recognised as unsustainable due to global issues such as climate change [5–7], resource depletion, and pollution. Its dependence on the extraction of extensive quantities of raw materials, the generation of substantial waste, and its contribution to environmental deterioration render it a considerable impediment to attaining global sustainability objectives. The linear model emphasizes resource extraction while frequently overlooking the limited availability of several resources. These are the results of the exhaustion of natural resources, such as minerals, fossil fuels, and water, which can potentially lead to scarcity and disputes over access. Pollution and trash Generation: The "disposal" phase of the linear model culminates in substantial trash production, contaminating terrestrial, aquatic, and atmospheric environments. This pollution has a detrimental impact on ecosystems, human health, and biodiversity [8]. Climate Change Exacerbates the linear economy's dependence on fossil fuels, energy-intensive production methods, and waste disposal, which is significantly contribute to greenhouse gas emissions and intensify climate change. Increasing temperatures, severe weather phenomena, and elevated sea levels are but a few of the repercussions [9].

Unsustainable resource extraction, deforestation, and habitat destruction, often driven by the linear economy, lead to biodiversity loss. The decline in biodiversity jeopardizes the stability and resilience of ecosystems, affecting human welfare [6]. The linear economy can exacerbate social disparities, as marginalized areas often bear the adverse effects of pollution and resource depletion associated with industrial activity. The linear model fosters economic growth that is unsustainable and not dissociated from environmental consequences. This suggests that economic growth often results in environmental damage and resource depletion. A circular economy, which prioritizes the reduction, reuse, recycling, and regeneration of resources, presents a more sustainable alternative. This transition can mitigate the issues presented by the linear economy and foster a more sustainable future.

The objectives of the chapter is to introduce and clarify the fundamental principles of the circular economy, highlighting how it differs from the traditional linear model. It aims to explain the core concepts, such as resource efficiency, waste reduction, and product lifecycle extension, while emphasizing the economic, environmental, and societal value that the circular economy can generate.

1.3 Core Principles of the Circular Economy

1.3.1 Design Out Waste and Pollution

The circular economy concept aims to eliminate waste and pollution through intentional design, prolonging the use of products and resources, and restoring natural systems. This entails transitioning from a linear "take-make-dispose" paradigm to one that emphasises resource efficiency, product durability, and restorative methodologies. This chapter examines the fundamental ideas and their application within a circular economy framework, beginning with product design, which must be engineered for longevity, ease of disassembly, and recyclability, while minimizing the use of toxic materials and emphasizing renewable resources. The same principles apply to material selection these select materials are non-toxic, readily recyclable, and sustainably sourced. To explore alternatives to plastics and other products that significantly impacts the environment [10]. Process optimization enhances industrial processes to reduce waste production and energy usage. This ensures producers are liable for the complete lifecycle of their products; encompassing end-of-life management and disposal, as well as promoting the reuse of products and components through repair, refurbishment, and remanufacturing Recycling establish effective recycling systems to reclaim valuable resources from waste streams [11]. Sharing economy advocates promote frameworks of sharing, leasing, and collaborative consumption to reduce the need for individual ownership of goods.

Product-as-a-Service transitions from product sales to service provision, motivating enterprises to uphold the quality and durability of their solutions. Restore natural systems to employ biodegradable materials that can be safely reintegrated into the biosphere, enhancing soil quality and facilitating natural processes [12]. Utilising renewable energy sources are to promote circular economy initiatives, thereby diminishing dependence on fossil fuels and their corresponding emissions. Invest in efforts that restore and enhance natural ecosystems, thereby fostering biodiversity and food security and resilience [13]. Policy and regulation can significantly influence governments by enacting laws that promote circular practices, including extended producer responsibility, waste reduction objectives, and support for circular enterprises. Business Innovation as organisations can develop pioneering business models, products, and services that align with the principles of the circular economy. Consumer awareness aims to educate consumers about the benefits of circularity and encourage them to adopt more sustainable consumption habits along with proper policy frame work thereby promoting long-term economic growth and sustainability [14].

1.3.2 Regenerate Natural Systems

It includes replanting native trees and vegetation, restoring wetlands, and controlling erosion to recreate healthy habitats for wildlife, promoting the growth and proliferation of native plants and animals, while managing invasive species, which is crucial for a thriving ecosystem [15, 16]. Regenerative agriculture practices focus on enhancing soil health through techniques such as no-till farming, cover cropping, and composting, which improve carbon sequestration, water retention, and nutrient cycling [17]. Moving away from a linear "take-make-dispose" model to a circular economy that eliminates waste, circulates materials, and are regenerates natural systems. These practices actively work with natural processes to create positive impacts. For example, regenerative agriculture not only reduces harmful emissions but also removes them from the atmosphere through carbon sequestration in the soil. Instead of just minimizing harm, the goal is to create a net positive impact on the environment, like restoring ecosystems to a healthier state. Understanding and utilizing natural processes, such as photosynthesis, nutrient cycling, and water purification are to enhance ecosystem health. Building systems that can withstand disturbances and recover quickly, ensuring long-term sustainability. Creating systems that can sustain themselves over time, reducing reliance on external inputs recognizing the interconnectedness of all living things and their environments. Promoting practices are that benefit the entire system, soil health, biodiversity, and carbon sequestration, while producing food and other resources [18, 19]. Planting trees are to restore forests and improve carbon sequestration. Implementing practices that enhance the quality of water, recharge aquifers, and reduce water scarcity. Engaging local communities in the regeneration process to ensure long-term sustainability and equitable benefits. By embracing these principles and practices, we can move beyond simply minimizing our negative impact on the environment and actively work towards regenerating natural systems, creating a healthier and more sustainable future for all [15, 16].

1.3.3 The "3R" and "9R" Frameworks (Refuse, Rethink, Reduce, Reuse, Repair, Refurbish, Remanufacture, Repurpose, Recycle)

The 3R and 9R frameworks serve as tools for advancing a circular economy, emphasizing resource efficiency and waste minimization. The 3Rs (Reduce, Reuse, Recycle) constitute fundamental concepts, while the 9Rs (Refuse, Rethink, Reduce, Reuse, Repair, Refurbish, Remanufacture, Repurpose, Recycle) offer a more comprehensive framework. The 9Rs offer a comprehensive range of solutions for extending product lifecycles and minimising waste, aiming to optimise resource value and mitigate environmental impact [20]. The "3R" approach minimizes waste, which in turn reduces the initial quantity of trash produced. Reuse the act of repurposing goods by utilizing them again, whether for their original use or an alternative one. The conversion of utilized materials into new goods, eschewing superfluous consumption or items that generate waste, analyzing existing consumption trends, and investigating new, more sustainable solutions. Reduce the diminishing quantity of resources utilized in production and consumption. Reuse prolongs the longevity of products through repeated utilization and restores damaged objects rather than substituting them. Refurbish and revitalize pre-owned items to a condition akin to new. Remanufacturing is the process of reconstructing goods to their original specifications, utilizing either recycled or new components. Repurpose by assigning innovative and functional applications to abandoned items. Recycling is the conversion of used materials into new goods. The 9Rs framework represents a comprehensive strategy for waste management and resource efficiency, promoting a transition from a linear "take- make-dispose" model to a circular economy that maximizes the utilization of resources over extended periods [21].

1.4 Systems Thinking in Circular Economy

1.4.1 Interconnectedness of Production, Consumption, and Waste

Production, consumption, and waste are intricately linked operations that constitute a linear system with considerable environmental repercussions. Unsustainable production and consumption habits lead to resource depletion and pollution, while insufficient waste management worsens these issues by permitting garbage accumulation that further damages the ecosystem. Transitioning to a circular economy, which prioritises waste reduction, material reuse, and resource recycling, is essential for alleviating these adverse effects [22]. Consumption trends significantly impact production choices. Elevated demand for commodities often leads to increased production, which can potentially strain resources and generate additional waste. In contrast, unsustainable manufacturing methods, such as the application of toxic chemicals or excessive resource depletion, can result in environmental degradation and subsequently influence consuming behaviours. Waste is a direct consequence of consumption and production. When products are disposed of after use, they become waste. Unsustainable consumption leads to excessive waste production, while inadequate waste management can result in pollution and resource depletion. The management of waste can impact manufacturing processes. A deficiency in recycling facilities may promote the ongoing extraction of virgin resources for new products, hence sustaining a linear system. The circular economy seeks to dismantle the linear paradigm by emphasizing waste reduction, material reuse, and resource recycling. By reevaluating production and consumption and enhancing waste management procedures, a more sustainable system can be established [23].

1.4.2 Feedback Loops and Lifecycle Thinking

Feedback loops and lifecycle thinking are interrelated concepts that highlight the cyclical nature of processes and the significance of evaluating the long-term consequences of decisions. Feedback loops fundamentally entail utilizing a system's output as input for subsequent activities, facilitating modifications and enhancements. Lifecycle thinking is a holistic approach that evaluates the environmental, social, and economic effects of a product or process from inception to disposal. The significance of lifecycle thinking lies in its capacity to pinpoint opportunities for enhancement, minimize waste, and foster sustainability by acknowledging the interrelatedness of various stages [24]. The relationship between feedback loops and lifecycle thinking, implementing feedback loops throughout lifecycle stages (Fig. 4).

Feedback loops can be utilized at every stage of a product's lifecycle to enhance its performance. Feedback regarding the environmental impact of a manufacturing process can be utilized to improve energy efficiency. Lifecycle thinking provides a framework for understanding the potential consequences of decisions made at various stages of development. Feedback loops can be employed to enhance decisions informed by insights derived from lifecycle evaluations. Through the integration of feedback loops and lifecycle thinking, organizations can pursue ongoing enhancement in all facets of their operations, resulting in more sustainable goods and processes [25].

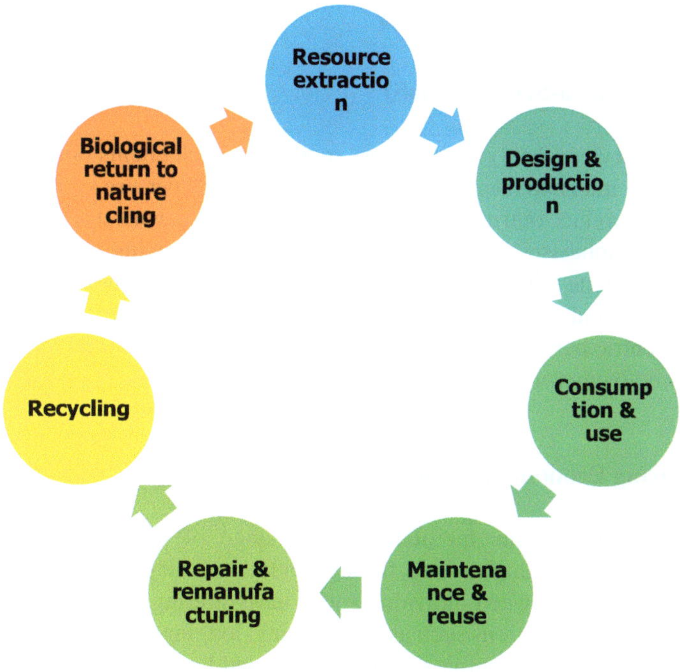

Fig. 4 Circular economy feedback loop

1.4.3 Circular Economy vs. Traditional Sustainability Models

The circular economy and conventional sustainability models, although both striving for a more sustainable future, diverge in their fundamental methodologies. The circular economy seeks to minimise waste and enhance resource efficiency through methods such as reuse, repair, and recycling, thereby establishing a closed-loop system. Conventional sustainability, typically characterised by the linear "take-make-dispose" paradigm, emphasises the reduction of adverse effects associated with resource extraction and waste production, rather than their complete elimination. The fundamental distinction between a circular economy and conventional sustainability models lies in their methodologies for resource utilisation and waste management [26, 27]. Conventional sustainability models often emphasise reducing the adverse effects of a linear "take-make-dispose" framework. In contrast, a circular economy proactively seeks to redesign systems and products to eliminate waste and pollution, emulating natural cycles.

1.5 Circular Business Models

1.5.1 Product-as-a-Service (PaaS)

Product as a Service (PaaS) is a business model in which clients obtain a product's functionality or intended results via a subscription or pay-per-use system, rather than acquiring the product outright. The provider maintains ownership of the tangible product and emphasises value through continuous servicing and support. This strategy often aligns with circular economy principles by promoting product durability, refurbishment, and recycling. PaaS signifies a transition from a transactional, ownership-centric paradigm to a more sustainable, service-oriented framework that advantages both enterprises and consumers [28].

1.5.2 Sharing Economy Platforms

Sharing economy systems facilitate the exchange of underutilized assets and resources among individuals, often through online marketplaces. These platforms provide connections between individuals who have offerings and those who require them, facilitating transactions such as ride-sharing, lodging rentals, and peer-to-peer financing. Illustrations encompass Airbnb, Uber, and LendingClub [29].

1.5.3 Remanufacturing and Closed-Loop Supply Chains

Remanufacturing is a crucial component of a closed-loop supply chain (CLSC), designed to reclaim value from used items and materials, thereby reducing waste

Table 1 Comparing design for durability, disassembly, and reuse

Design principle	Description	Benefits
Durability	Creating products that last longer and resist wear and tear	Reduces need for frequent replacement, lowers waste, conserves resources
Disassembly	Designing products to be easily taken apart at end-of-life	Facilitates repair, refurbishing, recycling; preserves material quality
Reuse	Enabling components or products to be reused in original or new forms	Extends product lifecycle, reduces demand for virgin materials, cuts emissions

Source: Compiled by the Author

and optimising resource efficiency. CLSCs amalgamate forward (conventional) and reverse logistics, encompassing remanufacturing processes, to accomplish this objective. This methodology is crucial for achieving sustainable development, preserving resources, and safeguarding the environment [30].

1.5.4 Design for Durability, Disassembly, and Reuse

Designing for durability, disassembly, and reuse, referred to as Design for Disassembly (DfD), is a sustainable methodology that emphasizes the creation of products and buildings that can be readily disassembled after their functional lifespan for recycling, repurposing, or additional use. This technique minimizes waste, preserves resources, and fosters a circular economy [31] (Table 1).

1.6 Value Creation in the Circular Economy

1.6.1 Economic Value: Resource Efficiency, Cost Savings, Innovation

Resource efficiency, cost reduction, and innovation are interrelated ideas that provide economic value. By optimizing resource utilization, companies can save expenses and create novel, inventive products and processes. This results are in heightened competitiveness and the potential to access new markets, ultimately fostering a more sustainable and lucrative economy. A focus on resource efficiency, cost reduction, and innovation fosters a virtuous cycle that leads to a more sustainable and economically viable future.

1.6.2 Environmental Value: Reduced Emissions, Pollution Control, Biodiversity Conservation

Pollution, originating from industrial sources, agriculture, or various human activities, can significantly impair ecosystems. It can pollute water and soil, adversely affecting flora and fauna, hence diminishing biodiversity. Conserving biodiversity requires the protection and restoration of natural habitats, controlling pollution, mitigating climate change, and managing invasive species. This encompasses the preservation of genetic, species, and environmental diversity. The three ideals are inherently interconnected. For instance, reducing emissions helps alleviate climate change, a significant factor contributing to biodiversity loss [32]. Efficient pollution management can avert additional harm to ecosystems and safeguard the species they sustain. In contrast, healthy ecosystems characterized by elevated biodiversity are more resilient to the effects of pollution and climate change. Biodiversity is crucial for delivering key ecosystem services, including clean air and water, pollination, nutrient cycling, and climate regulation [18, 19]. These services are essential to human welfare and economic success (Fig. 5).

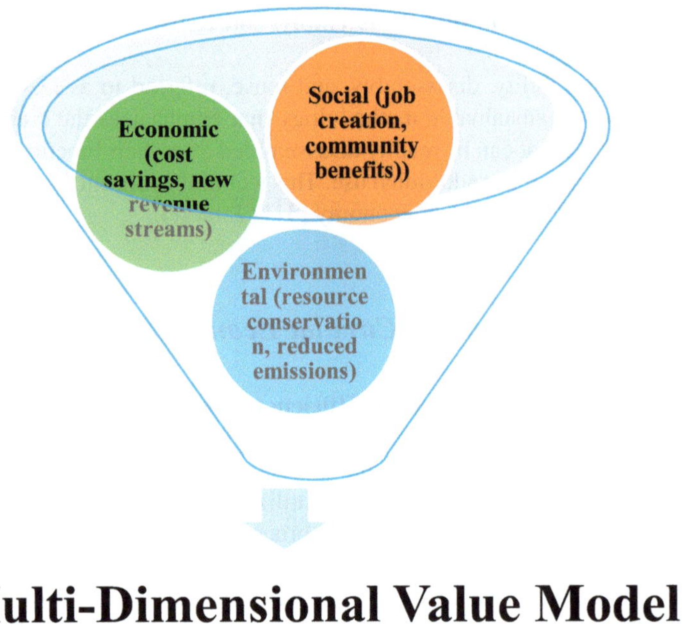

Fig. 5 Value pyramid

1.6.3 Social Value: Job Creation, Community Resilience, Ethical Supply Chains

It denotes the beneficial influence that businesses and organizations exert on society beyond their fundamental economic functions. It underscores the creation of employment opportunities, the fortification of communities, and the assurance of ethical and equitable behaviours across the supply chain. Generating work opportunities is primarily aimed at individuals who are most distant from the labour market, such as those with disabilities or those who face employment barriers. Investing in training and apprenticeships equips individuals with the necessary skills for current and future jobs. Promoting the development of local enterprises and entrepreneurs enhances a diverse and prosperous local economy, fosters social relationships, encourages community involvement, and supports local causes. Essentially, enterprises that emphasise social value priorities not only profit but also generate beneficial and enduring effects on the people and environments in which they operate. This strategy may lead to increased consumer loyalty, enhanced employee morale, and a more robust and sustainable business model [33].

1.7 *Enablers and Tools*

1.7.1 Digital Technologies: IoT, Blockchain, AI, and Digital Twins

The successful transition to a circular economy relies heavily on a range of enabling tools and frameworks that support innovation, traceability, and system-level thinking. Among the most impactful enablers are digital technologies, such as the Internet of Things (IoT), blockchain, artificial intelligence (AI), and digital twins [34]. These technologies improve transparency, track material flows, and optimize resource use in real time. For example, blockchain can ensure traceability and authenticity in supply chains, while digital twins—virtual replicas of physical assets—allow for predictive maintenance and enhanced product lifecycle management [35].

1.7.2 Life Cycle Assessment (LCA) and Material Flow Analysis

Critical enabler is Life Cycle Assessment (LCA) and Material Flow Analysis (MFA), which provide systematic methods to evaluate environmental impacts and resource use across the entire lifecycle of products and systems. These tools help identify hotspots of inefficiency and inform data-driven decision-making to design more sustainable solutions [36].

1.7.3 Design Thinking and Biomimicry

Design thinking and biomimicry also play a foundational role in circular innovation. Design thinking encourages user-centered, iterative problem-solving that aligns business needs with sustainability goals. Biomimicry, on the other hand, draws inspiration from nature to design products and systems that are inherently regenerative and resource-efficient [37]. These approaches foster creativity in developing circular products that minimize waste and mimic natural ecosystems.

1.7.4 Role of Standards and Certifications (e.g., Cradle to Cradle, ISO)

The implementation and credibility of circular practices are strengthened by the use of standards and certifications. Certifications like Cradle to Cradle, ISO 14040 (LCA), and ISO 14001 (environmental management systems) provide structured frameworks for evaluating and validating circular performance. These standards not only help organizations benchmark their progress but also enhance trust among consumers and stakeholders [38].

1.8 Challenges and Critiques

1.8.1 Technological and Financial Limitations

One of the primary challenges facing the circular economy is the technological and financial barrier to implementation. Many circular solutions—such as advanced recycling methods, product redesign for disassembly, or reverse logistics systems require significant investment and technical expertise. While large corporations may have the resources to experiment and adapt, small and medium-sized enterprises (SMEs) often struggle to afford these transitions. In addition, the lack of supportive infrastructure—like collection systems, repair hubs, or material recovery facilities—can limit the feasibility of circular models, particularly in less developed regions.

1.8.2 Cultural and Behavioral Resistance

Another significant obstacle is cultural and behavioral resistance. Transitioning from a traditional linear economy to a circular one requires not only technical change but also a fundamental shift in mindset among consumers, businesses, and policymakers. Many consumers are used to the convenience and low cost of disposable goods and may resist practices like renting, repairing, or buying second-hand. On the business side, circular models can challenge long-standing revenue models that depend on high sales volumes. Overcoming this resistance involves not just awareness campaigns but also systemic changes in incentives and education to promote long-term thinking.

1.8.3 Risk of "Circular Washing" and Lack of Accountability

A growing critique of the circular economy is the risk of "circular washing", a phenomenon where companies exaggerate or misrepresent their circular initiatives to appear more sustainable than they truly are. Without clear regulatory standards or verification systems, it is difficult to distinguish between genuinely circular efforts and superficial marketing. This lack of accountability can undermine consumer trust and allows companies to benefit reputationally without making meaningful environmental contributions. To address this, there is a need for transparent reporting, standardized certification, and stronger policy enforcement.

1.8.4 Measuring Circularity and Impact

A persistent technical challenge lies in measuring circularity and its impact. Unlike carbon emissions or energy use, circularity lacks universally accepted indicators. Companies and governments often struggle to track and quantify metrics like resource retention, material efficiency, or lifecycle impacts. Without reliable data, it becomes difficult to evaluate progress, compare initiatives, or optimize performance. Developing standardized, science-based tools to measure circular performance is essential for setting goals, ensuring accountability, and scaling successful practices.

1.9 Future Directions

1.9.1 Scaling Circular Models Across Sectors and Regions

As the circular economy continues to gain momentum, a key priority for the future is the scaling of circular models across sectors and regions. While many successful examples exist in industries such as fashion, electronics, and construction, these remain fragmented and localized. To achieve a broader impact, circular principles must be integrated into global supply chains, public procurement policies, and mainstream business practices. This includes supporting both industrialized and developing economies in adopting circular solutions tailored to their specific resources, infrastructures, and socio-economic conditions (Table 2).

1.9.2 Role of Innovation and Cross-Sector Collaboration

Innovation and cross-sector collaboration will be critical drivers of this scaling process. Emerging technologies such as AI, blockchain, and digital product passports can improve resource tracking, enable predictive maintenance, and support transparent material flows. At the same time, partnerships between businesses, governments, researchers, and civil society can foster system-wide change by combining

Table 2 Future directions in the circular economy

Focus area	Key strategies
Scaling Circular Models	Expand adoption across sectors (e.g., agriculture, construction, electronics); tailor to local/regional contexts
Innovation and Collaboration	Foster partnerships between industries, governments, and academia; invest in R&D
Policy and Regulation	Implement supportive policies, incentives, and harmonized international standards
Global Integration	Promote circular trade, technology transfer, and cross-border resource management
Metrics and Impact Measurement	Develop standardized tools to track circularity (e.g., material flow, lifecycle impact)

Source: Compiled by the Author

expertise, aligning incentives, and sharing best practices. For example, collaborations between manufacturers and recyclers can ensure products are designed for reuse or easier material recovery, enhancing the efficiency of circular loops.

1.9.3 Vision for a Global Circular Economy

The vision for a global circular economy requires more than isolated initiatives it calls for a unified, strategic approach rooted in international cooperation. This includes harmonizing standards, investing in circular infrastructure, and building global governance frameworks that promote sustainable trade, innovation transfer, and equitable access to circular opportunities. A truly global circular economy would not only reduce waste and resource consumption but also create inclusive economic growth, protect ecosystems, and build resilience against climate change.

1.10 Conclusion

The circular economy presents a transformative vision for how societies can produce, consume, and manage resources more sustainably. By shifting away from the traditional linear model of "take-make-dispose," it offers pathways to reduce environmental impact, conserve natural resources, and create long-term economic and social value. However, realizing this vision is not without significant challenges. Technological and financial barriers, cultural resistance, risks of circular washing, and difficulties in measuring progress all pose obstacles to widespread adoption. These issues highlight the need for stronger policy support, innovation, standardized metrics, and greater public awareness. Despite these hurdles, the potential of the circular economy remains substantial, especially if supported by coordinated efforts across sectors and regions. Looking forward, the focus must be on scaling successful circular practices, fostering cross-sector collaboration, and working toward a global, inclusive circular economy. With the right investments, governance, and shared commitment, the circular economy can evolve from a promising concept into a practical and impactful global system for sustainable development.

References

1. Singh S, Singh P, Sharma A, Choudhury M (eds) (2022) Agriculture waste management and bioresource: the circular economy perspective. Wiley. https://doi.org/10.1002/9781119808428
2. Corvellec H, Stowell AF, Johansson N (2022) Critiques of the circular economy. J Ind Ecol 26(2):421–432. https://doi.org/10.1111/jiec.13187
3. Tuladhar A, Iatridis K, Dimov D (2022) History and evolution of the circular economy and circular economy business models. In: Circular economy and sustainability. Elsevier, pp 87–106. https://doi.org/10.1016/B978-0-12-819817-9.00031-4
4. Dabija D, Năstase C (2024) Exploring the evolution of sustainability paradigms: from linear to circular economy models. Proc Int Conf Bus Excell 18(1):754–769
5. Beck D, Pal A, Zaveri P, Zheng A, Tiwari AK, Majumdar S (2025) Ecosystem resilience for climate resilience, strategies and innovations: a case study from developing countries. In: Choudhury M, Dixit G, Majumdar S (eds) Sustainable synergy: harnessing ecosystems for climate resilience. Climate change management. Springer, Cham. https://doi.org/10.1007/978-3-031-77957-2_4
6. Dixit G, Majumdar S, Choudhury M (2025) Introduction to ecosystem services for climate resilience: a theoretical framework. In: Choudhury M, Dixit G, Majumdar S (eds) Sustainable synergy: harnessing ecosystems for climate resilience. Climate change management. Springer, Cham. https://doi.org/10.1007/978-3-031-77957-2_1
7. Kumaraswamy TR, Choudhury M, Ponappa S, Majumdar S (2025) A significance of climate resilience: strategies for local, regional, and national levels. In: Choudhury M, Dixit G, Majumdar S (eds) Sustainable synergy: harnessing ecosystems for climate resilience. Climate change management. Springer, Cham. https://doi.org/10.1007/978-3-031-77957-2_13
8. Majumdar S (2023) Uses of artificial intelligence and GIS technologies in healthcare services. In: Tomar RS, Chauhan RK, Nayak SK (eds) Artificial intelligence and data science in contemporary healthcare (Chapter 18). CRC Press. https://doi.org/10.1201/9781003363606-18
9. Chowdhury BR et al (2025) Building a resilient tomorrow: ecosystem-based approaches to measurement, assessment, prediction, and future perspectives in climate resilience. In: Choudhury M, Dixit G, Majumdar S (eds) Sustainable synergy: harnessing ecosystems for climate resilience. Climate change management. Springer, Cham. https://doi.org/10.1007/978-3-031-77957-2_14
10. Roy and Choudhury (2022) Bioeconomy and green plastic production. In: Baskar C, Ramakrishna S, Daniela La Rosa A (eds) Encyclopedia of green materials. Springer, Singapore. https://doi.org/10.1007/978-981-16-4921-9_189-
11. Chowdhury R, Barman S, Choudhury M, Kim KH, Bhattacharya SS (2024) Earthworm modifies microbial community and functional genes for lignocellulosic waste valorization: isolating plant-growth-promoting bacteria via next generation sequencing. Int Biodeterior Biodegradation 193:105854. https://doi.org/10.1016/j.ibiod.2024.105854
12. Paul S, Choudhury M, Deb U, Pegu R, Das S, Bhattacharya SS (2019) Assessing the ecological impacts of ageing on hazard potential of solid waste landfills: a green approach through vermitechnology. J Clean Prod 236:117643. https://doi.org/10.1016/j.jclepro.2019.117643
13. Dey S, Santra M, Choudhury M, Ghosh AR, Samanta P (2024) Food waste generation and its industrial utilization: an overview. Environ Sci Pollut Res:1–20. https://doi.org/10.1007/s11356-024-34252-3

14. Choudhury M, Roy P (2025) Challenges with microplastic pollution in the regime of UN sustainable development goals. World Dev Sustain 2025:100216. https://doi.org/10.1016/j.wds.2025.100216
15. Bhardwaj LK, Dhanaraj K, Rath P, Singh V, Choudhury M (2025) Urban wetlands and climate resilience: a case study of Surajpur Wetland, Greater Noida, Uttar Pradesh, India. In: Impact of environmental degradation on ecosystems and preventive measures. IGI Global Scientific Publishing, pp 161–194. https://doi.org/10.4018/979-8-3373-1549-2.ch007
16. Bhardwaj LK, Rath P, Jain H, Choudhury M (2025) A comprehensive review of wetlands in India and their conservation. In: Nature-based technologies for wastewater treatment and bioenergy production, vol 43. IWA Publishing. https://doi.org/10.2166/9781789064100_0043
17. Kumar D, Choudhury M (2021) Recognizing economic values of wetland ecosystem services: a study of emerging role of monetary evaluation of Chandubi ecosystem and biodiversity. In: Wetlands conservation: current challenges and future strategies, pp 87–110. https://doi.org/10.1002/9781119692621.ch5
18. Gogoi N, Choudhury M, Ul Hasan MS, Changmai B, Baruah D, Samanta P (2024) Soil organic carbon pool in diverse land utilization patterns in North-East India: an implication for carbon sequestration. Environ Dev Sustain:1–27. https://doi.org/10.1007/s10668-024-05037-y
19. Gogoi N, Sarma A, Choudhury M, Samanta P, Faizan M, Sadhak S (2024) Soil heavy metal pollution and ecological risk assessment in disturbed and undisturbed soil of Morigaon, Assam. Discover Environ 2(1):13. https://doi.org/10.1007/s44274-024-00039-z
20. Alivojvodic V, Kokalj F (2024) Redefining waste: R-strategies and metrics as a framework for driving progress of circular economy performance. In: International conference of experimental and numerical investigations and new technologies. Springer Nature Switzerland, Cham, pp 75–91. https://doi.org/10.1007/978-3-031-78635-8_9
21. Pedrini S, Tartari M (2025) Cultural and creative practices as drivers of circular economy: a 9R framework perspective. In: International conference on computational science and its applications. Springer Nature Switzerland, Cham, pp 283–298. https://doi.org/10.1007/978-3-031-97654-4_18
22. Rajpal A, Choudhury M, Goswami S, Chakravorty A, Raghavan V (eds) (2024) Waste management and treatment: advances and innovations, 1st edn. CRC Press. https://doi.org/10.1201/9781003258377
23. Castiglione A, Cimmino L, Di Nardo M, Murino T (2023) A framework for achieving a circular economy using the blockchain technology in a sustainable waste management system. Comput Ind Eng 180:109263. https://doi.org/10.1016/j.cie.2023.109263
24. Cecere G, Bottausci S, Degli Esposti A, Magrini C, Mazzi A, Camana D et al (2024) The role of life cycle thinking-based methodologies in the development of waste management plans. Waste Manag 173:109–117. https://doi.org/10.1016/j.wasman.2023.11.005
25. Choudhary R, Kumar A, Narayan S, Mishchenko I, Choudhury M, Bhardwaj LK, Ahmad A, Andrienko M, Gaman P, Boiko O (2025) Integrating fuzzy logic into life cycle assessment for sustainable municipal solid waste management in metro cities. Ecol Eng Environ Technol 26(7):316–331. https://doi.org/10.12912/27197050/205500
26. Gallardo-Vázquez D, Scarpellini S, Aranda-Usón A, Fernández-Bandera C (2024) How does the circular economy achieve social change? Assessment in terms of sustainable development goals. Human Soc Sci Commun 11(1):1–18. https://doi.org/10.1057/s41599-024-03217-9
27. Santibanez Gonzalez EDR, Kandpal V, Machado M, Martens ML, Majumdar S (2023) A bibliometric analysis of circular economies through sustainable smart cities. Sustainability 15(22):15892. https://doi.org/10.3390/su152215892
28. Vogt Duberg J, Sakao T (2024) How can manufacturers identify the conditions for financially viable product-as-a-service? Front Manuf Technol 4:1498189. https://doi.org/10.3389/fmtec.2024.1498189
29. Zuhroh D, Jermias J, Ratnasari SL, Nurjanah E, Fahlevi M (2025) The impact of sharing economy platforms, management accounting systems, and demographic factors on financial performance: exploring the role of formal and informal education in MSMEs. J Open Innov Technol Mark Complex 11(1):100447. https://doi.org/10.1016/j.joitmc.2024.100447

30. Wang Z, Ye C, Guo J (2024) Game analysis on two-tier remanufacturing green closed-loop supply chain considering incentive compatibility under uncertain demand. Int J Syst Sci: Oper Logist 11(1):2368132. https://doi.org/10.1080/23302674.2024.2368132
31. Roskladka N, Bressanelli G, Saccani N, Miragliotta G (2025) Repairable electronic products for the circular economy: a review of design for repair features practices and measures to contrast obsolescence. Discov Sustain 6(1):66. https://doi.org/10.1007/s43621-024-00753-x
32. Choudhury M, Dixit G, Majumdar S (eds) (2025) Sustainable synergy: harnessing ecosystems for climate resilience. Springer Nature, Switzerland. https://doi.org/10.1007/978-3-031-77957-2
33. Gualandris J, Branzei O, Wilhelm M, Lazzarini S, Linnenluecke M, Hamann R et al (2024) Unchaining supply chains: transformative leaps toward regenerating social–ecological systems. J Supply Chain Manag 60(1):53–67. https://doi.org/10.1111/jscm.12314
34. Liao Y, Liao W, Wu D, Zhang C, Xu J, Majumdar S (2024) Inspired applications of artificial intelligence in disaster-resilient regional construction for climate risk. In: Ghosh S, Majumdar S, Cheshmehzangi A (eds) Cities of tomorrow: urban resilience and climate change preparedness. Urban sustainability. Springer, Singapore. https://doi.org/10.1007/978-981-97-9658-8_16
35. Antikainen M, Uusitalo T, Kivikytö-Reponen P (2018) Digitalisation as an enabler of circular economy. Proc CIRP 73:45–49. https://doi.org/10.1016/j.procir.2018.04.027
36. Hauschild MZ, Rosenbaum RK, Olsen SI (eds) (2018) Life cycle assessment: theory and practice. Springer. https://doi.org/10.1007/978-3-319-56475-3
37. Benyus JM (2002) Biomimicry: innovation inspired by nature. Harper Perennial
38. Ellen MacArthur Foundation (2013) Towards the circular economy: economic and business rationale for an accelerated transition. Ellen MacArthur Foundation. https://ellenmacarthurfoundation.org

From Past to Present: The Evolution of Waste Management and the Rise of Circular Economy Principles

Saurav Narayan, Deepesh Yadav, Shikha Jha, Basona Khatun, Rakesh Choudhary, Ujjwal Kumar Verma, Moharana Choudhury, and Deepak Kumar

Abstract This chapter examines the historical trajectory and future potential of waste management, tracing its evolution from rudimentary disposal methods to sophisticated, sustainability-oriented circular systems. It begins by surveying ancient and industrial-era waste practices, illustrating the emergence of the linear economy and its associated environmental costs. A bibliometric analysis maps the global research trends from 2015 to 2024, revealing an exponential rise in academic interest in circular economy models. The chapter then presents influential frameworks such as Cradle-to-Cradle, Industrial Symbiosis, and Doughnut Economies that are reshaping policy and industrial practice. Real-world applications are illustrated through case studies from Amsterdam, San Francisco, IKEA, and Philips, demonstrating successful strategies for municipal, corporate, and technological circular economies. The chapter further extends to global challenges, best practices, and governance frameworks, concluding with a phased roadmap to mainstream circularity by 2050. This chapter situates the circular economy not just as an environmental imperative, but as a transformative principle for global economic and social systems.

S. Narayan (✉) · D. Yadav · B. Khatun
Department of Law, School of Legal Studies, Central University of Punjab, Bathinda, Punjab, India
e-mail: s4srvnarayan@gmail.com

S. Jha
Advocacy Forum Nepal, Kathmandu, Nepal

R. Choudhary
Department of Civil Engineering, National Institute of Technology Delhi, New Delhi, India

U. K. Verma · D. Kumar
Dnipro State Medical University, Dnipro, Ukraine

M. Choudhury
Environmental Research and Management Division, Voice of Environment (VoE), Guwahati, Assam, India

S. Majumdar et al. (eds.), *Sustainable Urban Future*, Urban Sustainability,
https://doi.org/10.1007/978-981-95-2976-6_2

Keywords Circular economy · Waste management · Municipal solid waste · Sustainability · Policy governance

1 Introduction

1.1 Background and Significance

Global waste has dramatically increased over the years. Approximately 2 billion metric tons of municipal solid waste (MSW) is generated annually. This is expected to grow by approximately 70% by 2050 [1]. With the tremendous increase in waste comes the need to manage it systematically. With time, the need for systematic waste management has been realised from the shift from a linear to a circular model. Linear economy, also critiqued as the '*take-make-dispose*' approach of production, is a traditional economic model that is based on the linear process, which is extracting raw materials, transforming them into products, and disposing of them at the end of their life [2, 3]. The concept of a linear waste management model is contrary to the idea of sustainability.

Sustainability is the practice of utilizing available resources without sacrificing the ability of future generations to meet their own needs. Given the current generation's management of the depletion of raw materials and resources, it is necessary to switch from the unsustainable linear model of the economy and waste management to the circular model. The circular economy (CE) is predicated on the principles of 'reducing, reusing, and recycling' waste resources [4]. The circular economy framework treats waste as a source of new resources and materials, in contrast to the linear economy worldview. CE encourages conserving raw material and recycling waste as a means to reduce carbon emissions [5]. Taken together, changing the perception of waste from a "disposable resource" in a linear economy model to a "raw material" used in the creation of another resource is evidence that sustainable practices are receiving growing attention.

Different degrees of sustainability are represented by waste avoidance, minimization, reuse, recycling, recovery, and disposal [6]. Sustainability has been considered the basis for waste management. The three Rs of sustainability—reduce, reuse, and recycle—have become the foundation of waste management [7]. Therefore, sustainability and waste management are closely intertwined. To this end, laws and legislation can be an effective tool for ensuring proper waste management and achieving sustainability.

Figure 1 illustrates that the global champions of recycling are setting a powerful example in sustainability, with Austria and Wales leading the charge, both achieving an impressive 59% recycling rate. Taiwan follows closely at 53%, showcasing how technology and smart policy can transform waste management. Germany and Belgium, long recognized for their environmental leadership, each recycle 52% of their waste. In comparison, the Netherlands and Denmark maintain a strong 51%

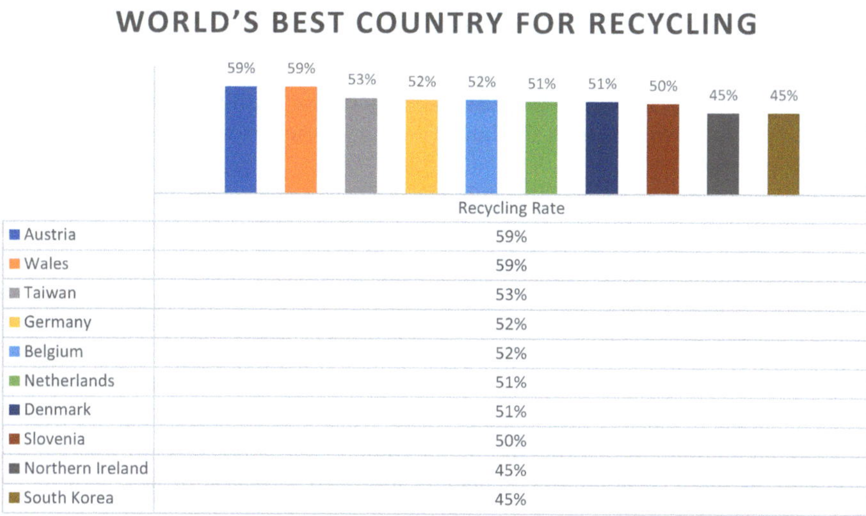

Fig. 1 World best country for recycling. (Source [8])

rate, thanks to innovative infrastructure and circular economy practices. Slovenia, with 50%, shines as a green leader in the Balkans, while Northern Ireland and South Korea round out the top ten with 45% each. South Korea particularly excels in food waste recycling, with rates as high as 95%. These nations demonstrate that, with the right combination of government action, citizen engagement, and innovation, achieving high recycling rates is not only possible but also essential for a sustainable future [9].

This chapter is important because it thoroughly addresses waste generation, collection, and transportation, which make up a foundational node in modern solid waste management and are often, overlooked aspects of sustainability transitions. Cumulatively, urbanization, increased consumerism, and diminishing landfill capacities have strained the capacity for good and sustainable waste logistics. The implications of failure in waste collection and transit not only create public health and environmental safety hazards (e.g., incremental greenhouse gas emissions) but also challenge the systems thinking that is more prevalent in circular economies. By identifying waste typologies, their generation trends, and established collection and routing practices, this chapter serves to inform local municipalities, urban planners, and researchers, thereby helping to bridge a substantial on-the-ground implementation gap with higher-level circular economy policies.

The timing of this chapter is also timely as countries work to implement SDG targets (specifically SDG 11 & 12) focused on sustainable cities and sustainable production and consumption, respectively.

1.2 Historical Overview of Waste Management

The human history brought along itself the need for waste management. Waste management has evolved with the human history; brief overview of the evolution has been provided below.

1.2.1 Ancient and Pre-industrial Practices

Each era in human history saw unique approaches to waste management. In the early stages of civilization, there was virtually no concept of waste management. Waste was typically discarded into nearby water sources or open spaces. Nomadic communities relied on nature to manage their organic waste, which decomposed naturally without human intervention [10]. During the hunter-gatherer era, waste was left where it was produced, as these communities frequently migrated in search of food and shelter [11].

Systematic approaches began to emerge in ancient civilizations. The Mesopotamians designated specific areas for waste disposal and developed storm drainage and sanitary sewer systems [12]. Cities like Ur and Babylon had well-designed stormwater drainage systems [13]. In Egypt, urban centers allocated separate zones for waste, much like modern practices. The Romans constructed sewer systems, such as the Cloaca Maxima, and established public services for street cleaning and waste collection [12]. The Roman Empire introduced organized waste collection systems, managed by dedicated workers, and constructed early sewers to manage human waste [14]. The Harappan and Minoan civilizations also developed advanced drainage and sewerage systems [15]. Notably, pits used for waste disposal in 3000 B.C. in Knossos, Crete, are considered the earliest form of landfill (Fig. 2).

In 1388, the English Parliament prohibited the dumping of waste into water bodies and ditches. By 1657, New Amsterdam (modern-day Manhattan, NY) passed the first anti-littering law prohibiting street littering [16]. As urban populations grew, so did the volume of waste. London employed "rakers" to remove street waste, marking the first organized waste collection service [17]. With each era, waste management evolved, and so did the volume of refuse.

1.2.2 Industrial Revolution and the Rise of the Linear Economy

The Industrial Revolution marked the growth of industries, the introduction of mechanization, and the generation of industrial waste, laying the groundwork for the linear economy. '*Toshers*,' or sewer hunters, scavenged valuable items from sewers while other waste was discarded [18]. Roles like mud-larks and dust-men emerged to manage street waste [19]. Waste separation practices began, with women sorting reusable materials, such as glass and metal, by hand [20]; however, this was far from a structured model.

Fig. 2 Large-scale waste segregation at a landfill site using heavy machinery and conveyor systems to separate recyclable materials from garbage, Delhi, India. (Photo By: Saurav Narayan)

Benjamin Franklin introduced public waste disposal pits in 1757, aligning with linear disposal [21]. The UK's Public Health Act of 1875 led to the formalization of waste collection and the introduction of movable garbage bins. In 1885, New York developed the first garbage incinerator, a significant milestone in waste management.

1.2.3 Twentieth Century Developments

The twentieth century, ushering in the Fourth Industrial Revolution, marked a significant shift in waste management. With growing environmental concerns, the circular economy began to replace the linear '*take-make-dispose*' model. This new approach emphasized the principles of reuse, recycling, and sustainable design. Theories such as cradle-to-cradle and industrial ecology promoted regenerative, closed-loop systems as alternatives to wasteful industrial models [22].

The chapter explores the logic of waste in current systems and introduces circular alternatives. It outlines a bibliometric review and key concepts like cradle-to-cradle design and industrial symbiosis. Case studies from Amsterdam, Kalundborg, and leading firms illustrate practical applications. A gap analysis highlights governance and equity challenges. The chapter closes with a phased roadmap toward circularity and zero waste by 2050.

2 Mapping the Evolution of Waste Management and Circular Economy Research: A Scopus-Based Bibliometric Study

A bibliometric analysis was conducted using the Scopus database to examine the evolving research landscape surrounding waste management and the circular economy. The search query applied was TITLE-ABS-KEY (evolution AND of AND waste AND management OR circular AND economy) AND PUBYEAR > 2014 AND PUBYEAR < 2025, targeting literature published between 2015 and 2024. This 10-year timeframe was chosen to capture recent developments, innovations, and scholarly interest in sustainable resource use and waste reduction strategies. The analysis focuses on trends in publication output, geographic distribution of contributions, and prominent academic institutions contributing to the field via Source citations by year. Figures 3, 4, and 5 illustrate these dimensions, including yearly publication trends, leading countries by research volume, and the performance metrics of publishing entities involved in circular economy and waste management studies.

Between 2015 and 2024, a total of **528 research documents** were published on the topic of the **evolution of waste management or circular economy**, based on the provided Scopus data. The field has shown a steady increase in scholarly attention over the decade. Starting with just 5 publications in 2015, the number grew gradually to 8 in 2016 and 10 in 2017. This upward trend continued through 2018 (16) and 2019 (25), followed by more significant growth in 2020 (45) and 2021 (73). The acceleration continued in 2022 (86) and 2023 (109), culminating in a peak of 151 publications in 2024—the highest annual output recorded. This consistent rise highlights a growing global interest in sustainable practices, particularly the circular economy and the need for effective waste management solutions.

Fig. 3 Year wise Publication (2015–2024)

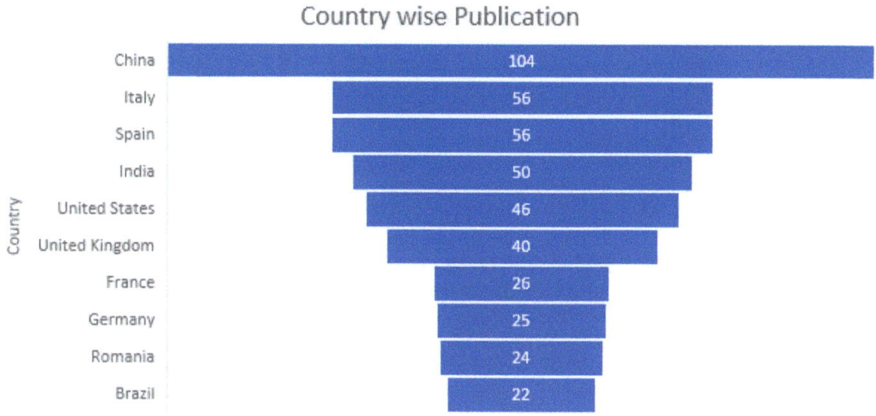

Fig. 4 Country wise Publication

China is at the forefront of research output in the area concerning the advancement of waste management or circular economy, producing a total of 104 publications from 2015 to 2024. Italy and Spain are tied for second place, having 56 publications each, followed by India with 50 publications. The United States (46), the United Kingdom (40), and France (26) also made valuable contributions. Germany (25), Romania (24) and Brazil (22) also contributed equally notable science outputs. This indicates that China continues to play a dominant role in research output on evolution of waste management or circular economy, along with high outputs from several European nations as well as emerging economies like India and Brazil.

A review of publication trends in leading academic journals from 2015 to 2024 indicates a distinct increase in research assignments centered on the ecological implications of ineffective waste management. Specifically, the Journal of Cleaner Production has had the greatest impact on the literature across the decade, by increasing citation volume from 25,750 citations in 2015 to 462,886 citations in 2024 and consequently demonstrating its robust influence on sustainability in industrial design. Additionally, Sustainability (Switzerland) illustrated a stunning escalation from 3584 citations in 2015 to 394,629 citations in 2024, identifying it as a primary conduit for the global conversation of sustainable practices serving as a global discussion of sustainability and circular economy principles. Following a similar citation trajectory, the Journal of Environmental Management expands from 22,735 citations in 2015 to 165,105 citations in 2024 illustrating the consistent academic commitment to waste management and the governance and policy aspects impacting waste management. Lastly, Environmental Science and Pollution Research increased citation frequency from 12,642 in 2015 to 225,630 in 2024 and had the most significant citation impact of the decade reinforcing its essential role in pollution-based investigations. Similarly, Resources, Conservation and Recycling also showcased growth, but with moderate increases from 8584 to 67,934 citations for a similar timeframe. Overall, the citation

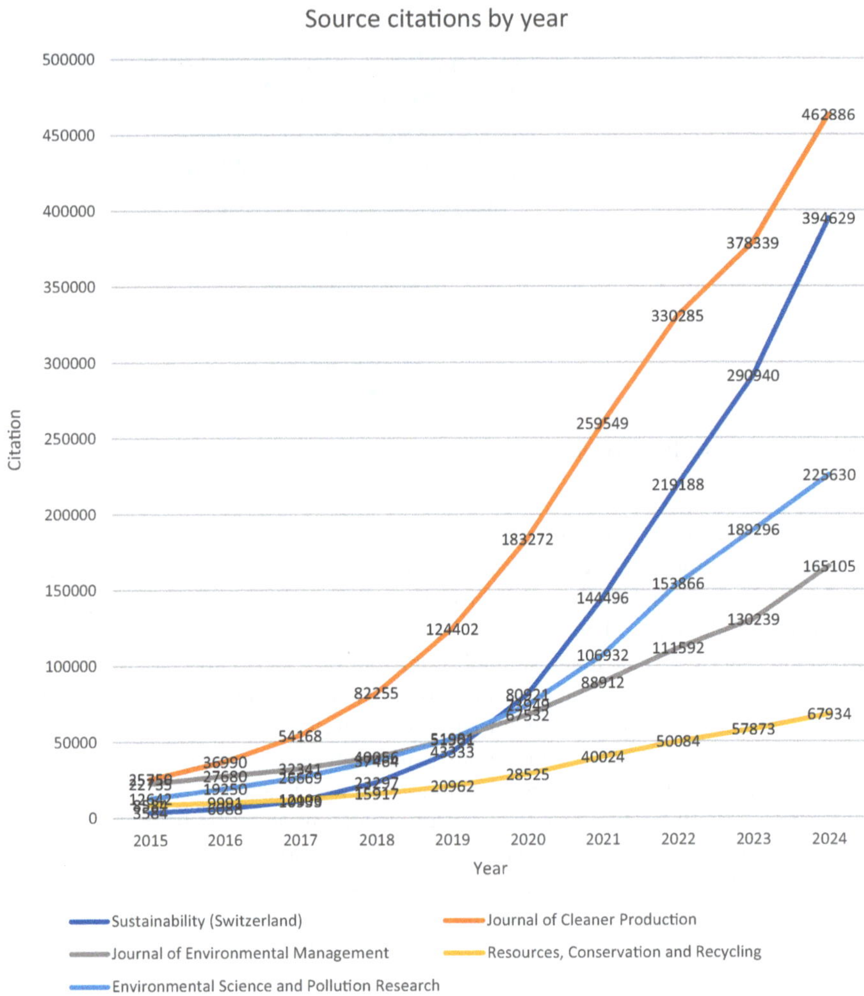

Fig. 5 Source citation by year

growth across the five journals illustrates a burgeoning interdisciplinary commitment to environmental impacts of waste and examining comprehensive solutions for sustainability in society.

As well, a bibliometric assessment of literature published from 2015 to 2024 reflects a continuous increase in academic research focus in the development of the waste management theory and the circular economy, wherein publication counts increased from just 5 documents published in 2015 to 151 published in 2024 resulting in a total of 528 documents published over the decade. This publication growth demonstrates clear interest worldwide in the efficacious use of resources and important waste minimization methods. In this decade publication analysis, China was the most prolific country with 104 published documents, followed by Italy, Spain, and India. The Journal of Cleaner Production and Sustainability (Switzerland) again demonstrated citation growth, and the pattern therefore it is inferred that these journals were essential measures of the dissemination of circular economy frameworks and environmental governance.

3 Influential Models and Theories in Circular Economy and Sustainability

The transition toward a circular and sustainable economy has been shaped by several influential theories and frameworks that reimagine the relationship between production, consumption, and ecological boundaries. Among these, the Cradle-to-Cradle design framework, the Performance Economy, Industrial Symbiosis, and Doughnut Economics have provided foundational concepts that guide policy-making, industrial practice, and academic research. This section discusses these models and their theoretical underpinnings [23].

3.1 Cradle-to-Cradle Design Framework

The Cradle to Cradle (C2C) concept, developed by William McDonough and Michael Braungart, represents a radical departure from the traditional linear economy model, often summarized as '*take, make, dispose.*' Instead, C2C envisions products designed to return safely to either biological or technical nutrient cycles without generating waste [24]. It emphasizes material health, reutilization, renewable energy, and social fairness.

In this framework, all materials are classified as either biological nutrients, which can safely return to the environment, or technical nutrients, which can be infinitely cycled within industrial systems [25]. The approach goes beyond minimizing harm; it advocates for eco-effectiveness, aiming to create systems that are regenerative by design. Has had a significant influence on sustainable product design, leading to the development of third-party certification schemes, as illustrated by Report Yak [26], which outlines the principles, benefits, and real-world applications of cradle-to-cradle design in promoting regenerative production systems.

3.2 Performance Economy

Proposed by Walter R. Stahel, the Performance Economy is a business model focused on delivering the performance or utility of products rather than the products themselves. This model extends the life of goods through repair, maintenance, reuse, and remanufacturing, thereby retaining value and reducing raw material consumption [27].

The Performance Economy introduces concepts such as product life extension, product-service systems, and functional sales, aligning with the goals of the circular economy [28]. It seeks to decouple economic growth from resource throughput, and is particularly relevant in sectors such as manufacturing, mobility, and electronics. The model enhances local employment and reduces environmental impact by embedding closed-loop service systems at its core.

3.3 Industrial Symbiosis

Industrial symbiosis is a systemic approach in which traditionally separate industries engage in a collective exchange of materials, energy, water, and byproducts. The aim is to achieve mutual benefits while reducing waste and environmental burdens [29]. This model is one of the pillars of eco-industrial development, emphasizing cross-sector collaboration and resource efficiency [30].

A well-known practical example is the Kalundborg Symbiosis in Denmark, where excess heat, steam, and byproducts from one facility serve as inputs for others in the same network [31]. Industrial symbiosis promotes the valorization of underutilized resources and fosters innovation in resource management.

3.4 Doughnut Economics

Developed by economist Kate Raworth, Doughnut Economics presents a visual and conceptual framework for sustainable development. The model defines a safe and just space for humanity bounded by a social foundation (covering basic human needs) and an ecological ceiling (planetary boundaries) [32]. Economic activity should occur within this '*doughnut*,' ensuring human well-being without overshooting Earth's ecological limits [33].

This model challenges traditional economic indicators, such as GDP growth, and instead promotes regenerative and distributive systems [34]. It has been adopted in policy frameworks by cities and governments (e.g., the Amsterdam Doughnut City Model). It has stimulated interdisciplinary research across economics, environmental science, and urban planning [35].

4 The Emergence of Circular Economy Principles

Circular-economy thinking did not spring fully formed, but evolved through a lineage of systems-minded scholars who questioned the 'take-make-waste' model. Kenneth Boulding's 1966 '*Spaceship Earth*' metaphor framed the planet as a closed vessel whose finite stocks must circulate [36]. Building on that insight, environmental economists David Pearce and R. Kerry Turner in 1989 recast economic activity as a restorative loop rather than a linear pipeline [37]. During the same era, Walter Stahel's '*Product-Life Factor*' argued that value lies in extending product lifespans [38], inspiring industrial ecology, Cradle-to-Cradle design, the Blue Economy, and biomimicry. Together, these ideas provide the conceptual foundation for today's circular economy agenda.

4.1 Conceptual Foundations

- **Early thought leaders and systems thinking**: In 1966, Kenneth Boulding introduced the idea of a '*closed economy*'—'*Spaceship Earth*'—where resources are finite and must circulate, laying a conceptual groundwork that's central to CE [39]. Environmental economists Pearce & Turner (1989/1990) formalized the transition from a linear to a '*circular economic system*' in their *Economics of Natural Resources and the Environment* [40].
- **Industrial ecology and closing resource loops**: In 1976, Walter Stahel and Geneviève Reday introduced the concept of extending product life and replacing energy use with labor [41]. Stahel's concept of the '*Product-Life Factor*' (1982) won a Mitchell Prize and helped coin the notion of a circular economy [42]. Concepts such as Industrial Ecology, Cradle-to-Cradle, Performance Economy, Blue Economy, and Biomimicry have further shaped the theoretical foundation by integrating environmental, economic, and design principles [43].
- **Definitional clarity and frameworks**: Studies such as '*Conceptualizing the Circular Economy*' highlighted the need to converge diverse definitions and anchor CE in competitiveness and profitability [44]. The three core principles, as outlined by the Ellen MacArthur Foundation, are: Eliminate waste and pollution; Keep products and materials at their highest value; Regenerate natural systems [45].

4.2 Key Milestones and Global Adoption

Table 1 outlines key historical milestones and global developments that have shaped the adoption of circular economy principles. It highlights significant events, policy initiatives, and influential contributions that collectively advanced the concept

Table 1 Milestones in the development and global adoption of Circular Economy Principles

Year	Milestone	Impact
1976–1982	Stahel's report/book and Product-Life Factor [46]	Introduced product life extension, recycling, and reuse, catalysing CE discourse
1989–1990	Pearce and Turner's economic framing [47]	Positioned CE within mainstream environmental economics
2002–2004	China embeds CE in policy post-16th CPC Congress [48]	CE enters national strategies under '*reduce-reuse-recycle*' framework
2010s	Rise of '*3R*', '*10R*' hierarchies and preventive environmental management [49]	Progressive models and emphasis on eco-design and prevention
2013	Ellen MacArthur + McKinsey '*Towards the Circular Economy*' report [50]	Landmark study estimating ~$630 billion EU savings and catalysed private-sector interest
2014–2020	EU Circular Economy Action Plan(s) and legislation [51, 52]	Structured policy with a 54-measure Action Plan (2015–2018), followed by a robust 2020 CE Action Plan
2017	Standards: BSI's BS 8001 and ISO TC 323 launched [53]	Created frameworks aiding organizational implementation of CE
2018	Launch of PACE (WEF & partners)	Elevated CE to global agenda; coalition of CEOs &governments.
2020s	COP, global treaties [54]	Plastic pollution treaties, GHG goals tied to circularity (e.g., COP29 Azerbaijan), and binding commitments [55]
Today	Country-level strategies	Netherlands aims for 50% material reuse by 2030 and full circularity by 2050; China enforces CE via laws and 11th Five-Year Plan; EU continually updates directives [56]

from early theoretical foundations to widespread implementation across nations, industries, and international sustainability frameworks.

5 Circular Economy in Practice

5.1 Redesigning Products and Systems

A key principle of the CE is designing out waste and inefficiency from the very beginning of a product's life. Unlike linear models that prioritize throughput and short-term utility, CE emphasizes redesigning products to extend their lifespan, improve repairability, and ensure material recoverability at end-of-life. This shift often involves eco-design principles, which incorporate modularity and disassembly into product architecture [57]. For example, the Fairphone is a modular smartphone developed in the Netherlands that allows users to easily replace individual components, thereby extending the device's usable lifespan [58]. Furthermore, the Cradle to Cradle design philosophy, pioneered by William McDonough and Michael Braungart, encourages the use of non-toxic, biodegradable materials that can safely return to nature or industry [59]. The integration of digital technologies, such as the Internet of Things (IoT), artificial intelligence, and digital twins, further enables circular product innovation by optimizing production processes and predicting maintenance needs [60]. In parallel, biomimicry and regenerative design approaches draw inspiration from natural systems to develop closed-loop, zero-waste models, reflecting a more holistic view of sustainability that integrates ecology, engineering, and economics.

5.2 *Circular Business Models*

The adoption of circular economy principles requires transformative changes not only in design but also in how businesses create and deliver value [61]. Traditional business models based on ownership and volume-driven sales are increasingly being replaced by circular business models that focus on longevity, reuse, and resource efficiency [62]. One of the most prominent approaches is the Product-as-a-Service (PaaS) model, where companies retain ownership of a product and provide it as a service [63]. This model incentivizes manufacturers to design longer-lasting products since they bear the responsibility for maintenance and end-of-life management. Notable examples include Rolls-Royce's '*Power by the Hour*' model for aircraft engines [64] and Philips' lighting-as-a-service offerings for buildings and cities [65]. Another evolving model is the sharing economy, where digital platforms facilitate the collective use of underutilized assets such as in the cases of Airbnb and Uberor B2B sharing of equipment in construction and agriculture [66].

Additionally, reverse logistics and remanufacturing have become integral to circular operations. Companies like Caterpillar and Renault have developed systems for collecting used equipment or auto parts, refurbishing them to as-new condition, and reintegrating them into the market [67]. Lastly, industrial symbiosis, where waste or by-products of one firm become input resources for another, has gained traction. Kalundborg in Denmark serves as a global benchmark for such cooperative networks, demonstrating reduced emissions and enhanced economic resilience through inter-firm material and energy exchanges [68].

5.3 Policy and Governance

Policies and governance structures are typically crucial for facilitating the implementation of circular economy practices across diverse industries and countries [69]. There is a growing recognition among governments of the need for strong policy measures that not only stimulate circularity but also eliminate systemic barriers [70]. Similarly, the inclusion of disaster-resilient systems in wastewater governance is of equal importance; emergency preparedness and adaptive wastewater systems can facilitate the continuity and sustainability of wastewater during climate events [71]. Various countries have initiated national circular economy roadmaps to enable this transition. For example, the Netherlands aims to halve the use of virgin raw materials by 2030 and achieve full circularity by 2050, through both targeted investments and regulatory policies [72].

Meanwhile, in China, where the circular economy was first institutionalized through the Circular Economy Promotion Law in 2008, the commitments to a circular economy are epitomized through the inclusion of circular economy measures in national Five-Year Plans [73]. At the regional level, the European Union has been a leader in launching its Circular Economy Action Plan in 2015, followed by an updated action plan as part of the European Green Deal in 2020 [51, 52]. These plans include the right to repair, the prohibition of single-use plastics, and extended producer responsibility (EPR), which holds producers accountable for waste management costs, as examples of governance action plans [74]. Additionally, standardization initiatives, such as BS 8001 in the UK, as well as the ISO TC 323 committee, are increasing the transparency and fidelity of circular economy implementation at an international level [53, 75]. Globally, the United Nations Environment Programme (UNEP) and multi-stakeholder platforms, such as the Platform for Accelerating the Circular Economy (PACE), are working to foster international cooperation, build aligned climate goals with circular strategies, and develop governance frameworks regarding plastics, e-waste and resource governance [76, 77]. Robust governance systems facilitate innovation, equity, and accountability in transitioning to a more circular and sustainable economy.

In addition to these well-established initiatives, several emerging global and regional alliances are contributing to the diffusion of circular economy principles. The African Circular Economy Alliance (ACEA), a government-led coalition, supports the development of CE policy and investment in infrastructure across the continent, with a focus on inclusive growth, local job creation, and resilience [78]. Similarly, the Latin America and the Caribbean Circular Economy Coalition, established under the auspices of the UN Economic Commission for Latin America and the Caribbean (ECLAC), fosters policy harmonization and capacity-building among member states [79]. In Asia, India's Resource Efficiency and Circular Economy Framework, led by the Ministry of Environment, Forests, and Climate Change, emphasizes sector-specific roadmaps, especially in construction, electronics, and mobility [80].

At the international level, the Organisation for Economic Cooperation and Development (OECD) has launched a comprehensive programme to assist countries in measuring circular performance, designing policy packages, and promoting innovation [81]. The OECD's analytical reports and peer reviews provide critical insights for cross-border learning and the harmonization of CE practices [82]. Additionally, the in-progress Global Plastics Treaty, being negotiated under the leadership of UNEP, represents a landmark attempt to establish a binding international agreement targeting the full life cycle of plastics, from production to disposal, through circular solutions [83].

These initiatives, in conjunction with national strategies and multilateral dialogues, demonstrate the growing globalisation of circular economy governance and the diversification of approaches to suit diverse development contexts. Together, they reinforce the importance of multilevel coordination, knowledge exchange, and global solidarity in driving a prosperous and equitable circular transition.

6 Case Studies of Circular Economy in Practice

As the circular economy (CE) concept evolves from theory to implementation, cities, corporations, and innovation ecosystems are beginning to translate circular principles into operational models. These case studies provide practical evidence of how CE can be applied across diverse sectors and geographies, showcasing both systemic transformations and targeted interventions. This section categorizes such efforts into three domains: municipal programs, corporate circular models, and enabling technologies and innovations [5, 23].

6.1 Municipal Circular Economy Programs

Urban centers are playing an increasingly vital role in advancing circular economy objectives due to their high population densities, concentrated resource flows, and governance capabilities. Several cities around the world have emerged as pioneers in embedding circularity into municipal policies and infrastructure.

6.1.1 Amsterdam: A Circular City Strategy

Amsterdam is widely recognized for institutionalizing circularity within its municipal development agenda. The Amsterdam Circular Strategy 2020–2025 aims to halve the use of new raw materials by 2030 and achieve full circularity by 2050 [84]. The strategy builds upon earlier pilot projects and employs the *Doughnut Economics* framework developed by Raworth [85], positioning social foundations and ecological ceilings as guiding parameters for urban development.

The city has implemented circular procurement protocols, promoted reuse and modular construction practices in the built environment, and created innovation districts such as Buiksloterham and Marineterrein, which serve as test beds for urban circularity [86]. The integration of circularity into spatial planning and housing design exemplifies a multi-sectoral approach.

6.1.2 San Francisco: Toward Zero Waste

A strong policy orientation toward waste minimization has driven San Francisco's circular ambitions. In 2018, San Francisco updated its zero-waste commitments to reduce solid waste generation by 15% and landfill or incineration disposal by 50% by 2030. The city supports local circular entrepreneurship, including repair services, reuse stores, and community-scale resource exchange initiatives [87].

Public-private partnerships have played a critical role, with Recology and other waste service providers offering closed-loop solutions in textile and e-waste recovery [88]. These efforts underscore the importance of behavioral change, regulatory enforcement, and civic engagement in achieving municipal circular goals.

6.2 Corporate Circular Economy Models

Private sector engagement in circular economy practices has accelerated as companies recognize the strategic advantages of resource efficiency, brand differentiation, and regulatory preparedness. The following case examples illustrate diverse approaches to circularity across industries.

6.2.1 IKEA: Furniture and Material Circularity

IKEA has committed to becoming a fully circular and climate-positive business by 2030, as outlined in its *People and Planet Positive Strategy* [89]. The company's circularity agenda focuses on product design for modularity and disassembly, furniture leasing and take-back programs, and the integration of recycled and renewable materials into product lines [90].

Examples include the KUNGSBACKA kitchen front, manufactured from recycled PET bottles and industrial wood waste, as well as refurbishment services piloted across European markets [91]. These initiatives reflect an operationalization of circular design principles and supply chain integration.

6.2.2 Philips: Circular Health Technology

Philips has adopted circularity in its healthcare division by implementing *product-as-a-service* models and designing for equipment refurbishment. The company's **circular propositions** include leasing MRI and CT scanners, which are returned, remanufactured, and redeployed, thereby extending the lifecycle and reducing material inputs [92].

Design for disassembly and remote monitoring technologies support circular operations, enabling predictive maintenance and optimizing product use. Philips exemplifies how high-value, regulated sectors can integrate circularity without compromising on safety and innovation [93].

6.2.3 Patagonia: Responsible Apparel and Consumer Engagement

Patagonia's circular model focuses on **durability, repairability, and reuse**, aligning with the company's environmental mission. Its *Worn Wear* program facilitates the repair, resale, and recycling of used garments, while also educating consumers on mindful consumption [94]. The company uses recycled polyester, organic cotton, and regenerative wool to minimize lifecycle impacts.

Patagonia's model emphasizes the socio-cultural dimension of circularity, leveraging customer engagement and transparency to drive behavioural change and brand loyalty.

6.3 Innovation and Technology Enablers

Technological and material innovations increasingly support the implementation and scaling of circular economy practices. Digital technologies and advanced materials are crucial in enabling circular flows, enhancing product traceability, and supporting the development of new business models.

6.3.1 Digital Technologies

Technological enablers, such as **artificial intelligence (AI), the Internet of Things (IoT)**, and **blockchain**, are facilitating a data-driven circular economy [95]. AI is employed in predictive maintenance and intelligent sorting systems for recycling [96]. More recently, explainable AI models, such as stacked ensemble regression integrated with SHAP analysis, have been used to predict water quality indices in Indian river systems, demonstrating scalable applications of machine learning in environmental circularity [97–100]. Advanced spatial intelligence systems integrated with

AI are also enabling intelligent sewer networks in urban areas, contributing to optimized flow management, early warning systems, and sustainable wastewater infrastructure [101].AI tools have also been successfully deployed for air quality prediction and environmental impact assessment in industrial cities such as Kryvyi Rih, Ukraine, demonstrating the broader applicability of intelligent systems in circular urban transitions [102]. Spatial intelligence tools have also been incorporated into urban sewer networks to enhance wastewater management and sustainability through smart system integration [97–100]. IoT devices enhance the real-time tracking of products and materials, thereby enabling *product-as-a-service* offerings and optimized reverse logistics [103]. Geospatial analysis is also emerging as a vital tool for optimizing wastewater networks, providing location-based insights for efficient planning and achieving sustainable outcomes [97–100]. Blockchain contributes to transparency and trust in supply chains, particularly in the fashion and electronics sectors, by documenting material provenance, compliance, and lifecycle transactions.

6.3.2 Material Innovation and Circular Design

Advancements in material science have expanded the potential for circular product development. Innovations include biodegradable bioplastics, cradle-to-cradle certified materials, and mycelium-based packaging solutions [104]. Product designs that incorporate modularity, disassembly, and recyclability are increasingly supported by *digital twins* and *lifecycle simulation tools* [105].

Start-ups and R&D centers are developing alternative materials, such as **ECOR** fibreboard and **Mushroom Packaging**, which offer both functional performance and end-of-life biodegradability [106, 107]. These innovations are critical to closing material loops and aligning with CE principles from the design phase.

7 Challenges and Best Practices

7.1 Challenges from a Global Perspective

The implementation of the circular economy varies significantly across continents due to diverse socio-economic, political, and environmental contexts. Table 2 outlines the key challenges each region faces in adopting and sustaining circular practices:

7.2 Best Practices from Global Perspective

Across continents, diverse and innovative circular economy practices have emerged, reflecting localized solutions to global sustainability challenges. The following

Table 2 Challenges—continent wise

Continent	Challenges	References
Africa	Infrastructure Gaps, Informal Sector, Policy Enforcement, Financial Constraints	[108–111]
Asia	Population pressure, policy fragmentation, industrial growth, and economic diversity often complicate effective waste management. Additionally, the lack of integrated analytical tools for decision-making under uncertainty remains a major barrier, particularly in rapidly urbanizing cities. The unchecked rise of informal electronic waste processing further aggravates environmental risks, leading to hazardous exposure and serious biodiversity loss across terrestrial and aquatic ecosystems	[97–100, 112–119]
Europe	Patchy Frameworks, Cross-Border Waste, Consumer Behaviour, Legacy Infrastructure	[120–124]
North America	Regulatory Diversity, Consumer Culture, Corporate Influence, Awareness Gaps	[125–128]
South America	Informal Economy, Policy Gaps, Socioeconomic Disparities, Urban-Rural Divide	[129–133]
Oceania	Transportation Cost, Small Markets, Resource Dependence, Awareness vs. Cost	[134–137]

best practices highlight region-specific strategies that offer valuable insights and inspiration for nurturing circularity worldwide (Table 3).

8 The Path forward

To accelerate the transition from waste management to a genuinely circular economy over the next decade, research must move beyond simple publication counts to incorporate social-justice metrics, behavioral economics, Indigenous knowledge, and stronger **collaboration with the Global South**, all underpinned by dynamic, AI-enabled resource-flow models that share open, standardized data. Policy makers should weave circularity into climate and trade frameworks, expand right-to-repair and product-passport legislation across regions, and mainstream the protective integration of informal-sector actors, such as waste-picker cooperatives. Businesses and technologists need to scale IoT- and blockchain-based traceability, design for modularity and disassembly in line with emerging ISO 59000 and BS 8001 standards, and enlarge eco-industrial parks that exchange energy, water, and materials across sectors. Financial instruments must internalise externalities through landfill levies and carbon-adjusted material taxes, channel green-bond capital toward circular assets, nurture SME and start-up ecosystems, and deploy public-sector procurement as a mass-market signal. Society at large should embed circular-economy principles from primary school to MBA curricula, normalise a repair culture through sharing

Table 3 Best practices—continent wise

Region	Circular Economy—best practices	Examples
Africa	1. Informal Sector Integration	*Kenya*: Plastic recycling uses informal waste collectors, ensuring fair wages [138] *Nigeria*: Social enterprises collaborate with informal e-waste collectors for safe collection and refurbishing [139]
	2. Public-Private Partnerships	*Rwanda*: Ban on single-use plastics through government-private partnerships; support for biodegradable alternatives [140]
	3. Community-Based Waste Management	*Africa*: Grassroots NGOs train communities to compost organic waste for urban agriculture. Composting hubs act as local resource models [141]. In Eastern Europe, particularly in Ukraine's Poltava Region, the implementation of community bin systems has shown effectiveness in rural waste segregation and public engagement [142]
Asia	1. Extended Producer Responsibility (EPR)	*Japan, India and South Korea*: Laws requiring manufacturers to design recyclable products and manage take-back programs [143–145]. In addition to regulatory strategies, countries like India are also adopting green engineering approaches and systemic waste valorization strategies that support circular economy transitions [146]
	2. Circular Agriculture and Aquaculture	*China*: Circular farming integrates livestock and crop systems, reducing waste and chemical inputs [147]
	3. Industrial Symbiosis	*Singapore*: Eco-industrial parks share resources and reuse byproducts, reducing costs and waste [148]
Europe	1. Comprehensive Policy Frameworks	*EU*: Circular Economy Action Plan sets waste reduction targets and product design regulations [149]
	2. Eco-Design and Product Life Extension	*Scandinavia*: Tax incentives support repair services over replacement for electronics and furniture [150]
	3. Circular Procurement	*Netherlands*: Public contracts favor suppliers with circular products; use of circularity scores [151]
North America	1. City-Level Zero-Waste Programs	*San Francisco (USA)*: Zero-waste mandates, high diversion rates from landfills [152]
	2. Corporate-Led Innovation	*USA and Canada*: Companies invest in closed-loop manufacturing and take-back programs [153]

(continued)

Table 3 (continued)

Region	Circular Economy—best practices	Examples
	3. Collaboration with Indigenous Communities	*Canada*: Integration of traditional stewardship in land/resource management for circular outcomes [154]
South America	1. Circular Startups and Social Enterprises	*Brazil*: Upcycling plastic waste into durable goods, creating green jobs [155] *Chile*: '*Circular Chile*' supports startups in circular solutions [156]
	2. Municipal Recycling and Composting	*Colombia and Peru*: Cities work with waste picker cooperatives to expand recycling [157, 158]
	3. Collaboration on Regional Policies	*Regional alliances*: Framework harmonization and knowledge-sharing across Latin America [159]
Oceania	1. Extended Producer Responsibility and Deposit Schemes	*Australia*: Container deposit systems and EPR for electronics and packaging [160] *New Zealand*: Product stewardship and buy-back initiatives [161]
	2. Sustainable Procurement and Construction	Local agencies prioritize circular materials in public infrastructure projects [162]
	3. Circular Community Initiatives	Repair cafés, swap meets, and sharing platforms reduce neighbourhood-level waste [163]
Antarctica (Special Mention)	Waste Minimization Protocols	*Environmental Protocol of the Antarctic Treaty*: Requires treatment/removal of waste, enforcing strict environmental standards [164].

and upcycling hubs, and foreground gender-responsive health and livelihood safeguards for the workers who keep **secondary material loops flowing**. Finally, governments, firms, and civil society must converge **today's competing** '*3R*', '*10R*', and ESG scorecards into a harmonized dashboard tracking material circularity, carbon intensity, and social value added, while subjecting strategies to adaptive five-year reviews that align with the SDG reporting cycle. Phasing these actions through a foundation period (2025–2027), an acceleration phase (2028–2030), and a mainstreaming horizon (2031–2035) offers a realistic yet ambitious trajectory toward a 2050 economy in which materials remain in perpetual, value-generating use and the concept of waste is effectively designed out.

9 Conclusion

Bridging centuries of practice with cutting-edge scholarship, this chapter demonstrates that the circular economy is no longer an abstract ambition but an emerging organising principle for science, policy, and enterprise. A chapter bibliometric scan confirms an exponential rise in cross-disciplinary output since 2015, yet also reveals persistent geographic and thematic gaps that must **be addressed if** benefits are to be shared equitably. Historical analysis reveals how each technological leap—from Roman sewers to AI-enabled digital twins—reshapes both the scale of waste and the logic of resource stewardship. Case studies reveal that cities, corporations, and start-ups are already **translating theory** into profitable, **low-carbon practices**, while global alliances **are beginning to** align standards, finance, and trade. Nevertheless, entrenched linear subsidies, fragmented governance, and social inequities still slow momentum. By adopting the phased roadmap outlined in this chapter, stakeholders can mainstream circularity by 2035 and, by mid-century, **consign the concept** of waste to history for all communities worldwide.

References

1. Valavanidis A (2023) Global municipal solid waste (MSW) in crisis: two billion tonnes of MSW every year, a worrying worldwide environmental problem. Sci Rev, 17 January 2023
2. Rashid S, Malik SH (2023) Transition from a linear to a circular economy. In: Bandh SA, Malla FA, Hoang AT (eds) Renewable energy in circular economy. Springer, pp 1–20. https://doi.org/10.1007/978-3-031-42220-1_1
3. Rome Business School (2023) Why is the linear economic model no longer sustainable? https://romebusinessschool.com/blog/linear-economy/
4. Ekins P, Domenech Aparisi T, Drummond P, Bleischwitz R, Hughes N, Lotti L (2020) The circular economy: what, why, how and where. The OECD Centre for Entrepreneurship. https://discovery.ucl.ac.uk/id/eprint/10093965/
5. Rajpal A, Choudhury M, Goswami S, Chakravorty A, Raghavan V (eds) (2024) Waste management and treatment: advances and innovations, 1st edn. CRC Press. https://doi.org/10.1201/9781003258377
6. Mostaghimi K, Behnamian J (2023) Waste minimization towards waste management and cleaner production strategies: a literature review. Environ Dev Sustain 25:12119–12166. https://doi.org/10.1007/s10668-022-02599-7
7. Mohanty CRC (2011) Synergizing resource efficiency with informal sector towards sustainable waste management. Conference presentation, CSD-19 Learning Centre, New York, NY, United States. https://sustainabledevelopment.un.org/content/dsd/csd/csd_pdfs/csd-19/learningcentre/presentations/May%209%20am/1%20-%20Learning_Centre_9May_ppt_Mohanty.pdf
8. TOMRA Systems ASA (2024) Austria, Wales and Taiwan leading the world when it comes to rates of recycling [News release]. TOMRA Systems ASA. https://www.tomra.com/news-and-media/news/2024/austria-wales-and-taiwan-leading-the-world-when-it-comes-to-rates-of-recycling
9. TOMRA, & Eunomia (2024) Global recycling league table: phase 1 report. https://www.tomra.com/-/media/project/tomra/tomra/about-tomra/documentation/global-recycling-league-table-phase-1-report.pdf

10. Borjigin GON (2025) North Asian Mongolian nomads' traditional ecological wisdom: modern relevance in land management & enhancement. Working paper. https://doi.org/10.2139/ssrn.5258633
11. Musariri L, Moyer E (2022) Hunting treasure, gathering trash: politics and precarity in the plastic recycling industry. Etnofoor 34(2):49–66. https://www.jstor.org/stable/27191495
12. De Feo G, Antoniou G, Fardin HF, El-Gohary F, Zheng XY, Reklaityte I et al (2014) The historical development of sewers worldwide. Sustainability 6(6):3936–3974. https://doi.org/10.3390/su6063936
13. Jones DE (1967) Urban hydrology—a redirection. Civil Eng 37:58–62
14. Lofrano G, Brown J (2010) Wastewater management through the ages: a history of mankind. Sci Total Environ 408(22):5254–5264. https://doi.org/10.1016/j.scitotenv.2010.07.062
15. Khan S, Dialynas E, Kasaraneni VK, Angelakis AN (2020) Similarities of Minoan and Indus Valley hydro-technologies. Sustainability 12(12):4897. https://doi.org/10.3390/su12124897
16. Commercial Zone Products (2020) A brief history of waste management [Blog post]. https://www.commercialzone.com/a-brief-history-of-waste-management/. Accessed 26 Jan 2025
17. Craig HC (2019) Rags, ragpickers, and early modern papermaking. Lit Compass 16(5):e12523. https://doi.org/10.1111/lic3.12523
18. Merab M (2024) This is the worst job in history: the toshers of Victorian London [Vocal Media article]. Vocal.media. https://vocal.media/history/this-is-the-worst-job-in-history-the-toshers-of-victorian-london
19. Waste Advantage Magazine (2017) Things you didn't know about garbage men [Blog post]. https://wasteadvantagemag.com/things-you-didnt-know-about-garbage-men/
20. Barles S (2014) History of waste management and the social and cultural representations of waste. In: Agnoletti M, Neri Serneri S (eds) The basic environmental history, vol 4. Springer, pp 199–226. https://doi.org/10.1007/978-3-319-09180-8_7
21. Kamita RY (2002) Talking trash: the impact of geographic differentiation and municipal ownership on competition in solid waste disposal. Doctoral dissertation, University of California, Berkeley
22. Geissdoerfer M, Savaget P, Bocken NMP, Hultink EJ (2017) The circular economy—a new sustainability paradigm? J Clean Product 143:757–768. https://doi.org/10.1016/j.jclepro.2016.12.048
23. Singh S, Singh P, Sharma A, Choudhury M (eds) (2022) Agriculture waste management and bioresource: the circular economy perspective. Wiley. https://doi.org/10.1002/9781119808428
24. Meyer A, Schneider P (2019) Cradle-to-Cradle for sustainable development: from ecodesign to circular economy. In: Leal Filho W (ed) Encyclopedia of sustainability in higher education. Springer, Cham, pp 321–330. https://doi.org/10.1007/978-3-030-11352-0_273
25. Bjørn A, Hauschild MZ (2018) Cradle to cradle and LCA. In: Hauschild MZ, Rosenbaum RK, Olsen SI (eds) Life cycle assessment: theory and practice. Springer, pp 605–631. https://doi.org/10.1007/978-3-319-56475-3_25
26. Report Yak (2024) What is Cradle-to-Cradle design? Report Yak. https://reportyak.com/blog/cradle-to-cradle-design/. Accessed 5 Jul 2025
27. Stahel WR (2010) The performance economy, 2nd edn. Springer
28. Stahel WR, Clift R (2016) Stocks and flows in the performance economy. In: Clift R, Druckman A (eds) Taking stock of industrial ecology. Springer, pp 137–158. https://doi.org/10.1007/978-3-319-20571-7_7
29. Genc O (2024) Harmony in industry and nature: exploring the intersection of industrial symbiosis and food webs. Circ Econ Sustain 5:79–100. https://doi.org/10.1007/s43615-024-00421-2
30. Norberg B (2024) A case study of maritime clusters in Western Norway: exploring the role of industrial symbiosis in advancing the green transition. Master's thesis, Norwegian School of Economics. NHH Brage. https://hdl.handle.net/11250/3158933
31. Ulusoy K, Doğan-sağlamtimur N, Çiner F, Sternik A, Sekula P (2025) Industrial symbiosis practices: a case study of Türkiye and Denmark. Environ Res Technol 8(1):224–233. https://doi.org/10.35208/ert.1271679

32. Shao Q (2025) Sustainability evaluation of the Doughnut Economics: a bibliometric analysis. Earths Fut 13(4):e2024EF004638. https://doi.org/10.1029/2024EF004638
33. Hasanov RI, Azizli NN (2025) Doughnut economics: exploring a new alternative sustainable model. Econ Altern 1(1):149–165. https://doi.org/10.37075/EA.2025.1.08
34. Wahlund M, Hansen T (2022) Exploring alternative economic pathways: a comparison of foundational economy and Doughnut economics. Sustain Sci Pract Pol 18(1):171–186. https://doi.org/10.1080/15487733.2022.2030280
35. Van Espen S (2024) Doughnut Economics within urban climate action: a comparative study between Amsterdam, Copenhagen and Stockholm. Master's thesis, UiT—The Arctic University of Norway. https://hdl.handle.net/10037/33857
36. Grünsch J (2022) From recycling to recoupling: towards a philosophy of the circular economy in the EU. Master's essay, University of Twente. University of Twente Theses Repository. https://purl.utwente.nl/essays/89516
37. Costanza R, Cumberland JH, Daly HE, Goodland R, Norgaard RB (1997) An introduction to ecological economics. St. Lucie Press. https://library.uniteddiversity.coop/Measuring_Progress_and_Eco_Footprinting/An_Introduction_to_Ecological_Economics.pdf
38. Zhu J, Gao Y, Chen H, Li Y (2025) A comprehensive review of circular economy models in solid waste management: classification, challenges, and future perspectives. Sustain Product Consump 41:653–669. https://doi.org/10.1016/j.spc.2025.04.020
39. Blomsma F, Brennan G (2017) The emergence of circular economy: a new framing around prolonging resource productivity. J Ind Ecol 21(3):603–614. https://doi.org/10.1111/jiec.12603
40. Wautelet T (2018) The concept of circular economy: Its origins and its evolution [Working paper]. Positive ImpaKT. https://doi.org/10.13140/RG.2.2.17021.87523
41. Upadhayay S, Alqassimi O, Khashadourian E, Sherm A, Prajapati D (2024) Development in the circular economy concept: systematic review in context of an umbrella framework. Sustainability 16(4):1500. https://doi.org/10.3390/su16041500
42. Rasmussen TS, Esclusa R, Petrova EA, Bohnstedt KD (2021) From linear to circular: Circular Economy in the Danish construction industry [Conference paper]. In: ECPPM 2021—eWork and eBusiness in architecture, engineering and construction. Aalborg University, pp 423–430. https://projekter.aau.dk/projekter/files/319999750/From_Linear_to_Circular.pdf
43. Saidani M, Yannou B, Leroy Y, Cluzel F, Kim H (2020) How circular economy and industrial ecology concepts are intertwined? A bibliometric and text mining analysis [Preprint]. arXiv. https://doi.org/10.48550/arXiv.2007.00927
44. Kirchherr J, Reike D, Hekkert M (2017) Conceptualizing the circular economy: an analysis of 114 definitions. Resour Conserv Recycl 127:221–232. https://doi.org/10.1016/j.resconrec.2017.09.005
45. Ellen MacArthur Foundation (2015) Towards a circular economy: business rationale for an accelerated transition. https://www.ellenmacarthurfoundation.org/publications/towards-a-circular-economy-business-rationale-for-an-accelerated-transition
46. Circular (2021) Circular motion [Interview with Dr Walter Stahel]. Circular Online. https://www.circularonline.co.uk/interviews/circular-motion/
47. Pearce DW, Turner RK (1991) Economics of natural resources and the environment. Am J Agric Econ 73(1):227–228. https://doi.org/10.2307/1242904
48. Chen R (2023) China's circular economy policies: review and reflection. Circular Economy Stakeholder Platform. https://circulareconomy.europa.eu/platform/sites/default/files/2023-08/China%E2%80%99s%20Circular%20Economy%20Policies_%20Review%20and%20Reflection.pdf
49. Reike D, Vermeulen WJV, Witjes S (2022) Conceptualization of Circular Economy 3.0: synthesizing the 10R hierarchy of value retention options. In: Alvarez-Risco A, Rosen MA, Del-Aguila-Arcentales S (eds) Towards a circular economy: transdisciplinary approach for business, CSR, sustainability, ethics & governance. Springer Nature, pp 47–69. https://doi.org/10.1007/978-3-030-94293-9_3

50. World Economic Forum, Ellen MacArthur Foundation, McKinsey & Company (2014) Towards the circular economy: accelerating the scale-up across global supply chains. World Economic Forum. https://www3.weforum.org/docs/WEF_ENV_TowardsCircularEconomy_Report_2014.pdf
51. European Commission (2020a) Circular economy action plan. European Commission—Environment. https://environment.ec.europa.eu/strategy/circular-economy-action-plan_en
52. European Commission (2020b) Circular economy action plan: for a cleaner and more competitive Europe (EC Publication No. 9789276190707). https://rcbc.ca/wp-content/uploads/2022/08/EU-European-Union-Circular-Economy-Action-Plan-2020.pdf
53. Arana Landín G, Sigüenza W, Landeta Manzano B, Laskurain Iturbe I (2023) Circular economy: on the road to ISO 59000 family of standards. Corporate Soc Respons Environ Manag 31(3):1977–2009. https://doi.org/10.1002/csr.2665
54. Kar J, Brandes MS, Helstroffer A (2024) Extended producer responsibility and a global plastics treaty—what do the experts say? World Economic Forum. https://www.weforum.org/stories/2024/11/extended-producer-responsibility-global-plastics-treaty-experts/
55. Hornstein DT (2025) Plastics, carbon, politics, and experimentation in environmental governance. William Mary Environ Law Policy Rev 49(3):555–600. https://scholarship.law.wm.edu/wmelpr/vol49/iss3/4
56. Gorokhova T, Kozlova A, Laguta Y, Skrashchuk L, Kaminskyi S (2024) The role of the circular economy in advancing environmentally sustainable industrial production. Grassroots J Nat Resour 7(3):s191–s211. https://doi.org/10.33002/nr2581.6853.0703ukr10
57. Boix Rodríguez N, Favi C (2023) Disassembly and repairability of mechatronic products: insight for engineering design. J Mech Des 146(2):020906. https://doi.org/10.1115/1.4064075
58. Thomas Gomez R (2024) From disposability to conviviality: investigating planned obsolescence and conviviality in smartphones. Master's thesis, Aalborg University. https://essay.utwente.nl/102884/
59. Mohajan HK (2021) Cradle to cradle is a sustainable economic policy for the better future [Working paper]. Ann Spiru Haret Univ Econ Ser 21(4):559–572. https://doi.org/10.26458/21433
60. Rejeb A, Zailani S, Rejeb K, Seuring S, Treiblmaier H (2022) The Internet of Things and the circular economy: a systematic literature review and research agenda. J Clean Product 350:131439. https://doi.org/10.1016/j.jclepro.2022.131439
61. Asgari A, Asgari R (2021) How circular economy transforms business models in a transition towards circular ecosystem: the barriers and incentives. Sustain Product Consump 28:566–579. https://doi.org/10.1016/j.spc.2021.06.020
62. Timergaleeva D (2025) Pathways of business model innovation: a process perspective on theory, industry evolution, and firm-level transformation. Doctoral dissertation, University of Milano Bicocca. https://hdl.handle.net/10281/543861
63. Hidalgo Crespo J, Riel A, Vogt Duberg J, Bunodiere A, Golińska Dawson P (2024) An exploratory study for product-as-a-service (PaaS) offers development for electrical and electronic equipment. Proc CIRP 122:521–526. https://doi.org/10.1016/j.procir.2024.01.076
64. Rolls Royce plc (2012) Rolls Royce celebrates 50th anniversary of Power by the Hour [Press release]. https://www.rolls-royce.com/media/pressreleasesarchive/yr2012/121030thehour.aspx
65. Smart Cities World (2016) Philips' light-as-a-service offering [News article]. https://www.smartcitiesworld.net/news/philips-light-as-a-service-offering-1137
66. Ganapati S, Reddick CG (2018) Prospects and challenges of sharing economy for the public sector. Gov Inf Q 35(1):77–87. https://doi.org/10.1016/j.giq.2018.01.001
67. World Economic Forum (2014) Towards the circular economy: accelerating the scale-up across global supply chains. https://www3.weforum.org/docs/WEF_ENV_TowardsCircularEconomy_Report_2014.pdf
68. Hunt OB (2025) Repurposing the grid: territorialized assets and reconfiguring ties in gas production networks in Denmark. Energy Res Soc Sci 120:103925. https://doi.org/10.1016/j.erss.2025.103925

69. Kandpal V, Jaswal A, Santibanez Gonzalez EDR, Agarwal N (2024) Policy framework for sustainable energy transition and the circular economy. In: Sustainable energy transition: circular economy and sustainability. Springer, Cham, pp 217–238. https://doi.org/10.1007/978-3-031-52943-6_7
70. Tan J, Tan FJ, Ramakrishna S (2022) Transitioning to a circular economy: a systematic review of its drivers and barriers. Sustainability 14(3):1757. https://doi.org/10.3390/su14031757
71. Tripathi S, Agarwal S, Choudhary R, Choudhury M, Padoin N, van Hullebusch ED et al (2025) Strengthening disaster resilience: emergency response strategies in wastewater management. In: Choudhury M, Majumdar S, Goswami S, Sillanpää M (eds) Smart wastewater systems and climate change: innovations through spatial intelligence, Chapter 10, vol 15. Royal Society of Chemistry, pp 135–154. https://doi.org/10.1039/9781837676880-00135
72. Abbas Q, Imran M, Sattar A (2025) From linear to circular: the impact of economic policies and technological innovations on greenhouse gas emissions in the Netherlands. Carbon Balance Manag 20:Article 11. https://doi.org/10.1186/s13021-025-00297-1
73. Luo A (2022) Cooperating on new sustainable visions with old global politics? Doctoral dissertation, University of Freiburg, Germany. https://freidok.uni-freiburg.de/files/233589/
74. Choudhury M, Roy P (2025) Challenges with microplastic pollution in the regime of UN sustainable development goals. World Dev Sustain 6:100216. https://doi.org/10.1016/j.wds.2025.100216
75. Pauliuk S (2018) Critical appraisal of the circular economy standard BS 8001:2017 and a dashboard of quantitative system indicators for its implementation in organizations. Resour Conser Recycl 129:81–92. https://doi.org/10.1016/j.resconrec.2017.10.019
76. Fadeeva Z, Van Berkel R (2021) Unlocking circular economy for prevention of marine plastic pollution: an exploration of G20 policy and initiatives. J Environ Manag 277:111457. https://doi.org/10.1016/j.jenvman.2020.111457
77. Pedrazzoli P, Sorlini M, Rovere D, Lazaro O, Malò P, Fiorello M (2022) Challenges and founding pillars for a manufacturing platform to support value networks operating in a circular economy framework. Appl Sci 12(6):2995. https://doi.org/10.3390/app12062995
78. Godfrey L, Roman H, Smout S, Maserumule R, Mpofu A, Ryan G, Mokoena K (2021) Unlocking the opportunities of a circular economy in South Africa. In: Ghosh SK, Ghosh SK (eds) Circular economy: recent trends in global perspective. Springer, pp 145–180. https://doi.org/10.1007/978-981-16-0913-8_5
79. United Nations Environment Programme (2021) Circular Economy Coalition launched for Latin America and the Caribbean [Press release]. UNEP News and Stories. https://www.unep.org/news-and-stories/press-release/circular-economy-coalition-launched-latin-america-and-caribbean
80. Jena S, Tyagi A, Kale A, Gupta S, Bassi N, Chaturvedi K, Kashyap A, Kovacs C, Khan A, Singh P, Das R, Mishra S, Aggarwal A, McNamara M, Ghate A, Garg T, Pallerlamudi SSH, Zakai Z, Kumaraswamy SK, Lad V et al (2024) Unlocking India's circular waste economy potential for sustainability. Discussion Paper. Council on Energy, Environment and Water (CEEW); WRI India; RMI. https://www.ceew.in/sites/default/files/india-circular-economy-potentialwebceew16dec.pdf
81. Johnstone I (2023) Implementing circular economy principles in environmental law: insights from OECD guidance. Yearb Int Environ Law 34(1):yvae026. https://doi.org/10.1093/yiel/yvae026
82. Yamaguchi S (2021) International trade and circular economy—policy alignment (OECD Trade and Environment Working Papers No. 2021/02). OECD Publishing. https://doi.org/10.1787/ae4a2176-en
83. United Nations (2025) Global push to end plastic pollution gains ground in Nice [News article]. UN News. https://news.un.org/en/story/2025/06/1164301
84. City of Amsterdam (2020) Amsterdam Circular 2020–2025 strategy [PDF]. Circle Economy & City of Amsterdam. https://carbonneutralcities.org/wp-content/uploads/2020/06/Amsterdam-Circular-2020-2025_Strategy_HighRes.pdf

85. Turner RA, Wills J (2022) Downscaling doughnut economics for sustainability governance. Curr Opin Environ Sustain 56:101180. https://doi.org/10.1016/j.cosust.2022.101180
86. Croatian Circular Economy Committee & World Bank (2022) Circular economy approaches in solid waste management: study visit by the Croatian Circular Economy Committee to Amsterdam, 3–6 July 2022 (Summary Report). World Bank Group. https://thedocs.worldbank.org/en/doc/c5d60443a89bd228d3f45023aa9223e1-0080012022/original/CE-RAS-NL-Study-Trip-Report.pdf
87. U.S. Environmental Protection Agency (2025) Zero Waste case study: San Francisco, California. U.S. EPA—Managing and Transforming Waste Streams. https://www.epa.gov/transforming-waste-tool/case-study-san-francisco
88. Sarkhoshkalat MM, Afkham A, Bonyadi Manesh M, Sarkhosh M (2024) Circular economy and the recycling of e-waste. In: New technologies for energy transition based on sustainable development goals. Springer, pp 319–354. https://doi.org/10.1007/978-981-97-2527-4_16
89. Inter IKEA Group (2018) IKEA commits to becoming people and planet positive by 2030 [News release]. IKEA Newsroom. https://www.ikea.com/global/en/newsroom/sustainability/ikea-takes-sustainable-living-to-a-new-level-with-new-commitments-to-become-people-and-planet-positive-by-2030-180607/
90. Mahalakshmi S, Nallasivam A, Kumar H, Kautish S, Madan S (2024) From assembly to reassembly: Ikea's circular design for a sustainable future. In: Singh K, Dubey R, Renwick D, Crichton R (eds) Utilizing technology for sustainable resource management solutions. IGI Global Scientific Publishing, pp 261–280. https://doi.org/10.4018/979-8-3693-2346-5.ch017
91. Kock A, Bengtsson M (2018) Catalyzing transition of the Swedish waste regime through niche market initiatives: a case study of a planned recirculation facility in Kungsbacka. Master's thesis, Chalmers University of Technology, Gothenburg, Sweden. https://hdl.handle.net/20.500.12380/254941
92. Philips (2025) Innovation Spotlight: how a circular economy is driving sustainable healthcare [Feature article]. Philips News & Insights. https://www.philips.com/a-w/about/news/archive/features/2025/innovation-spotlight-how-a-circular-economy-is-driving-sustainable-healthcare.html
93. Aimaro M (2025) The circular economy in the European Union: the corporates' responses through innovative strategies. Doctoral dissertation, Università degli Studi di Torino. https://hdl.handle.net/20.500.14242/199358
94. Ram A (2021) Our quest for circularity [Feature article]. Patagonia stories. https://www.patagonia.com/stories/our-quest-for-circularity/story-96496.html. Accessed 5 Jul 2025
95. Turskis Z, Šniokienė V (2024) IoT-driven transformation of circular economy efficiency: an overview. Math Computat Applic 29(4):49. https://doi.org/10.3390/mca29040049
96. Lakhouit A (2025) Revolutionizing urban solid waste management with AI and IoT: a review of smart solutions for waste collection, sorting, and recycling. Results Eng 25:104018. https://doi.org/10.1016/j.rineng.2025.104018
97. Choudhary R, Bharadwaj LK, Tan TN, Choudhury M, Sharma D, Rath P et al (2025) Spatial intelligence integration in smart wastewater systems: advancing efficiency and sustainability in urban sewer networks. In: Choudhury M, Majumdar S, Goswami S, Sillanpää M (eds) Smart wastewater systems and climate change: innovations through spatial intelligence, Chapter 3, vol 15. Royal Society of Chemistry, pp 28–41. https://doi.org/10.1039/9781837676880-00028
98. Choudhary R, Kumar A, Priyadharsini C, Mude M, Naik M, Moharana C, Khan NA (2025) Predicting water quality index using stacked ensemble regression and SHAP based explainable artificial intelligence. Sci Rep 15(1):9463. https://doi.org/10.1038/s41598-025-09463-4
99. Choudhary R, Kumari S, Kumar A, Kumar P, Choudhury M, Sharma D et al (2025) Optimizing wastewater management through geospatial analysis. In: Choudhury M, Majumdar S, Goswami S, Sillanpää M (eds) Smart wastewater systems and climate change: innovations through spatial intelligence, Chapter 6, vol 15. Royal Society of Chemistry, pp 72–87. https://doi.org/10.1039/9781837676880-00072
100. Choudhary R, Kumar A, Narayan S, Mishchenko I, Choudhury M, Bharadwaj LK et al (2025) Integrating fuzzy logic into life cycle assessment for sustainable municipal solid waste

management in metro cities. Ecol Eng Environ Technol 26(7):316–331. https://doi.org/10.12912/27197050/205500
101. Choudhury M, Choudhary R, Sharma D, Bhattacharjee B, Majumdar S, Chegini S et al (2025) Spatial intelligence integration in smart wastewater systems: advancing efficiency and sustainability in urban sewer networks. In: Choudhury M, Majumdar S, Goswami S, Sillanpää M (eds) Smart wastewater systems and climate change: innovations through spatial intelligence, Chapter 11, vol 15. Royal Society of Chemistry, pp 155–169. https://doi.org/10.1039/9781837676880-00155
102. Halaktionov M, Bredun V, Choudhary R, Goroneskul M, Kumar A, Ouiya F, Sydorenko V, Markina L (2025) AI-enhanced air quality assessment and prediction in industrial cities: a case study of Kryvyi Rih, Ukraine. Ecol Eng Environ Technol 26(6):45–56. https://doi.org/10.12912/27197050/203725
103. Purushothaman R, Alamelu R, Sudha M (2024) Leveraging solutions for advancing circular economy strategies: an overview. In: Sustainability-focused digital innovations. IGI Global, pp 283–304. https://doi.org/10.4018/979-8-3693-8492-3.ch014
104. Rayonier Advanced Materials (2025) Discover the future of packaging with sustainable materials [Feature article]. RYAM Newsroom. https://www.ryam.com/newsroom/discover-the-future-of-packaging-with-sustainable-materials/
105. Abanda FH, Jian N, Adukpo S, Tuhaise VV, Manjia MB (2024) Digital twin for product versus project lifecycles' development in manufacturing and construction industries. J Intell Manuf 36(2):801–831. https://doi.org/10.1007/s10845-023-02301-2
106. Giordanengo G (2024) Development of innovative materials and processes for cellulose-based food packaging in a circular economy perspective. Doctoral dissertation, Università degli Studi di Torino. https://hdl.handle.net/2318/2030836
107. Kim Y, Ruedy DR (2023) Mushroom packages: an Ecovative approach in packaging industry. In: Dhiman S (ed) Sustainable development and environmental stewardship. Springer, pp 199–223. https://doi.org/10.1007/978-3-031-28885-2_9
108. Debrah JK, Teye GK, Dinis MAP (2022) Barriers and challenges to waste management hindering the circular economy in Sub-Saharan Africa. Urban Sci 6(3):57. https://doi.org/10.3390/urbansci6030057
109. Muriithi JK, Ngare IO (2023) Transitioning circular economy from policy to practice in Kenya. Front Sustain 4:1190470. https://doi.org/10.3389/frsus.2023.1190470
110. Omar HM (2018) Sustainability of informal solid waste collection livelihood in urban areas: a case of Kinondoni Municipality, Dar es Salaam City, Tanzania. Doctoral dissertation, Open University of Tanzania
111. Thorleifsson T (2020) Barriers for private sector companies operating in waste management and recycling in Sub-Saharan Africa: a qualitative exploration. Master's thesis
112. Bleischwitz R, Yang M, Huang B, Xu X, Zhou J, McDowall W et al (2022) The circular economy in China: achievements, challenges and potential implications for decarbonisation. Resour Conserv Recycl 183:106350. https://doi.org/10.1016/j.resconrec.2022.106350
113. Chowdhury PR, Goswami S, Choudhary R, Choudhury M (2025) Electronic wastes, its impact on wildlife and biodiversity. In: Electronic waste (Chapter 8). CRC Press. https://doi.org/10.1201/9781003582311-8
114. Kumar S, Smith SR, Fowler G, Velis C, Kumar SJ, Arya S et al (2017) Challenges and opportunities associated with waste management in India. R Soc Open Sci 4(3):160764. https://doi.org/10.1098/rsos.160764
115. Lehmann S (2019) Conceptualizing the urban nexus framework for a circular economy: linking energy, water, food, and waste (EWFW) in Southeast-Asian cities. In: Droege P (ed) Urban energy transition, 2nd edn. Elsevier, pp 371–398. https://doi.org/10.1016/B978-0-08-102074-6.00032-2
116. Liu K, Tan Q, Yu J, Wang M (2023) A global perspective on e-waste recycling. Circ Econ 2(1):100028. https://doi.org/10.1016/j.cec.2023.100028
117. Srivastava V, Ismail SA, Singh P, Singh RP (2015) Urban solid waste management in the developing world with emphasis on India: challenges and opportunities. Rev Environ Sci Biotechnol 14(2):317–337. https://doi.org/10.1007/s11157-014-9352-4

118. Zhang D, Keat TS, Gersberg RM (2019) A comparison of municipal solid waste management in Berlin and Singapore. Waste Manag 78:800–810. https://doi.org/10.1016/j.wasman.2018.10.044
119. Zhang Z, Chen Z, Zhang J, Liu Y, Chen L, Yang M et al (2024) Municipal solid waste management challenges in developing regions: a comprehensive review and future perspectives for Asia and Africa. Sci Total Environ 930:172794. https://doi.org/10.1016/j.scitotenv.2024.172794
120. Boermans DD, Jagoda A, Lemiski D, Wegener J, Krzywonos M (2024) Environmental awareness and sustainable behavior of respondents in Germany, the Netherlands and Poland: a qualitative focus group study. J Environ Manag 370:122515. https://doi.org/10.1016/j.jenvman.2024.122515
121. Domenech T, Bahn-Walkowiak B (2019) Transition towards a resource efficient circular economy in Europe: policy lessons from the EU and the member states. Ecol Econ 155:7–19. https://doi.org/10.1016/j.ecolecon.2017.11.001
122. Haliskou A (2020) Transboundary movements of hazardous waste by sea in Europe and practice in Hellas
123. Leitão FO, Silva ME, Salim OO (2023) Transition from linear to circular economy of electrical and electronic equipment. A review. Bus Strat Dev 6(2). https://doi.org/10.1002/bsd2.249
124. Rizos V, Bryhn J (2022) Implementation of circular economy approaches in the electrical and electronic equipment (EEE) sector: barriers, enablers and policy insights. J Clean Product 338:130617. https://doi.org/10.1016/j.jclepro.2022.130617
125. Alzaydi A (2024) Balancing creativity and longevity: the ambiguous role of obsolescence in product design. J Clean Product 445:141239. https://doi.org/10.1016/j.jclepro.2024.141239
126. Andeobu L, Wibowo S, Grandhi S (2021) An assessment of e-waste generation and environmental management of selected countries in Africa, Europe and North America: a systematic review. Sci Total Environ 792:148078. https://doi.org/10.1016/j.scitotenv.2021.148078
127. Thyberg KL, Tonjes DJ (2016) Drivers of food waste and their implications for sustainable policy development. Resour Conserv Recycl 106:110–123. https://doi.org/10.1016/j.resconrec.2015.11.016
128. Veleva V, Bodkin G (2018) Corporate-entrepreneur collaborations to advance a circular economy. J Clean Product 188:20–37. https://doi.org/10.1016/j.jclepro.2018.03.196
129. Economist Intelligence Unit (2017) Progress and challenges for inclusive recycling: an assessment of 12 Latin American and Caribbean cities. https://doi.org/10.18235/0006489
130. Ferronato N, Rada EC, Gorritty Portillo MA, Cioca LI, Ragazzi M, Torretta V (2019) Introduction of the circular economy within developing regions: a comparative analysis of advantages and opportunities for waste valorization. J Environ Manag 230:366–378. https://doi.org/10.1016/j.jenvman.2018.09.095
131. Gallego-Schmid A, Vásquez-Ibarra L, Guerrero AB, Henninger CE, Rebolledo-Leiva R (2025) Circular economy in a recently transitioned high-income country in Latin America and the Caribbean: barriers, drivers, strengths, opportunities, key stakeholders and priorities in Chile. J Clean Product 486:144429. https://doi.org/10.1016/j.jclepro.2024.144429
132. Mancini SD, de Medeiros GA, Paes MX, dos Santos MP, da Silva RM (2021) Circular economy and solid waste management: Challenges and opportunities in Brazil. Circ Econ Sustain 1:261–282. https://doi.org/10.1007/s43615-021-00031-2
133. Pereira A, Ribeiro F d M, Jeffrey R, Doron A (2020) Waste policy reforms in developing countries: a comparative study of India and Brazil. Waste Manag Res 38(9):987–994. https://doi.org/10.1177/0734242X20938435
134. Hossain R, Islam MT, Ghose A, Sahajwalla V (2022) Full circle: challenges and prospects for plastic waste management in Australia to achieve circular economy. J Clean Product 368:133127. https://doi.org/10.1016/j.jclepro.2022.133127
135. Lee W (2019) Australia's waste crisis: circular economy responses to the contradictions of capitalism. Doctoral dissertation, The University of Sydney
136. Oliver DL (2021) The Pacific Islands. University of Hawaii Press

137. Zaman A, Caceres Ruiz AM, Shooshtarian S, Ryley T, Caldera S, Maqsood T (2023) Development of the circular economy design guidelines for the Australian built environment sector. Sustainability 15(3):2500. https://doi.org/10.3390/su15032500
138. Gall M, Wiener M, Chagas de Oliveira C, Lang R, Hansen EG (2020) Building a circular plastics economy with informal waste pickers: recyclate quality, business model, and societal impacts. Resour Conserv Recycl 156:104685. https://doi.org/10.1016/j.resconrec.2020.104685
139. Abogunrin-Olafisoye OB, Adeyi O, Adeyi AJ, Oke EO, Ojediran GO (2025) Sustainable waste electrical and electronics equipment recycling in Nigeria: policies and emerging technologies. Environ Qual Manag 34(3):e70020. https://doi.org/10.1002/tqem.70020
140. Karangwa C (2025) Plastic waste: debate about law banning plastic bags in Angola: the advisory policy to a Member of Parliament (MP) of RMLA (Revolutionary Movement for Liberation of Angola) in Angola. J Liberal Arts Human 2(1):23–39. https://jlahnet.com/wp-content/uploads/2025/06/2-1.pdf
141. Training and Research Support Centre, ADFT, ACTogether Uganda, CMPD, FRA, GCC, Kwekwe City Council, Municipality Kariba, Rikoloto, TalkAB[M]R, UAAU, YMCA (2024) Promoting sustainable wellbeing in urban food and waste management systems in East and Southern Africa: learning from case studies of promising practice (EQUINET Discussion Paper No. 132). EQUINET. https://equinetafrica.org/sites/default/files/uploads/documents/Diss%20132%20Integrated%20urban%20health%20May2024.pdf
142. Bredun V, Choudhary R, Kumar A (2024) Regional specifics of using community bins in waste management: a case study of rural communities in Poltava Region (Ukraine). Trends Ecol Indoor Environ Eng 2(4):10–17. https://doi.org/10.62622/TEIEE.024.2.4.10-17
143. Ghosh SK, Baidya R (2021) Circular economy in India: reduce, reuse, and recycle through policy framework. In: Ghosh SK (ed) Circular economy: recent trends in global perspective. Springer, pp 183–217. https://doi.org/10.1007/978-981-16-0913-8_6
144. Lee D-H, Park E-H (2023) Decision making in a closed-loop supply chain with a waste management program: manufacturers' take-back activity and governmental subsidies for remanufacturing. Processes 11(11):3132. https://doi.org/10.3390/pr11113132
145. Yolin C (2015) Waste management and recycling in Japan: opportunities for European companies (Minerva Fellowship Report). EU–Japan Centre for Industrial Cooperation. https://www.eu-japan.eu/sites/default/files/publications/docs/waste_management_recycling_japan.pdf
146. Sreenath AB, Dhanaraj K, Rath P, Bharadwaj LK, Choudhary R (2025) Waste management and circular economy. In: Sustainable chemistry and pioneering green engineering solutions (Chapter 11). IGI Global. https://doi.org/10.4018/979-8-3373-1409-9.ch011
147. Han Z, Han C, Shi Z, Li J, Luo E (2023) Rebuilding the crop-livestock integration system in China: based on the perspective of circular economy. J Clean Product 393:136347. https://doi.org/10.1016/j.jclepro.2023.136347
148. Ecomondo (2025) Eco-industrial parks: 7 examples of sustainable industrial ecosystem in Italy and around the world. Ecomondo. https://www.ecomondo.com/en/news-detail/eco-industrial-parks-7-examples?newsId=2549063
149. Chioatto E, Sospiro P (2023) Transition from waste management to circular economy: the European Union roadmap. Environ Dev Sustain 25:249–276. https://doi.org/10.1007/s10668-021-02050-3
150. Almén J, Anehagen M, Enström E, Hartman C, Jonsson C, Lindbladh F, Ohlsson J (2020) In: Dalhammar C, Richter JL (eds) Promoting the repair sector in Sweden. International Institute for Industrial Environmental Economics (IIIEE), Lund University. https://lucris.lub.lu.se/ws/portalfiles/portal/77933910/Promoting_the_repair_sector_in_Sweden_2020_IIIEE.pdf
151. Puente-Rodríguez D, van Laar H, Veraart M (2022) A circularity evaluation of new feed categories in The Netherlands—squaring the circle: a review. Sustainability 14(4):2352. https://doi.org/10.3390/su14042352
152. Abruzzo S (2019) Best practices to encourage landfill diversion in waste management programs. Senior thesis, Scripps College, Scripps Senior Theses, 1329. https://scholarship.claremont.edu/scripps_theses/1329

153. Wang J, He S (2023) Government interventions in closed-loop supply chains with modularity design. Int J Product Econ 264:108965. https://doi.org/10.1016/j.ijpe.2023.108965
154. Atabaki Fard Tehrani Y, Sarkar A, Shetranjiwalla S (2025) Integrating traditional ecological knowledge into a circular economy framework using the example of consumer durable goods for northern rural communities. RSC Sustain 3(5):2048–2063. https://doi.org/10.1039/D4SU00623B
155. Sanches RA, Mendes FD, Duarte AYS, de Campos Araújo M, Issac MG (2025) Application of upcycling on discarded clothing in Vila Maria (Brazil): a case of study on jeans articles. In: Raposo D, Neves J, Silva R, Correia Castilho L, Dias R (eds) Advances in design, music and arts III: EIMAD 2024, vol 48. Springer, pp 98–110. https://doi.org/10.1007/978-3-031-74975-9_8
156. Circular Innovation Lab (2023) Chile is leading the way as an example of a regional circular economy. Circular Innovation Lab. https://www.circularinnovationlab.com/post/chile-is-leading-the-way-as-an-example-of-a-regional-circular-economy. Accessed 4 Jul 2025
157. Platform Cooperativism Consortium (2024) Waste pickers in Colombia: organization, rights and digitization. Platform Cooperativism Consortium. https://platform.coop/blog/waste-pickers-in-colombia-organization-rights-and-digitization/. Accessed 4 Jul 2025
158. Rateau M, Tovar L (2019) Formalization of wastepickers in Bogota and Lima: recognize, regulate, and then integrate? EchoGéo 47. https://doi.org/10.4000/echogeo.16614
159. Ramos IA (2021) Assessing the incorporation of knowledge-sharing in the management of South–South cooperation projects. Master's thesis, KDI School of Public Policy and Management, KDI School Institutional Repository. https://archives.kdischool.ac.kr/handle/11125/42941
160. Laubinger F, Brown A, Dubois M, Börkey P (2022) Deposit-refund systems and the interplay with additional mandatory extended producer responsibility policies (OECD Environment Working Paper No. 208). OECD Publishing. https://doi.org/10.1787/a80f4b26-en
161. Ministry for the Environment (2024) About product stewardship in Aotearoa New Zealand. New Zealand Government. https://environment.govt.nz/what-government-is-doing/areas-of-work/waste/product-stewardship/about-product-stewardship-in-new-zealand/
162. Mathur D, Jayasinghe M, Zander KK (2025) Implementing a circular economy in regional Australia: who bears the economic costs in construction projects? Urban Res Prac:1–22. https://doi.org/10.1080/17535069.2025.2499633
163. Spekkink W, Rödl M, Charter M (2022) Repair Cafés and Precious Plastic as translocal networks for the circular economy. J Clean Product 380:135125. https://doi.org/10.1016/j.jclepro.2022.135125
164. Protocol on Environmental Protection to the Antarctic Treaty (1991/1998) Protocol on environmental protection to the Antarctic Treaty. Secretariat of the Antarctic Treaty. https://www.ats.aq/e/protocol.html

Circular Systems Unveiled: Essential Components and Their Impact on Recycling, Reuse, and Resource Recovery

P. Kowsalya, G. Jennifer Flora, A. Kousalya, Md. Abu Toha, Parvez Alam Khan, Moharana Choudhury, Saiyyeda Firdous, and T. Parthasarathi

Abstract Waste management is one of the serious problems resulting from industrial expansion globally, which is placing tremendous pressure on the Earth's ecosystem and natural resources. According to recent reports, India generates 157,000 tonnes of solid waste per day, from which only 20% is recycled and 80% is disposed of inappropriately into landfills. There is a need to recycle waste to promote a circular economy and protect the environment. Where today's economy is linear and "reduces and recycles and recovers," a circular economy seeks to eliminate the need for the disposal of products after use by the consumer. The circular economy extends the life of resources as far as possible, deriving as much value as possible from them in a way that directly benefits both the economy and society. This chapter discusses some of the barriers and prospects for more sustainable waste management in a circular economy and briefly introduces the concepts of reuse, recycling and resource recovery. This also highlights the economic and environmental considerations, including the impact of policy, regulation, and communities, in making the circular economy a reality.

P. Kowsalya · G. Jennifer Flora · A. Kousalya · S. Firdous (✉) · T. Parthasarathi (✉)
VIT School of Agricultural Innovations and Advanced Learning, Vellore Institute of Technology (VIT), Vellore, India
e-mail: Saiyyeda.firdous@vit.ac.in

T. Parthasarathi
e-mail: parthasarathi.t@vit.ac.in

Md. A. Toha
Faculty of Education, University of Cambridge, Cambridge, UK

P. A. Khan
Department of Management and Humanities, Universiti Teknologi PETRONAS, Seri Iskandar, Malaysia

M. Choudhury
Environmental Research and Management Division, Voice of Environment (VoE), Guwahati, Assam, India

S. Majumdar et al. (eds.), *Sustainable Urban Future*, Urban Sustainability,
https://doi.org/10.1007/978-981-95-2976-6_3

Keywords Circular systems · Material recovery · Resource efficiency · Waste minimization

1 Introduction

According to 2016 reports, the world's urbanisation and population growth generated 2.01 billion metric tons of municipal solid waste (MSW), and the accumulation of this MSW volume is expected to increase to 2.59 billion metric tons by 2030. The world consumed 90 billion tonnes of materials in 2017, and by 2060, this amount is projected to be around 165–195 billion tonnes [1]. Sustainable waste management is a key challenge that negatively impacts both healthcare and environmental quality. The global consumption of raw materials and products has increased year by year, driven by the rapid development of industrialisation, urbanisation, and population growth [2]. The increasing demand due to overconsumption has led to the overextraction and utilisation of natural resources, resulting in significant environmental damage, resource exhaustion, and an increased generation of waste [3]. Especially considering the issues with waste on a planetary level, the urgency of transitioning to a Circular Economy (CE) and greener, efficient environmental practices is becoming increasingly relevant [4, 5]. The Circular Economy is an "alternative to our extractive linear economy," in which a take-make-consume-dispose cycle has been replaced by a cycle that is restorative and regenerative, both by design and intention, thereby reducing waste and keeping resources in use [6]. The concept of the Circular Economy is inspired by Earth's natural systems, where waste is eliminated, energy is continually regenerated, and innovation drives creation [7].

According to the Ellen McArthur Foundation, the concept of Circular Economy is "an industrial system that is restorative or regenerative by intention and design," a 'closed loop' system of production and consumption that uses resources more efficiently by recycling them back into the manufacturing process rather than allowing them to become waste. The ultimate goal of integrating waste processing with the circular economy is to achieve a closed-loop system within the circular economy, where materials no longer leave the economy after their single use [8, 9]. It is a cycle or philosophy where products are designed for recycling, manufacturers are incentivised to produce products from recycled materials, and customers are educated to consume responsibly, thereby wasting less [10]. In this circular model, the core concepts of loop closure, narrowing, delay, intensification, and dematerialisation are deployed to reduce the level of resources, byproducts, and discharge emanating from the organisation [11].

A circular economy relies on a holistic and sustainable approach to resources and waste management. The reduce, reuse, recycle, and recover cycle, as described by Adefemi et al. [12], offers a model through which society can transition from a linear and wasteful system to a regenerative, circular one. Incorporating them into waste management efforts sets an environmental precedent, but also moulds a more resilient or sustainable future. Hence, the concept of a circular economy implies waste minimisation, extension of product and material lives, restoration of natural cycles, and ultimately, degrowth to encompass the system of production and consumption [13]. Encouraging this economy can, in turn, foster not only a variety of interventions beginning with a reduction of waste itself, but also with the unlocking of economic opportunity to recycle, compost, and recover resources [1].

1.1 Bibliographic Analysis

Figure 1 shows bibliometric data on research output between 2015 and 2025 in the "Main Information" area. This interest is evident in the research documents, which comprise a total of 761 documents and 250 sources produced during this period, with an annual growth rate of 16.06%. Only 21 single-authored documents were found out of a total of 3227 authors, which suggests a collaborative culture. Co-authorship averaged 5.16 per document, and 34.56% of the publications were internationally collaborative, indicating significant international participation in the field. The scholarly impact is measurable by 28.2 citations per document and a total of 46,423 cited references. Additionally, unique author keywords suggest a high diversity of themes, and a mean document age of 3.64 years indicates relatively recent literature that is still in development. Such patterns also reflect the bibliometric results of the related papers, which indicate a dynamic and interdisciplinary field of research.

The trends in key terms used in the research over time are in Fig. 2. Terms such as "removal," "reuse," and "resource recovery" are now more frequently used keywords, indicating a shift toward the implementation and adoption of recycling technologies. This supports the analysis in this paper, which indicates that research is starting to branch out towards specific topics, such as e-waste, food waste, and behaviour change. The fact that the terms "management" and "performance" are on the rise also points to an increasing interest in efficacy and policy implementation regarding recycling practices.

The trend represented in Fig. 3 reflects the growth in recycling behaviour, as indicated by the number of publications since 2015, with projections extending until 2025. The surge in the number of studies after 2018, with a peak in 2020, also

Fig. 1 Main information

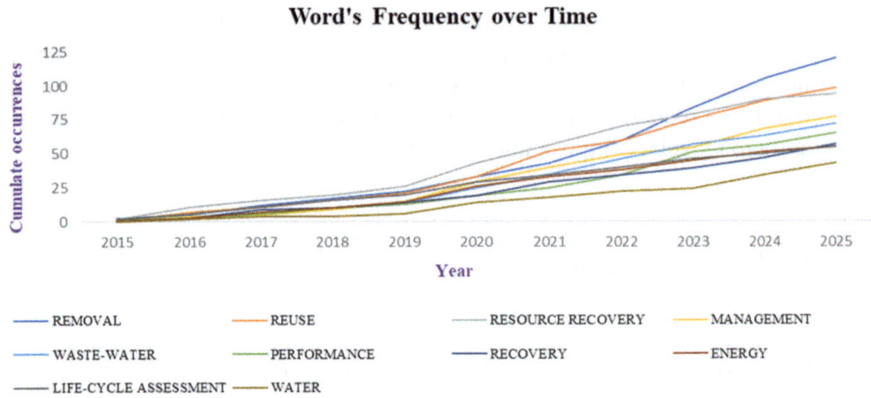

Fig. 2 Word's Frequency over Time

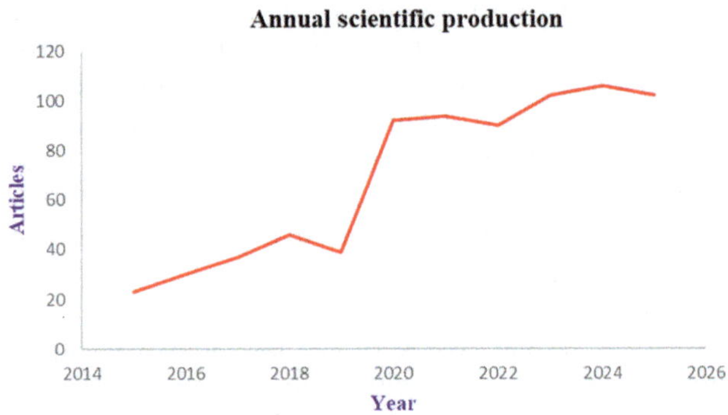

Fig. 3 Word's Frequency over Time

correlates with global awareness of sustainability, which was triggered particularly by the UN's Sustainable Development Goals. There is a minor downward trend in recent years; however, the data indicate continued interest in this area within academia.

2 Design of Circularity

Circular economy marks a shift from a linear economy as it focuses on the system's "flow of material, energy, and information in and within the system". This vision encapsulates four major concepts: reduce, re-use, recycle, and recover (Figs. 4, 5) [14].

2.1 Recycling Systems and Remanufacturing

The need to conserve resources and protect the environment has recently brought attention to recycling and remanufacturing technologies. In a circular economy, it is vital to reprocess waste into valuable resources to minimize the extraction of virgin raw materials and mitigate environmental damage [3].

2.2 Reuse

Extending the lifespan of products and materials through reuse is a key element of the circular economy [15]. This approach fosters a culture of design and fabrication for durability and repair. Through fostering reuse, the circular economy aims to mitigate the environmental impacts resulting from the constant production and consumption of goods [16].

2.3 Recovery of Resources

The circular economy model considers waste as a resource. In addition to recycling, waste-to-energy processes constitute another important approach within the circular economy paradigm [12]. Techniques such as waste-to-energy processes and anaerobic digestion help extract energy from organic waste, significantly reducing the environmental impact. These are important concepts to consider in sustainable waste management policy, helping to achieve a more sustainable and resilient approach [17]. Ideally, resource recovery is a valuable resource from these waste streams, which are recirculated back into the production cycle [18]. Successful types of resource recovery focus on electronic waste, also known as e-waste recycling, where valuable metals are reclaimed and reintegrated into the material circulation cycle. Improved recycling techniques reduce the need for raw materials and minimise the environmental impact of manufacturing.

2.4 Product Lifecycle Management

Extending the lifespan of products through a circular economy strategy can be achieved through repair, refurbishment, and remanufacturing. For example, businesses create incentives for consumers to return their used items for repair or refurbishment instead of disposing of them. The opportunity to provide such services means that these companies will not only avoid waste but also maintain closer relationships with their customers, potentially creating additional revenue-generating

Fig. 4 Waste Reduce, Reuse, Recycling, and Resource Recovery are for embracing the Circular economy. (The authors didn't need permission from any copyright holder because it was originally made by the authors)

business models [19]. A logical extension of this is the movement towards "right to repair" laws, which are gaining popularity and create benchmarks for consumers looking to be more ethical and earth-conscious [1] (Fig. 4).

3 Impact of Recycling on Urban Sustainability

Recycling offers numerous environmental, economic, and social benefits. Recycling projects can foster community cohesion through participation and sustainability. Recycled spaces, when utilised as multipurpose urban areas, can offer a better quality of life and enhanced recreational opportunities for the city and its residents [1, 20].

3.1 Environmental Benefits

3.1.1 Landfill Waste Reduction

This use of urban space also has the effect of reducing the amount of waste sent to landfills, which can have positive implications for more sustainable urban waste management [21, 22]. This can lead to the preservation of scarce natural resources and subsequently reduce the need for additional virgin materials [13]. According to the EPA, if we recycle 37% of the waste we generate, it would save 65% of the landfill space. This is particularly important in the case of cities that are running out of space for waste dumping [23–25].

3.1.2 Natural Resource Conservation

Recycling also reduces the need, demand and exploitation of raw materials as well as saving natural resources and reducing damage to the environment. Recycling a ton of office paper saves 18 trees and 7400 gallons of water [26–28].

3.1.3 Decreased Greenhouse Gas Emissions

Compared with manufacturing new products from scratch, recycling consumes less energy. Modern forms of recycling strive for a 'closed loop system,' which has the potential to reduce greenhouse gas emissions. It was reported that recycled MSW can reduce 178 tons of greenhouse gases in 2017 [29, 30].

3.2 Economic Benefits

3.2.1 Job Opportunity

Recycling is a labour-intensive process in the collection, processing, and creation of recycled products, providing economic benefits throughout communities. For instance, the transition to the circular economy in Europe is expected to generate 1.2 million new jobs by 2030. It also creates multiple employment opportunities. In the United States, recycling generates 693,000 jobs and $38.6 billion in wages [3, 26].

3.2.2 Cost Savings

Recycling is ultimately advantageous to local governments because they have lower disposal costs and can generate revenue from recycled materials. This financial efficiency translates into lower costs for consumers [19].

3.2.3 Social Benefits

Recycling programs can elicit a sense of responsibility and environmental consciousness among the community. Because it reduces the buildup of waste and pollution created from "urban" landfills. Recycling promotes a healthier urban environment that in turn, can benefit the public health of urban residents. An increase in recycling inspires a progressive collective mindset surrounding sustainable practices. Through recyclers, the quality of urban life improves, making a better life for urban citizens, with cleaner and more sanitary living conditions [25, 31].

The new recycling technologies are fostering new opportunities to extract useful materials from waste that was formerly unrecyclable. New forms of recycling, such as chemical recycling, which breaks down complex plastics to their base chemical components, could be a game-changer in terms of recycling [32]. Through this process, these materials can be transformed into new things, thereby effectively utilising the "closed loop system". One such example is the methods that companies are developing to recycle plastic mixtures and multi-layer packaging that is not easily recoverable by traditional mechanical recycling processes [33]. In the same light, progress in the recovery of precious metals, such as gold, silver, and palladium, from electronic waste recycling has contributed to increasing resource efficiency and reducing dependence on virgin materials [13] and recycling rates while promoting a circular economy where waste is considered as a resource [34]. Recycling is vital for advancing the concept of fitting waste as a recurring resource into a circular economy. It is an approach that is both cost-saving and can add to economic gain through new recycling technology and techniques. This allows better use of municipal funds, promoting sustainable urban development by reducing waste management costs and promoting resource recovery [26, 35] (Fig. 5).

Fig. 5 4 R-Strategies for a Circular Economy. (The authors didn't need permission from any copyright holder because it was originally made by authors)

4 Effective Reuse Strategies

4.1 Reuse and Its Importance

The best way to protect the environment, conserve any natural resource and save money is to recycle and reuse. [36]. In the context of waste prevention measures, reuse is "any process using products or components that are no waste (altered or not) to perform the same function" [37]. This particular type of transition towards a circular economy could be one of the solutions to reduce industrial carbon emissions and to address global climate change [38]. "Reuse consists of the operations like checking, cleaning or repairing recovered products or components so that they can be put to the same use for which they were conceived without pre-processing" [37, 39, 40].

4.2 Practical Reuse Strategies and Applications

Before recycling, an item must be assessed for quality and may need to be repaired or slightly modified to meet the proper specifications. Recent estimates indicate that one-third of all materials arriving at recycling facilities and similar locations are still suitable for reuse; at least 25% of electronic waste retains considerable potential for reuse [41]. For instance, recycling a metric ton of paper can conserve around 4100 kWh of electricity, help save 17 trees, reduce oil consumption by 2.5 barrels, and protect 31,780 L of water. Recycling a metric ton of steel can reduce the demand for 287 L of fuel oil, 1.2 metric tons of iron ore, 0.5 metric tons of limestone, 0.7 metric tons of coal, and 2.3 cubic meters of landfill space. Additionally, it can lower water usage by as much as 40% and cut CO_2 emissions by 58% [29].

5 Resource Recovery Systems and Innovations

5.1 Overview of Resource Recovery

Resource Recovery from Waste (RRfW) is one of the stages of a sustainable Circular Economy [42]. All governments have the primary concern of tackling the growing quantities of municipal waste. From a circular economy perspective, it is a viable plan to obtain resources from waste, provided that it relies on effective waste management. Recycling and resource recovery have been regarded in recent years as a solution for better handling of wastes with greater resource efficiency and environmental concerns [13]. This can help reduce the environmental footprint of anthropogenic processes by reducing reliance on the extraction of virgin materials, which is crucial for conserving raw materials, energy, and water [43].

5.2 Techniques and Innovations

According to Nanda and Berruti [44], Resource recovery involves extracting valuable materials and energy from conventional waste streams through the application of various engineering solutions and technological innovations. Among these is automatic classification, where the use of sensors, combined with data analysis, enables the automatic identification of distinctions among various types of garbage. Thermal processing, specifically pyrolysis or gasification processes, can convert waste into heat or electricity, and is another form of energy recovery. The application of these techniques through aerobic or anaerobic composting may help speed the degradation of organic wastes. Advancements in modern technology, including automated sorting technology, have enabled us to extract secondaries from the waste stream, thereby reducing the need for mined raw materials, such as metals and plastics [45].

Artificial Intelligence technologies, such as machine learning algorithms, can generally be used to enhance the perception and accuracy of classification between different types of waste materials or objects, thereby improving waste management systems. The use of AI that can read trends in data and enables the machinery to classify waste materials, similar to how humans do. AI could also enhance the routes taken by garbage trucks, resulting in reduced fuel consumption and lower carbon emissions. The utilisation of AI to manage waste is a promising path that can help to increase the efficiency of waste management systems, also aiding the movement towards sustainable waste management (Fig. 6) [46].

6 Economic and Environmental Impacts

6.1 Economic Benefits and Barriers

Recycling paves the way for good economic opportunities and minimises the environmental impacts. Recycling saves on marginal production costs and reduces dependency on raw materials [47]. For instance, Alimi et al. [48] stated that producing aluminium from recycled aluminium scrap requires only 5% of the energy needed to produce an equivalent quantity of aluminium from bauxite, resulting in 95% energy savings through recycling. That limited the cost of production. Furthermore, Elroi et al. [49] indicated that the large quantity of waste sent to landfills was decreased as a result of the reduction in long-term expenses associated with recycling policies. The waste management market in India is projected to reach approximately 14 billion USD by 2025, with an annual growth rate of 7%. This growth in the sector is likely to create job opportunities, as reported by Sharma et al. [50]. However, investment in these products is lacking due to unestablished markets for recovered materials. For example, there is a well-established recycling market for metal scraps, which many plastics lack. This creates investment issues for recycling technologies and processes [51].

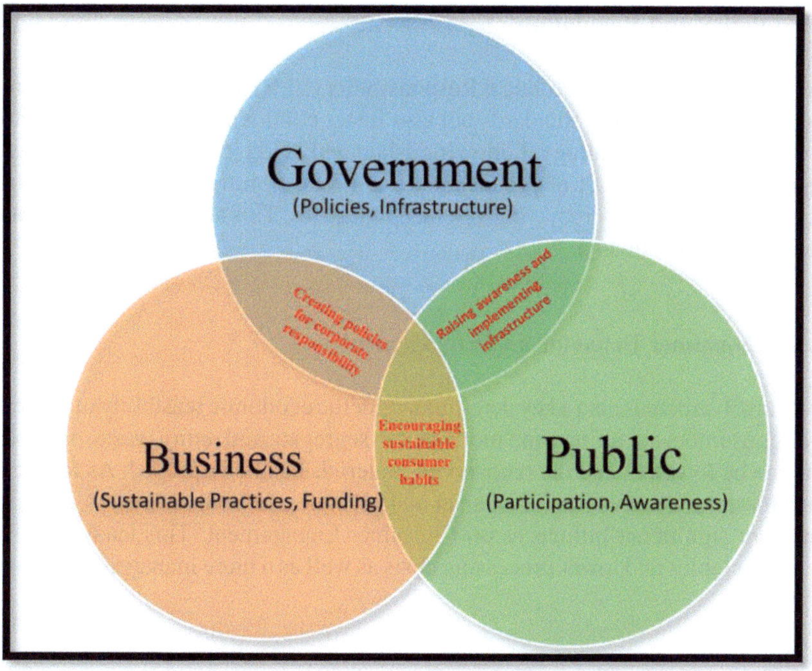

Fig. 6 Building a Sustainable Future: Policies, Infrastructure and Awareness in Circular Waste Management. (The authors didn't need permission from any copyright holder because it was originally made by the authors)

6.2 Enabling Mechanisms and Supporting Infrastructure

Though it is not much of a debate that recycling has economic and environmental advantages, the opportunities to achieve and expand on these impacts are highly dependent on a strong enabling environment. These are the logistical infrastructure, supportive policies, and consumer engagement that provide the scaffolding to good economic performance in waste management.

6.2.1 Reverse Logistics

Reverse logistics is a key economic activity that lowers the cost of production and helps recover materials by improving the process of returning and processing used goods. Julianelli et al. [52], Life cycle assessments, ICT, and planning are as having a major enabling role in the circular economy. These systems ultimately reduce drastically their operational costs by optimizing the recovery of resources and relying less and less on raw materials.

6.2.2 Policy and Regulation

Policies such as Extended Producer Responsibility (EPR) not only help shape compliance with environmental standards but also foster confidence in markets and investment in recycling technology. Following Islam and Huda [53], clear regulations and responsibility on the part of producers are to invest in infrastructure help to lower market entry barriers for recycling-oriented industries. This helps to extract economic value from waste streams.

6.2.3 Consumer Behavior and Engagement

Public involvement is also a key determinant for the economic feasibility of recycling. When consumers are trained and motivated to segregate and return waste, operational costs can be lowered, and the recovery of materials can be improved. As for the role of awareness and convenience, Khan et al. [54] state that both awareness and convenience are significant influences on Consumer Engagement. This leads directly to clean recyclables and lower processing costs as well as a more manageable economic cycle.

6.3 Environmental Advantage and Challenges

Recycling and reusing help conserve natural resources and mitigate environmental degradation by minimising the dependency on raw materials [55]. Additionally, Lamma [26] reported that the disposal of waste from landfills and incineration is diverted through recycling, thereby minimising air and water pollution and contributing to pollution mitigation. Furthermore, recycling requires less energy in production, leading to reduced greenhouse gas emissions [56].

According to Kumar et al. [57], challenges exist in recycling and reuse due to a lack of awareness among the public regarding participation in recycling programs and the adoption of relevant technologies. Additionally, the public's indifference to waste disposal contributes to poor segregation and an increase in landfill waste. Furthermore, poor infrastructure, the absence of strategic planning, and weak regulatory frameworks continue to pose significant challenges for the effective management of recycling.

6.4 Role of Policy, Regulation, and Public Participation in Overcoming Challenges

The processes of waste management, including recycling, reuse, and resource recovery, are primarily driven by the connection between policy, regulation, and public engagement [8, 9]. Extended producer responsibility represents an effective strategy that assigns substantial responsibility to producers for both the disposal and treatment of their products, encompassing financial and physical aspects. Additionally, it encourages the producers to manufacture products that can be easily recycled and disposed of responsibly [58]. Furthermore, India's Plastic Waste Management Rules enforce strict guidelines for controlling plastic pollution by regulating plastic usage and disposal, aiming to increase the rate of recycling [59]. For Instance, in China, to minimise resource consumption and maximise resource utilisation efficiency, the Circular Economy Promotion Law has been enforced to promote recycling [60].

In India, the Swachh Bharat Mission and the Solid Waste Management Rules, 2016, have encouraged people to compost, process waste, and generate biogas, which are effective strategies for recycling, reuse, and resource recovery [61]. Additionally, the Bhagidari system in Delhi is a community-level program that encourages the public to segregate waste at the source, thereby reducing waste generation. Municipal authorities provide collection bins and processing facilities to facilitate effective waste management [62]. Despite these guidelines and initiatives, the informal sector, which contributes a significant portion of recycling, mainly in Delhi, around 27% of waste, has not been integrated into formal waste management, which remains a challenge [61]. Moving a way forward, public engagement, community-level participation, and policy support will improve recycling, reuse, and resource recovery [61, 63].

6.5 Recommendations

Thus, to bolster waste management via policy and public engagement, a few targeted suggestions are made drawing from current literature. First, incorporating the informal waste sector into formal systems can significantly benefit recycling rates and resource recovery. As noted by Harfadli et al. [64], informal waste workers must be legally recognised, socially protected, and provided with training. They call for the creation of cooperatives and access to institutional support systems that can "scale up the efficiency of the informal workers, can dignify their work and help governments' national waste goals". Additionally, system dynamic modelling from Kala and Bolia [65] supports the creation of synergistic policies that enable cooperation between formal and informal actors. They find that "a good segregation, less dependence on landfills, and a better perception of the trust in public recycling programs come when local governments establish formalized partnerships and incentivize informal workers". Finally, awareness campaigns and nudging to raise public participation should be better designed. The quality and quantity of recyclables are also influenced by the education of citizens, as well as the availability of easily accessible segregation infrastructure, as noted by Harfadli et al. [64]. The government should strive to establish a sustainable waste culture at the household level through inclusive campaigns and waste education in schools and on e-platforms. These recommendations, taken together, then point the way towards a holistic, coherent, and integrated waste management process that aligns with a sustainable vision for India.

7 Future Prospective

To tackle the staggering amounts of waste produced, advanced technology in material recovery and waste-to-energy systems needs to be incorporated. Alongside proper planning and policy frameworks, public education and private-public partnership initiatives are required to curb waste. While these drivers may be overlooked, they are vital for establishing a circular economy that centres around reuse, recycling, and resource recovery for sustainability. Incorporating digital technologies such as AI and IoT can further enhance the efficiency of waste sorting, tracking, and processing. Developing sustainable product designs that emphasise recyclability and reduce waste generation strengthens circularity. Focus on education and community engagement can shift behaviours towards more widely embracing circular practices. Lastly, collaboration and sharing knowledge globally are crucial for tackling transboundary waste issues in resource management and environmental protection.

8 Conclusion

With the expected population and industry boom, the increase in waste is becoming more pressing and exacerbates the need for viable means to ensure a sustainable future. Reuse serves as a potent method for minimizing waste, conserving resources, and reducing greenhouse gas emissions. Numerous practical applications and case studies underscore the economic and environmental advantages of reusing, recycling, and recovering resources. They also highlight the need for support from policy and technological developments that will increase the efficiency in waste management. In partnership with local industries and governments, the community-based model can continue to push towards innovation and expansion of the circular economy. Increasing public awareness and education will mobilize people to participate proactively in waste reduction. Finally, achieving sustainable waste management as a part of economic planning can bring us closer to adopting a circular economy that provides resource and environmental security for future generations.

References

1. Kotyal K (2023) Sustainable waste management in the circular economy: challenges and opportunities. Environ Rep 05:1–5
2. Rajpal A, Choudhury M, Goswami S, Chakravorty A, Raghavan V (eds) (2024) Waste management and treatment: advances and innovations, 1st edn. CRC Press. https://doi.org/10.1201/9781003258377
3. Tanuja GG, Srimani PS, Rao ALN, Sharma S, Mashkour MS, Singh N (2023) RETRACTED: innovative technologies for sustainable recycling and re-manufacturing of materials and components. E3S Web Conf 430
4. Dey S, Santra M, Choudhury M, Ghosh AR, Samanta P (2024) Food waste generation and its industrial utilization: an overview. Environ Sci Pollut Res:1–20. https://doi.org/10.1007/s11356-024-34252-3
5. Singh S, Singh P, Sharma A, Choudhury M (eds) (2022) Agriculture waste management and bioresource: the circular economy perspective. Wiley. https://doi.org/10.1002/9781119808428
6. Kirchherr J, Yang NHN, Schulze-Spüntrup F et al (2023) Conceptualizing the circular economy (revisited): an analysis of 221 definitions. Resour Conserv Recycl 194:107001. https://doi.org/10.1016/j.resconrec.2023.107001
7. Reike D, Vermeulen WJV, Witjes S (2018) The circular economy: new or refurbished as CE 3.0?—exploring controversies in the conceptualization of the circular economy through a focus on history and resource value retention options. Resour Conserv Recycl 135:246–264
8. Aiguobarueghian I, Adanma UM, Ogunbiyi EO, Solomon NO (2024) Reviewing the effectiveness of plastic waste management in the USA. World J Adv Res Rev 22(2):1720–1733
9. Aiguobarueghian I, Adanma UM, Ogunbiyi EO, Solomon NO (2024) Waste management and circular economy: a review of sustainable practices and economic benefits. World J Adv Res Rev 22(2):1708–1719
10. Dumée LF (2022) Circular materials and circular design—review on challenges towards sustainable manufacturing and recycling. Circ Econ Sustain 2(1):9–23
11. Islam NF, Gogoi B, Saikia R, Yousaf B, Narayan M, Sarma H (2024) Encouraging circular economy and sustainable environmental practices by addressing waste management and biomass energy production. Reg Sustain 5(4):100174
12. Adefemi A, Daudu CD, Okoli CE, Ayorinde OB, Adekoya OO, Ibeh CV (2024) Reviewing the impact of LNG technology advancements on global energy markets. Eng Sci Technol J 5(1):128–151
13. Yu Z, Abdul S, Khan R, Ponce P, Zia-ul-haq HM, Ponce K (2022) Exploring essential factors to improve waste-to-resource recovery: a roadmap towards sustainability. J Clean Prod 350:131305
14. Chen TL, Kim H, Pan SY, Tseng PC, Lin YP, Chiang PC (2020) Implementation of green chemistry principles in circular economy system towards sustainable development goals: challenges and perspectives. Sci Total Environ 716:136998
15. Shevchenko T, Kronenberg J (2020) Management of material and product circularity potential as an approach to operationalise circular economy. Prog Ind Ecol 14(1):30–57
16. Corona B, Shen L, Reike D, Carreón JR, Worrell E (2019) Towards sustainable development through the circular economy—a review and critical assessment on current circularity metrics. Resour Conserv Recycl 151:104498
17. Fiksel J, Goodman I, Hecht A (2014) Resilience: navigating toward a sustainable future. Solutions 5(5):38–47
18. Murray A, Skene K, Haynes K (2017) The circular economy: an interdisciplinary exploration of the concept and application in a global context. J Bus Ethics 140(3):369–380. https://doi.org/10.1007/s10551-015-2693-2
19. McKinsey & Company (2020) The role of waste management in the circular economy
20. Lalbakhsh E (2012) The impact of recycling urban space in sustainable development in developing countries. APCBEE Procedia 1(January):331–334

21. Mondal T, Choudhury M, Kundu D, Dutta D, Samanta P (2023) Landfill: an eclectic review on structure, reactions and remediation approach. Waste Manag 164:127–142. https://doi.org/10.1016/j.wasman.2023.03.034
22. Paul S, Choudhury M, Deb U, Pegu R, Das S, Bhattacharya SS (2019) Assessing the ecological impacts of ageing on hazard potential of solid waste landfills: a green approach through vermitechnology. J Clean Prod 236:117643. https://doi.org/10.1016/j.jclepro.2019.117643
23. Amraoui S, Almi H (2024) Exploring circular economy contributes to sustainable development and green growth: case studies on circular economy practices and waste management in various nations. Manag Dev Res Stud 2(13):669–686
24. Lqliyti A-A, Qushish A (2020) Algeria's efforts in implementing the circular economy: a study of the experiences of some economic institutions. Horiz Manag Econ Sci J 4(2)
25. Xu J (2024) Research on innovative models of waste recycling in urban infrastructure. Acad J Sci Technol 9(2):245–248
26. Lamma OA (2021) The impact of recycling in preserving the environment. IJAR 7(11):297–302
27. Meric S, Selcuk H, Onat B, Ongen A (2018) Sustainable technologies for recycling and reuse: an overview. Environ Sci Pollut Res 25:2993–2995
28. Tabish M, Abdul S, Khan R, Yu Z, Tanveer M (2024) A thorough overview of the literature on waste recycling in the circular economy: current practices and future perspectives. Environ Sci Pollut Res 31(52):61377–61396
29. David A, Thangavel Y, Sankriti R (2019) Recover, recycle and reuse: an efficient way to reduce the waste. Int J Mech Product Eng Res Dev 09:31–42
30. Zaman K, Malik M, Awan U, Handayani W, Jabor MK (2022) Environmental effects of biowaste recycling on industrial circular economy and eco-sustainability. Recycling 7(4):60
31. Cansino JM (2020) Environmental and socioeconomic impacts of urban waste recycling as part of circular economy. The Case of Cuenca (Ecuador)
32. Choudhury M, Roy P (2025) Challenges with microplastic pollution in the regime of UN sustainable development goals. World Dev Sustain 6:100216. https://doi.org/10.1016/j.wds.2025.100216
33. Ragaert K, Delva L, Van Geem K (2017) Mechanical and chemical recycling of solid plastic waste. Waste Manag 69:24–58
34. Massoud M, Vega G, Subburaj A, Partheepan J (2023) Heliyon review on recycling energy resources and sustainability. Heliyon 9(4):e15107
35. Khosravani F, Abbasi E, Choobchian S, Ghazizade MJ (2023) OPEN a comprehensive study on criteria of sustainable urban waste management system: using content analysis. Sci Rep:1–15
36. Kabirifar K, Mojtahedi M, Wang C, Tam VW (2020) Construction and demolition waste management contributing factors coupled with reduce, reuse, and recycle strategies for effective waste management: a review. J Clean Prod 263:121265
37. EC (European Commission) (2008) Directive 2008/98/EC of the European Parliament and of the Council of 19 November 2008 on Waste and Repealing Certain Directives (Waste Framework Directive). EC, Brussels
38. Yang M, Chen L, Wang J et al (2023) Circular economy strategies for combating climate change and other environmental issues. Environ Chem Lett 21:55–80
39. Gharfalkar M, Court R, Campbell C, Ali Z, Hillier G (2015) Analysis of waste hierarchy in the European waste directive 2008/98/EC. Waste Manag 39:305–313
40. Gusmerotti NM, Corsini F, Borghini A, Frey M (2019) Assessing the role of preparation for reuse in waste-prevention strategies by analytical hierarchical process: suggestions for an optimal implementation in waste management supply chain. Environ Dev Sustain 21:2773–2792
41. Mehta D, Tege S, Sankhla VS (2018) Sustainable waste management: an approach towards sustainability. J Emerg Technol Innov Res 5(9):101–104
42. Lag-Brotons AJ, Velenturf AP, Crane R, Head IM, Purnell P, Semple KT (2020) Resource recovery from waste. Front Environ Sci 8:35
43. Shirazi MM (2023) Resource recovery for sustainable development and circular economy. J Resour Recov 1

44. Nanda S, Berruti F (2021) A technical review of bioenergy and resource recovery from municipal solid waste. J Hazard Mater 403:123970
45. Mashudi R, Sulistiowati R, Handoyo S, Mulyandari E, Hamzah N (2023) Innovative strategies and technologies in waste management in the modern era integration of sustainable principles, resource efficiency, and environmental impact. Int J Sci Soc 5:87–100
46. Hossain AA, Shohag JR, Ahamed N, Hossain S, Shuvo SA (2024) Innovative approaches in environmental waste management: merging technology with sustainability for a greener future. Adv Environ Waste Man Rec 7(3):01–05
47. Birat JP (2015) Life-cycle assessment, resource efficiency and recycling. Metallurg Res Technol 112(2):206
48. Al-Alimi S, Yusuf NK, Ghaleb AM, Lajis MA, Shamsudin S, Zhou W et al (2024) Recycling aluminium for sustainable development: a review of different processing technologies in green manufacturing. Results Eng 2024:102566
49. Elroi H, Zbigniew G, Agnieszka WC (2023) Enhancing waste resource efficiency: circular economy for sustainability and energy conversion. Front Environ Sci 11:1303792
50. Sharma S, Upadhyay S, Singh B (2019) Employment opportunities with promoting waste management in India. J Clean WAS 3(1):10–15
51. Rahimifard S, Coates G, Staikos T, Edwards C, Abu-Bakar M (2009) Barriers, drivers and challenges for sustainable product recovery and recycling. Int J Sustain Eng 2(2):80–90
52. Julianelli V, Caiado RGG, Scavarda LF, Cruz SPDMF (2020) Interplay between reverse logistics and circular economy: critical success factors-based taxonomy and framework. Resour Conserv Recycl 158:104784
53. Islam MT, Huda N (2018) Reverse logistics and closed-loop supply chain of Waste Electrical and Electronic Equipment (WEEE)/E-waste: a comprehensive literature review. Resour Conserv Recycl 137:48–75
54. Khan F, Ahmed W, Najmi A (2019) Understanding consumers' behavior intentions towards dealing with the plastic waste: perspective of a developing country. Resour Conserv Recycl 142:49–58
55. Cooper DR, Gutowski TG (2017) The environmental impacts of reuse: a review. J Ind Ecol 21(1):38–56
56. Jiang X, Bateer B (2025) A systematic review of plastic recycling: technology, environmental impact and economic evaluation. Waste Manag Res 0734242X241310658:1159–1178
57. Kumar S, Smith SR, Fowler G, Velis C, Kumar SJ, Arya S et al (2017) Challenges and opportunities associated with waste management in India. R Soc Open Sci 4(3):160764
58. Porwal P, Chatterjee S (2019) Extended producer responsibility on e-waste management in India: challenges and prospects. Int J Sci Res 8(5):1026–1034
59. Mani S, Singh S (2016) Sustainable municipal solid waste management in India: a policy agenda. Procedia Environ Sci 35:150–157
60. Bleischwitz R, Yang M, Huang B, Xiaozhen XU, Zhou J, McDowall W et al (2022) The circular economy in China: achievements, challenges and potential implications for decarbonization. Resour Conserv Recycl 183:106350
61. Sannabhadti RS (2020) Challenges in 'community organisation of scrap pickers for Swachh Bharat: reflections from Ahmedabad. In: Solid waste policies and strategies: issues, challenges and case studies. Springer Singapore, Singapore, pp 175–184
62. Chakrabarty B (2012) Citizen–government partnership: Bhagidari in Delhi. Int J Urban Sci 16(1):1–22
63. Yadav AK, Dash S (2020) ITC's community-based solid waste management strategy: a case study of Shivalik Nagar, Haridwar District, Uttarakhand. In: Solid waste policies and strategies: issues, challenges and case studies, pp 213–221
64. Harfadli MMA, Ramadan BS, Rachman I, Matsumoto T (2024) Challenges and characteristics of the informal waste sector in developing countries: an overview. J Mater Cycles Waste Manag 26(3):1294–1309
65. Kala K, Bolia NB (2022) Analysis of informal waste management using system dynamic modelling. Heliyon 8(8):e09993

Circular Economy Benefits: Analyzing Economic Advantages and Environmental Sustainability

Basona Khatun, Saurav Narayan, and D. D. M. I. L. Wijerathna

Abstract Beginning with a critical introduction to the circular economy, this chapter highlights *Reduce–Reuse–Recycle* principles as the cornerstone of a restorative economic model. Next, a conceptual framework synthesizes core concepts reuse, recycling, regeneration, and closed-loop systems underpinned by ecological economics and systems thinking, and aligned with stakeholder responsibilities. Economic advantages are analyzed through resource efficiency, innovation-driven business models, local job creation, and supply chain resilience. The environmental section assesses waste minimisation, greenhouse-gas mitigation, resource conservation, biodiversity protection, and life-cycle impacts. Global and sectoral case studies—from European Union policy mixes to construction, electronics, and textiles—demonstrate practical progress while exposing implementation gaps. The chapter further explores economic, regulatory, technological, data, and behavioral barriers that follow, paving the way for an integrated path forward that combines incentive reform, digital enablers, blended finance, and skills development through ambitious public-private partnership. The chapter concludes that circularity offers a pragmatic route to decouple prosperity from resource depletion and align economic systems with planetary boundaries.

Keywords Circular economy · Resource efficiency · Environmental sustainability · Closed-loop systems · Policy pathways

Abbreviation

AI	Artificial intelligence
BIM	Building information modelling

B. Khatun · S. Narayan (✉)
Department of Law, School of Legal Studies, Central University of Punjab, Bathinda, Punjab, India

D. D. M. I. L. Wijerathna
Faculty of Education, University of Colombo, Colombo, Sri Lanka

S. Majumdar et al. (eds.), *Sustainable Urban Future*, Urban Sustainability,
https://doi.org/10.1007/978-981-95-2976-6_4

CE	Circular economy
CO_2	Carbon dioxide
CSE	Centre for Sustainability and Excellence
EPR	Extended producer responsibility
EU	European Union
EY	Ernst & Young
GHG	Greenhouse gas
IoT	Internet of Things
LCA	Life cycle assessment
PaaS	Product-as-a-Service
PUBYEAR	Publication year
R&D	Research and development
SDG	Sustainable development goals
SMEs	Small and medium-sized enterprises
TITLE-ABS-KEY	Title, abstract, and keywords
UN	United Nations

1 Introduction

The Circular Economy (CE) represents a transformative economic model that prioritizes minimizing waste and maximizing the continual use of resources [1]. In contrast to the traditional linear '*take, make, use, and dispose*' approach [2], the CE is grounded in three core principles: *Reduce*, *Reuse*, and *Recycle* [3]. Reduction focuses on minimizing resource consumption and waste generation through the design of efficient products and processes [4, 5]. Reuse emphasizes extending the lifecycle of products and materials *via* maintenance, repair, refurbishment, and repurposing [6]. Recycling involves reintegrating materials into the production cycle, converting waste into valuable assets [7]. Together, these principles aim to eliminate pollution, maximize the value of materials, and restore natural systems. Restorative and regenerative by design, the circular economy seeks to decouple economic development from the depletion of finite resources, offering a promising framework for sustainable waste management practices globally [8–10].

Studying the circular economy is essential due to its potential to deliver substantial economic and environmental benefits [11]. Economically, extending the use of materials and products allows businesses to lower operational costs, unlock new revenue streams, and drive innovation. It also contributes to job creation, strengthens resource security, and enhances the resilience of supply chains. Environmentally, the circular model reduces greenhouse gas emissions, conserves finite natural resources, and minimizes pollution and waste [12, 13]. Furthermore, it supports biodiversity by decreasing the demand for raw material extraction and promoting the regeneration of natural systems. A clear understanding of these benefits equips policymakers,

Table 1 Linear Economy vs. Circular Economy

Aspect	Linear Economy	Circular Economy
Operational Model	Follows a '*take–make–dispose*' approach [2]	Emphasizes '*reduce–reuse–recycle*' principles [3]
Material Usage	Involves extraction, consumption, and discarding of raw materials [15]	Focuses on maximizing the Longevity and usability of resources [16]
Waste Management	Produces large amounts of waste and environmental pollutants [17]	Strives to minimize or eliminate waste through closed-loop systems [18, 19]
Product Longevity	Products are typically short-lived and not designed to last [20, 21]	Encourages product lifecycle extension through reuse, repair, and recycling [22]
Environmental Effect	Leads to natural resource depletion and ecological harm [16, 23]	Supports ecosystem renewal and helps lower carbon and resource impacts [1]
Economic Approach	Depends heavily on constant extraction of new resources [24]	Builds value by keeping materials and products in continuous circulation [25]

businesses, and society with the knowledge needed to adopt sustainable growth and strengthen in the face of environmental challenges [14].

Table 1 contrasts linear and CE. The linear model emphasizes consumption, waste, and resource depletion. In contrast, the CE promotes sustainability through reuse, longevity, and the renewal of ecosystems. It reduces environmental impact, supports closed-loop systems, and enhances economic value by maintaining material circulation instead of relying on constant extraction.

The chapter opens with a CE primer, follows with a bibliometric mapping of research trends, develops a Reduce–Reuse–Recycle conceptual framework, analyses economic opportunities and environmental life-cycle gains, illustrates sectoral case studies, diagnoses implementation barriers, proposes policy-digital-finance levers, and closes with future outlook, skills, stakeholder engagement, and metrics.

2 Global Trends and Influential Works in Circular Economy Research: Insights from a 2015–2024 Bibliometric Review

The applied search query—TITLE-ABS-KEY (circular AND economy OR economic AND advantage AND environmental AND sustainability) AND PUBYEAR > 2014 AND PUBYEAR < 2025—targeted scholarly literature published between 2015 and 2024. These 10 years were intentionally selected to capture recent developments in sustainable materials management, innovations in circular and closed-loop economic models, and the expanding influence of both policy-driven and industry-led sustainability initiatives. The bibliometric analysis explores several key dimensions: the frequency of key thematic terms, the global distribution of scholarly impact as measured by total citation counts, and the most influential publications as identified through co-citation analysis. These results are presented in Figs. 1, 2 and Table 2, which together provide an integrated perspective on the most frequently cited documents, country-level academic influence, and dominant research themes, offering a comprehensive overview of the scholarly discourse in the domains of circular economy and environmental sustainability.

The bibliometric analysis identified '*circular economy*' as the most frequently occurring term in the analysed body of literature, appearing 251 times. This finding reflects the growing centrality of circular economy frameworks in scholarly discourse, particularly in the context of sustainability transitions, resource efficiency, and systemic waste reduction. The second most prevalent term was

Fig. 1 Most frequently occurring terms in circular economy and sustainability literature (2015–2024)

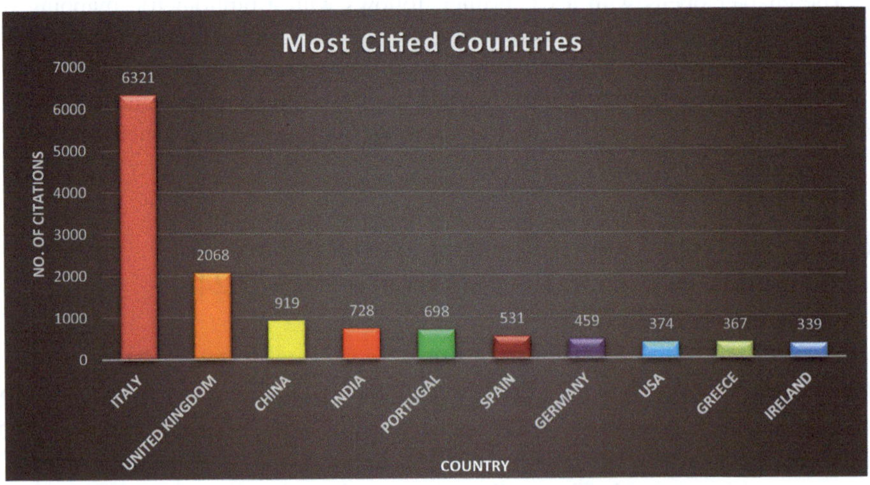

Fig. 2 Total citations by country in CE and sustainability research (2015–2024)

Table 2 Top co-cited documents and their network metrics in the circular economy and sustainability research field

Author(s) and year	Cluster	Betweenness centrality	Closeness centrality	PageRank
Kirchherr J. (2017-1)	2	119.989	0.003	0.071
Ghisellini P. (2016-1)	2	79.489	0.003	0.069
Our Common Future (1987)	1	67	0.002	0.047
Lieder M. (2016-1)	2	93.875	0.003	0.043
Kirchherr J. (undated/ variant entry)	2	214.221	0.003	0.042

'*sustainable development*', with 216 occurrences, signifying its foundational role in framing discussions around environmental policy, economic growth, and long-term ecological balance.

Closely aligned with these themes, the term '*sustainability*' appeared 196 times, further emphasizing its importance as a unifying concept that connects environmental, economic, and social dimensions of research. The prominence of '*environmental impact*' (79 occurrences) and '*recycling*' (62 occurrences) indicates sustained academic interest in evaluating ecological consequences and promoting material recovery strategies as integral components of circular systems. Together, these frequency patterns illustrate a tightly interlinked thematic landscape in which circular economy concepts are not only central but also embedded within a broader narrative of environmental sustainability and socio-economic advantage. This convergence of terminology demonstrates the interdisciplinary evolution of the field and its alignment with global sustainability agendas.

The country-level distribution of total citations offers valuable insights into the global academic influence within the field of CE and sustainability. Italy emerges as the most impactful contributor, with a remarkable 6321 total citations, indicating its leading role in shaping scholarly discourse and advancing research in this domain. This is followed by the United Kingdom with 2068 citations, reflecting a strong tradition of environmental and economic research, particularly in policy-oriented and interdisciplinary contexts.

China and India account for 919 and 728 citations, respectively, underscoring their growing prominence in sustainability research, likely driven by their large-scale environmental challenges and policy initiatives aimed at circular transformation. Southern European countries such as Portugal (698 citations) and Spain (531 citations) also exhibit notable academic engagement, reinforcing the regional focus on sustainability within the EU framework.

Other key contributors include Germany (459 citations), the United States (374 citations), Greece (367 citations), and Ireland (339 citations). Although these countries show relatively lower citation counts, their presence reflects a diverse and internationally distributed research effort. Collectively, the citation landscape demonstrates that while certain countries dominate in terms of academic influence, research on the CE has garnered global attention, reflecting its significance in addressing sustainability challenges across various economic and ecological contexts.

The co-citation network analysis provides critical insights into the intellectual structure and foundational literature shaping the discourse on CE and sustainability. The network, as represented in Table 2, reveals three distinct clusters, each representing a thematic grouping of frequently co-cited scholarly works. The analysis identifies Kirchherr et al. (2017-1) as the most central and influential node within the network, exhibiting the highest PageRank (0.071) and substantial betweenness centrality (119.99), suggesting its pivotal role in connecting various subfields within the domain.

Closely following is Ghisellini et al. (2016-1), another seminal work in the CE literature, with a Page Rank of 0.069, underscoring its sustained academic impact. Other influential contributors include Lieder and Rashid (2016-1) and a second record for Kirchherr et al., both of which demonstrate high levels of co-citation and network embeddedness, further affirming their scholarly relevance.

In a separate cluster, classic and policy-defining publications such as '*Our Common Future*' (1987) and the UN's '*Transforming Our World: The 2030 Agenda for Sustainable Development*' (2015) are prominently positioned. These works serve as theoretical and policy backbones for modern sustainability research, as reflected by their moderate to high centrality measures and foundational status.

The overall network topology suggests a well-integrated research field characterized by a core-periphery structure. Central nodes serve as bridges between emerging discourses and established frameworks, facilitating the interdisciplinary transfer of knowledge. The strong co-citation links between contemporary and foundational sources underscore the field's evolution from broad sustainability paradigms to more focused, actionable frameworks grounded in the concept of the CE.

3 Conceptual Framework

3.1 Core Concepts of the Circular Economy

The CE is founded on key concepts (Fig. 3) that redefine how resources are utilized and managed, distinguishing it from the traditional linear model [15, 26]. *Reuse* involves extending the life of products and materials by repurposing or finding new applications, keeping resources in circulation for as long as possible [6]. *Recycling* breaks down used products into raw materials that can be reintroduced into production, reducing reliance on virgin resources and minimizing waste [27]. *Regeneration* focuses on restoring and enhancing natural systems—such as improving soil through regenerative agriculture or adopting renewable energy—to generate positive environmental impacts [28, 29]. *Closed-loop* systems are designed to convert waste from one product or industry into a resource for another, thereby maintaining a continuous material flow within the economy and eliminating the concept of waste

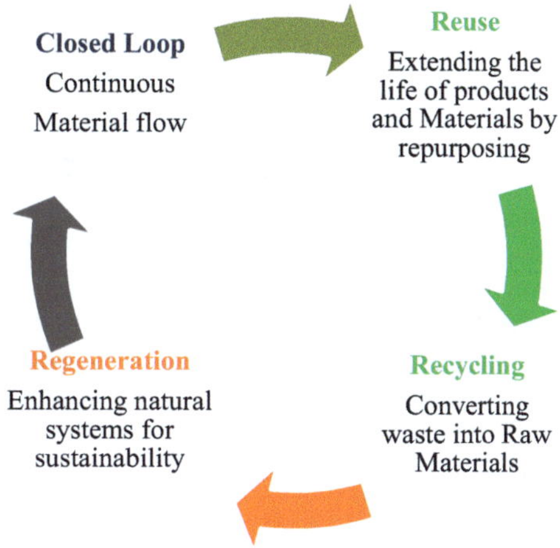

Fig. 3 Concepts of CE

[30, 31]. Nowadays composting, vermicompost and microbial community applications for better waste to resource creation technologies are for lignocellulosic waste valorization and waste resource conversion [32, 33]. Together, these concepts form the foundation of a circular economy that is restorative, efficient, and sustainable by design.

3.2 Theoretical Underpinnings

The CE is grounded in several interdisciplinary theoretical frameworks that guide its principles and practices. Ecological economics positions the economy within the broader context of ecological systems, advocating for sustainable development within planetary boundaries and challenging the notion of infinite economic growth [34]. Industrial symbiosis nurtures inter-industry collaboration, where the waste or byproducts of one process serve as inputs to enhance resource efficiency and drive innovation [35]. Systems are thinking provides a holistic lens for addressing economic and environmental challenges by examining the interconnectedness of production, consumption, and waste [36]. It supports the design of interventions that optimize entire systems rather than isolated components. These theoretical foundations collectively shape a circular economy that is systemic, regenerative, and aligned with ecological realities.

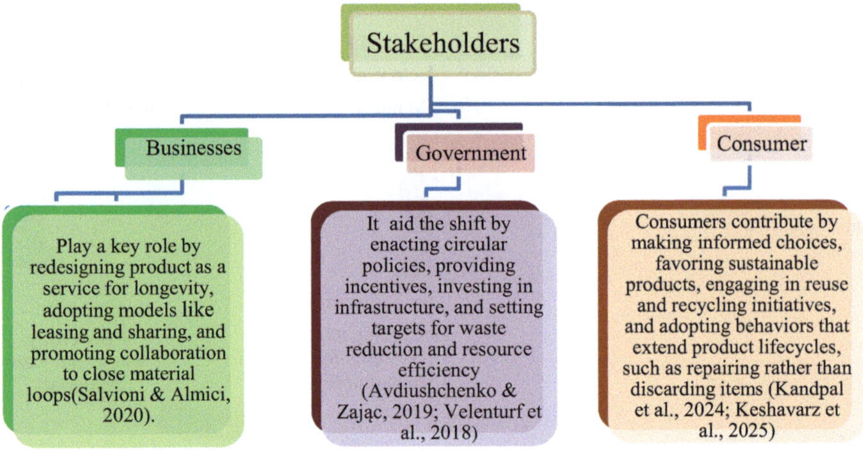

Fig. 4 Key Stakeholder of CE. *Source* (Salvioni & Almici, [38], Avdiushchenko &Zając, [39], Velenturf et al., [40], Keshavarz et al., [41]

3.3 Key Stakeholders

The successful transition to a CE relies on the active participation of multiple key stakeholders [37] (Fig. 4).

4 Economic Advantages of the Circular Economy

4.1 Resource Efficiency and Cost Savings

The CE promotes resource efficiency by emphasizing reuse, recycling, and the exchange of surplus materials, resulting in substantial cost reductions for businesses [42]. A notable example is the collaboration among 25 Danish companies engaged in industrial symbiosis, where the exchange of residual resources, such as water and energy, reduced waste management and procurement costs while generating additional revenue through product innovation and entry into new markets [43]. In manufacturing, practices such as remanufacturing and refurbishing can reduce the energy required to produce new items by up to 85%, thereby lowering both operational costs and environmental impacts [44]. Additionally, supply chain optimization and the implementation of closed-loop systems extend material lifecycles, further enhancing efficiency and reducing overall expenditure.

4.2 Innovation and New Business Models

CE principles boost innovation and support the development of transformative business models that decouple growth from resource consumption. *Product-as-a-Service (PaaS)* models [45], in which companies retain ownership and offer products as services, encourage the design of more durable, repairable, and upgradable goods, exemplified by Philips Lighting's '*light as a service*' offering [46]. The sharing economy promotes platforms that enable the collective use of products, extending their lifespan and reducing the need for new resources [47]. Remanufacturing initiatives by companies such as Caterpillar and Renault demonstrate how high-quality products can be restored and reused, resulting in significant cost savings while maintaining performance standards [48, 49]. In a notable case study, Dell Technologies has implemented a closed-loop recycling process that recovers materials from used electronics for incorporation into new products, effectively lowering material costs and minimizing environmental impact [50]. These innovative models demonstrate how circularity drives both economic and environmental value.

4.3 Job Creation and Local Economic Development

The CE stimulates job creation by expanding employment opportunities in sectors such as repair, refurbishment, and material recovery [51]. These '*circular jobs*' not only support local economies but also encourage entrepreneurship, particularly among small and medium-sized enterprises (SMEs) and social enterprises, which often pioneer repair, upcycling, and reuse initiatives [52]. By nurturing skill development and community engagement, the circular economy empowers local actors and strengthens social inclusion [53, 54]. Additionally, the establishment of regional recycling and remanufacturing hubs enhances local economic resilience, diversifies economic activity, and contributes to long-term, sustainable growth.

4.4 Risk Reduction and Supply Chain Resilience

Circular practices significantly reduce dependency on raw materials by maximizing the value derived from existing resources and minimizing waste [55]. By shifting reliance from virgin inputs to recovered or recycled materials, companies can mitigate risks associated with resource price volatility and supply chain disruptions [56]. This strategic shift enhances resilience, as circular supply chains are more adaptable to external shocks, just as disaster-resilient wastewater systems are being redesigned to ensure uninterrupted service and resource reuse during emergencies [57, 58]. Integrating circular principles not only stabilizes material sourcing but also supports long-term operational continuity and sustainability.

Fig. 5 Open dumping in nature (left) and Overflowing bins (right) reflect poor waste management—urgent need for proper disposal and recycling at Punjab, India. (Photo by: Saurav Narayan and Basona Khatun)

5 Environmental Sustainability Benefits

5.1 Reduction in Waste Generation

CE models play a critical role in minimizing waste by prioritizing reuse, recycling, and remanufacturing [59, 60]. A UN report projects that global adoption of circular practices could reduce annual municipal solid waste from over 4.5 billion tonnes to under 2 billion tonnes by 2050 [61]. This shift could lead to a more than 40% reduction in landfill waste and the elimination of uncontrolled dumping and burning [62], with regional case studies also showing how structured community bin use can enhance rural waste outcomes [63]. For instance, Australia's circular initiatives have resulted in 63% of waste being recovered for reuse, recycling, or energy, while per capita waste generation has stabilized or declined [64]. Additionally, waste-to-resource strategies—such as converting organic waste into biogas or compost—highlight the transition from traditional disposal methods to sustainable resource recovery, underscoring the circular economy's potential to transform waste into value (Fig. 5).

5.2 Lower Greenhouse Gas Emissions

CE strategies play a vital role in reducing greenhouse gas (GHG) emissions by decreasing reliance on virgin material extraction and energy-intensive production processes [65, 66]. According to the Circularity Gap Report, approximately 70% of global GHG emissions are associated with the production and use of goods, indicating that circular solutions have significant potential for climate mitigation [67]. In India, for example, processing 50% of wet waste through bio-methanation could reduce emissions by an estimated 10.36 million tonnes of CO_2 equivalent annually [68].

By decoupling economic growth from emissions and resource consumption, circular approaches make meaningful contributions to both national and international climate targets, thereby reinforcing the transition to a low-carbon, sustainable economy.

5.3 *Conservation of Natural Resources and Biodiversity*

The CE promotes the conservation of natural resources and biodiversity by enhancing resource efficiency and reducing environmental pressures associated with extraction, pollution, and waste [69]. By minimizing the need for raw material extraction, circular practices help preserve ecosystems, decrease habitat destruction, and limit water and land degradation [70]. For example, Europe has demonstrated that it is possible to achieve economic growth while stabilizing resource use and waste generation, indicating a modest but meaningful decoupling of growth from environmental harm. Circular models that prioritize material recovery and the regeneration of natural systems contribute to the protection of biodiversity by reducing landfill expansion, curbing harmful emissions, and safeguarding vital habitats [69, 71, 72].

5.4 *Life Cycle Thinking and Environmental Impact Assessment*

Life Cycle Assessment (LCA) is a fundamental tool in CE analysis, and recent advancements such as fuzzy-logic integrated LCA models offer more accurate and adaptable sustainability assessments in urban solid waste systems offering a comprehensive evaluation of environmental impacts throughout a product's entire life-cycle—from raw material extraction to end-of-life [28, 7374 7576]. By comparing circular and linear models, studies consistently demonstrate that circular approaches result in lower levels of waste, emissions, and resource depletion [77]. Integrating LCA into decision-making enables policymakers and businesses to identify the most effective strategies for sustainability, often enhanced by geospatial analysis techniques that optimize system-wide resource flows [31, 78]. This life cycle thinking supports more informed, impactful transitions toward circularity and long-term environmental resilience.

6 Case Studies

Table 3 outlines the progress of the CE across regions and sectors. The EU leads with strong policies and funding. China integrates circularity at multiple governance levels. Construction emphasizes design and material innovation. Electronics focus

Table 3 Key actions and outcomes of Circular Economy initiatives across regions and sectors

Region/sector	Key actions and outcomes
European Union (EU)	• Implementation of the CE Action Plan (CEAP) under the European Green Deal—Focus on high-impact sectors: electronics, textiles, construction—Initiatives include: Right to Repair, Digital Product Passports, Mandatory Recycled Content, Waste Reduction Targets—Outcomes: 6% growth in circular economy jobs (2012–2016), €10B+ public funding (2016–2020) [79, 80]
China	• National CE policy framework with Law for the Promotion of the CE—Three-tiered implementation: Micro (corporate), Meso (eco-industrial parks), Macro (eco-cities)—Suzhou New District as a model for industrial symbiosis—Policy tools include Five-Year Plans and performance tracking with material flow analysis [81]
Construction Sector	• Emphasis on modular, disassemblable design and lifecycle carbon assessment—Use of recycled/renewable materials like cross-laminated timber—Adoption of BIM tools for material efficiency—Regulatory reforms: UK requires CE assessments; China promotes waste-to-energy and recycled construction materials [82]
Electronics Sector	• Modular design for longevity and ease of repair—Advanced material recovery systems [83] • Circular business models: take-back schemes, product-as-a-service—Impact: on e-waste reduction, lower demand for virgin materials, reduction in raw material costs [84]
Textile Sector	• Strategic pillars: Reduce (sustainable materials), Reuse (rental/second-hand), Recycle (textile-to-textile technologies)—Notable programs: Circle Economy's Textile Programme—Policy instruments: EPR, tax incentives, waste targets—Challenges remain in technology and consumer adoption, but momentum is growing [85]

on repairability and recovery. Textiles adopt sustainable practices despite challenges, supported by policy tools and industry initiatives.

7 Challenges

Table 4 highlights the challenges to implementing a CE. Economic hurdles include high investment needs and limited demand. Regulatory systems favor linear models. Technological gaps persist in innovation and infrastructure. Data issues limit material tracking. Culturally, resistance from consumers and organizations delays adoption due to awareness gaps and short-term thinking.

Table 4 Barrier in implementation of Circular Economy

Category	Barrier	Description	Sources
Economic Barriers	High Upfront Investment	CE models require significant initial investments in R&D and scaling, often incompatible with linear-focused funding mechanisms	[86, 87]
	Limited Market Demand	Circular products face low consumer demand due to unfamiliarity, higher costs, and undervaluation in financial accounting systems	[88, 89]
Policy and Regulatory Gaps	Inadequate Regulatory Support	Current regulations favor linear models; circular initiatives face legal uncertainty, fragmented policies, and outdated standards	[87]
	Biased Subsidies and Incentives	Governmental support often prioritizes linear production, making circular alternatives financially less attractive	
Technological and Infrastructure Gaps	Lack of Advanced Technologies	Limited availability of innovative recycling and recovery technologies restricts circular adoption	[90]
	Insufficient Infrastructure	Many areas lack systems for efficient collection, sorting, and reuse of materials	[90]
	Data and Traceability Challenges	Poor data management and traceability hinder material flow monitoring and closed-loop systems	[91]
Cultural and Behavioural Resistance	Consumer Resistance	Preference for new products, low awareness, and misunderstanding of CE practices impede behavioural change	[92]
	Organizational Resistance	Lack of internal support and prioritization of short-term gains hinder circular economy implementation in businesses	[93]

8 Path Forward

To accelerate an economy-wide transition to circularity, policymakers should deploy complementary instruments, such as eco-modulated extended producer responsibility fees, green public procurement rules that prioritize products with digital passports and high recycled content, and fiscally neutral incentive mixes that reward repair, remanufacturing, and service-based business models while penalizing landfill and incineration. Public–private partnerships can de-risk scale-up by blending concessional finance with private capital, sharing physical and digital infrastructure, and tying disbursements to measurable key performance indicators for material recirculation, carbon abatement, and job creation. Parallel investments in education and awareness—from embedding life-cycle thinking in engineering and business curricula to citizen-facing '*right-to-repair*' campaigns—will build the skills and cultural norms needed to close intention–action gaps. Digital enablers amplify every lever: IoT sensors and AI optimize predictive maintenance and robotic sorting, and they are also being used for environmental monitoring such as AI-based air quality prediction in urban-industrial contexts [75, 94], helping cities align with CE targets; blockchain-secured product passports verify provenance and recycled content, facilitating both compliance and green procurement. Aligning these actions with the Sustainable Development Goals, the Paris Agreement, and emerging biodiversity frameworks unlocks additional climate-finance channels and ensures policy coherence. Finally, research and innovation should focus on the high-value recycling of complex composites and critical minerals, robust circular design metrics, socio-economic impact modeling, comparative policy mix analysis, and the development of bio-based, regenerative materials, all underpinned by transparent, ethically governed human–AI collaboration.

9 Conclusion

Drawing together the evidence presented in this chapter, the circular economy emerges as a pragmatic pathway for reconciling prosperity with planetary boundaries. By extending product lifetimes, valorizing secondary materials, and regenerating natural systems, CE strategies can unlock measurable cost savings, spur innovation-led growth, and create resilient local jobs while reducing waste and greenhouse-gas emissions. Case studies from the European Union, China, and key resource-intensive sectors demonstrate that targeted policy packages—EPR, green public procurement, and tax modulation—combined with digital enablers and public-private collaboration can translate theory into large-scale impact. However, realizing this potential requires closing technological, infrastructural, and behavioral gaps through sustained investment, skills development, and transparent data frameworks. Aligning national CE roadmaps with the SDG and the Paris Agreement will amplify climate and biodiversity co-benefits and mobilise new finance. Future research should prioritize high-value recycling, circular design metrics, and socio-economic modeling to guide this next phase of systemic transformation.

References

1. Kandpal V, Jaswal A, Santibanez Gonzalez EDR, Agarwal N (2024) Circular economy principles: shifting towards sustainable prosperity. In: Sustainable energy transition. Circular economy and sustainability. Springer, Cham, pp 125–165. https://doi.org/10.1007/978-3-031-52943-6_4
2. Aiguobarueghian I, Uwaga MA, Ogunbiyi EO, Solomon NO (2024) Waste management and circular economy: a review of sustainable practices and economic benefits. World J Adv Res Rev 22(2):1708–1719. https://doi.org/10.30574/wjarr.2024.22.2.1517
3. Madaan G, Singh A, Mittal A, Shahare P (2024) Reduce, reuse, recycle: circular economic principles, sustainability and entrepreneurship in developing ecosystems. J Small Bus Enterprise Dev 31(6):1041–1066. https://doi.org/10.1108/JSBED-01-2023-0009

4. Jawahir IS, Bradley RT (2016) Technological elements of circular economy and the principles of 6R-based closed-loop material flow in sustainable manufacturing. Procedia CIRP 40:103–108. https://doi.org/10.1016/j.procir.2016.01.067
5. Suárez-Eiroa B, Fernández E, Méndez-Martínez G, Soto-Oñate D (2019) Operational principles of circular economy for sustainable development: linking theory and practice. J Clean Prod 214:952–961. https://doi.org/10.1016/j.jclepro.2018.12.271
6. Psarommatis F, May G, Azamfirei V (2025) Product reuse and repurpose in circular manufacturing: a critical review of key challenges, shortcomings and future directions. J Remanuf. https://doi.org/10.1007/s13243-025-00153-y
7. Yu H, Zahidi I, Fai CM, Liang D, Madsen DØ (2024) Mineral waste recycling, sustainable chemical engineering, and circular economy. Results Eng 21:101865. https://doi.org/10.1016/j.rineng.2024.101865
8. Čegar S, Drezgić S, Čišić D (2024) Exploring the regenerative economy: a comprehensive review of literature. J Regen Econ 1(1):1–40. https://doi.org/10.5937/jre2401001C
9. Morseletto P (2020) Restorative and regenerative: exploring the concepts in the circular economy. J Ind Ecol 24(4):763–773. https://doi.org/10.1111/jiec.12987
10. Sreenath AB, Dhanaraj K, Rath P, Bharadwaj LK, Choudhary R (2025) Waste management and circular economy. In: Sustainable chemistry and pioneering green engineering solutions (Chapter 11). IGI Global. https://doi.org/10.4018/979-8-3373-1409-9.ch011
11. Nowicki P, Ćwiklicki M, Kafel P, Niezgoda J, Wojnarowska M (2023) The circular economy and its benefits for pro-environmental companies. Bus Strat Environ 32(7):4584–4599. https://doi.org/10.1002/bse.3382
12. Dey S, Santra M, Choudhury M, Ghosh AR, Samanta P (2024) Food waste generation and its industrial utilization: an overview. Environ Sci Pollut Res. https://doi.org/10.1007/s11356-024-34252-3
13. Rajpal A, Choudhury M, Goswami S, Chakravorty A, Raghavan V (eds) (2024) Waste management and treatment: advances and innovations, 1st edn. CRC Press. https://doi.org/10.1201/9781003258377
14. Singh S, Singh P, Sharma A, Choudhury M (eds) (2022) Agriculture waste management and bioresource: the circular economy perspective. Wiley. https://doi.org/10.1002/9781119808428
15. Velenturf APM, Archer SA, Gomes HI, Christgen B, Lag-Brotons AJ, Purnell P (2019) Circular economy and the matter of integrated resources. Sci Total Environ 689:963–969. https://doi.org/10.1016/j.scitotenv.2019.06.449
16. Figge F, Stevenson Thorpe A, Givry P, Canning L, Franklin-Johnson E (2018) Longevity and circularity as indicators of eco-efficient resource use in the circular economy. Ecol Econ 150:297–306. https://doi.org/10.1016/j.ecolecon.2018.04.030
17. Yang H, Ma M, Thompson JR, Flower RJ (2017) Waste management, informal recycling, environmental pollution and public health. J Epidemiol Commun Health 72(3):237–243. https://doi.org/10.1136/jech-2016-208597
18. Bharadwaj G, Chopde S, Taluja R, Lalitha G, Chandrashekar R, Dhahi HA (2023) Closing the loop: advances in materials, energy, and waste management. E3S Web Conf 453:01024. https://doi.org/10.1051/e3sconf/202345301024
19. Yang J (2022) From zero waste to material closed loop: The way towards circular economy (A. Zhang, Trans.). Springer, Singapore. https://doi.org/10.1007/978-981-16-7683-3
20. Bakker CA, Mugge R, Boks C, Oguchi M (2021) Understanding and managing product lifetimes in support of a circular economy. J Clean Prod 279:123764. https://doi.org/10.1016/j.jclepro.2020.123764
21. Luzzati T, Distefano T, Ialenti S, Andreoni V (2022) The circular economy and longer product lifetime: framing the effects on working time and waste. J Clean Prod 380:134836. https://doi.org/10.1016/j.jclepro.2022.134836
22. Oliveira BF, Cunha DP, Garcia R (2025) Organizational maturity and its influence on product life-cycle extension: from concept to circularity. In: Machado J, Trojanowska J, Antosz K, Leão CP, Knapčíková L, Sover A (eds) Innovations in industrial engineering IV. ICIENG 2025, Lecture notes in mechanical engineering. Springer. https://doi.org/10.1007/978-3-031-94484-0_27

23. Morseletto P (2023) Sometimes linear, sometimes circular: states of the economy and transitions to the future. J Clean Prod 390(136138):1–9. https://doi.org/10.1016/j.jclepro.2023.136138
24. Upadhayay S, Alqassimi O (2018) Transition from linear to circular economy. Westcliff Int J Appl Res 2(2):62–74. https://doi.org/10.47670/wuwijar201822OASU
25. Hopkinson P, De Angelis R, Zils M (2020) Systemic building blocks for creating and capturing value from circular economy. Resour Conserv Recycl 155:104672. https://doi.org/10.1016/j.resconrec.2019.104672
26. Korhonen J, Honkasalo A, Seppälä J (2018) Circular economy: the concept and its limitations. Ecol Econ 143:37–46. https://doi.org/10.1016/j.ecolecon.2017.06.041
27. Jensen HH (2025) 6th R—Recycle: circular closure. In: Circular economy opportunities and pathways for manufacturers. Springer, pp 201–237. https://doi.org/10.1007/978-3-031-75279-7_8
28. Choudhary R, Bharadwaj LK, Tan TN, Choudhury M, Sharma D, Rath P et al (2025a) Spatial intelligence integration in smart wastewater systems: advancing efficiency and sustainability in urban sewer networks. In: Choudhury M, Majumdar S, Goswami S, Sillanpää M (eds) Smart wastewater systems and climate change: innovations through spatial intelligence, Chapter 3, vol 15. Royal Society of Chemistry, pp 28–41. https://doi.org/10.1039/9781837676880-00028
29. Sher A, Li H, Ullah A, Hamid Y, Nasir B, Zhang J (2024) Importance of regenerative agriculture: climate, soil health, biodiversity and its socioecological impact. Discov Sustain 5:462. https://doi.org/10.1007/s43621-024-00662-z
30. Camilleri MA (2019) The circular economy's closed loop and product service systems for sustainable development: a review and appraisal. Sustain Dev 27(3):530–536. https://doi.org/10.1002/sd.1909
31. Choudhury M, Choudhary R, Sharma D, Bhattacharjee B, Majumdar S, Chegini S et al (2025) Spatial intelligence integration in smart wastewater systems: advancing efficiency and sustainability in urban sewer networks. In: Choudhury M, Majumdar S, Goswami S, Sillanpää M (eds) Smart wastewater systems and climate change: innovations through spatial intelligence, Chapter 11, vol 15. Royal Society of Chemistry, pp 155–169. https://doi.org/10.1039/9781837676880-00155
32. Chowdhury R, Barman S, Choudhury M, Kim KH, Bhattacharya SS (2024) Earthworm modifies microbial community and functional genes for lignocellulosic waste valorization: isolating plant-growth-promoting bacteria via next generation sequencing. Int Biodeterior Biodegrad 193:105854. https://doi.org/10.1016/j.ibiod.2024.105854
33. Paul S, Choudhury M, Deb U, Pegu R, Das S, Bhattacharya SS (2019) Assessing the ecological impacts of ageing on hazard potential of solid waste landfills: a green approach through vermitechnology. J Clean Prod 236:117643. https://doi.org/10.1016/j.jclepro.2019.117643
34. Sauvé S, Bernard S, Sloan P (2016) Environmental sciences, sustainable development and circular economy: alternative concepts for trans-disciplinary research. Environ Dev 17:48–56. https://doi.org/10.1016/j.envdev.2015.09.002
35. Ratsimandresy A, Miemczyk J (2023) Conceptualising collaborations beyond industrial boundaries: a literature review and a theoretical proposition to understand cross-industrial collaborations in the circular supply network. Sustainability 15(11):8850. https://doi.org/10.3390/su15118850
36. Iacovidou E, Hahladakis JN, Purnell P (2021) A systems thinking approach to understanding the challenges of achieving the circular economy. Environ Sci Pollut Res 28(19):24785–24806. https://doi.org/10.1007/s11356-020-11725-9
37. Kujala J, Heikkinen A, Blomberg A (eds) (2023) Stakeholder engagement in a sustainable circular economy: theoretical and practical perspectives, 1st edn. Palgrave Macmillan. https://doi.org/10.1007/978-3-031-31937-2
38. Salvioni DM, Almici A (2020) Transitioning toward a circular economy: the impact of stakeholder engagement on sustainability culture. Sustainability 12(20):8641. https://doi.org/10.3390/su12208641

39. Avdiushchenko A, Zając P (2019) Circular economy indicators as a supporting tool for European regional development policies. Sustainability 11(11):3025. https://doi.org/10.3390/su11113025
40. Velenturf APM, Purnell P, Tregent M, Ferguson J, Holmes A (2018) Co-producing a vision and approach for the transition towards a circular economy: perspectives from government partners. Sustainability 10(5):1401. https://doi.org/10.3390/su10051401
41. Keshavarz K, Álvarez MJ, García CJ (2025) The role of consumers in transitioning to a circular economy. Circ Econ Sustain. https://doi.org/10.1007/s43615-025-00617-0
42. de Sá P, Korinek J (2021) Resource efficiency, the circular economy, sustainable materials management and trade in metals and minerals (OECD Trade Policy Paper No. 245). OECD Publishing. https://doi.org/10.1787/69abc1bd-en
43. State of Green (2017) 10 examples of circular economy solutions. https://stateofgreen.com/en/news/10-examples-of-circular-economy-solutions/#:~:text=For%20example%2C%20an%20exchange%20of%20surplus%20resources,%E2%80%A6. Accessed 14 Jul 2025
44. Favi C, Germani M, Luzi A, Mandolini M, Marconi M (2017) A design for end-of-life approach and its application in a closed-loop product system. J Clean Prod 142(Part 4):2863–2876. https://doi.org/10.1016/j.jclepro.2017.05.083
45. Hidalgo Crespo J, Riel A, Vogt Duberg J, Bunodiere A, Golinska-Dawson P (2024) An exploratory study for product-as-a-service (PaaS) offers development for electrical and electronic equipment. Procedia CIRP 122:521–526. https://doi.org/10.1016/j.procir.2024.01.076
46. Kramer MR, Geradts T, Nadella B (2019) Philips lighting: light-as-a-service (Harvard Business School Case No. 719-446; Rev. May 2019). Harvard Business School Publishing
47. Chomachaei FI, Gal-Or E, Letizia P, Roma P (2024) The economic viability of the sharing economy business model and its environmental impact. Eur J Oper Res 315(3):1197–1209. https://doi.org/10.1016/j.ejor.2023.12.022
48. Renault Trucks (2025) The circular economy: Renault Trucks' approach to the environmental challenge. https://www.renault-trucks.com/en/circular-economy-renault-trucks-approach-environmental-challenge. Accessed 14 Jul 2025
49. Ridley SJ, Ijomah WL, Corney JR (2019) Improving the efficiency of remanufacture through enhanced pre-processing inspection: a comprehensive study of over 2,000 engines at Caterpillar remanufacturing, U.K. Prod Plann Contr 30(4):259–270. https://doi.org/10.1080/09537287.2018.1471750
50. Dell Technologies (2018) Design for environment: Dell's approach to circularity and sustainable product development [PDF]. Dell Technologies. https://www.delltechnologies.com/asset/en-sg/solutions/business-solutions/briefs-summaries/design-for-environment.pdf. Accessed 15 Jul 2025
51. Kirchherr J, Yang NHN, Schulze-Spüntrup F, Heerink MJ, Hartley K (2023) Conceptualizing the circular economy (revisited): an analysis of 221 definitions. Resour Conserv Recycl 194:107001. https://doi.org/10.1016/j.resconrec.2023.107001
52. OECD (2024) The circular economy in Central Macedonia, Greece (OECD Regional Development Paper No. 103). OECD Publishing. https://doi.org/10.1787/19f00380-en
53. Padilla Rivera A, Russo Garrido S, Merveille N (2020) Addressing the social aspects of a circular economy: a systematic literature review. Sustainability 12(19):7912. https://doi.org/10.3390/su12197912
54. Souza-Piao R, de Vincenzi TB, de Carvalho MM (2023) Strategies for social inclusion in circular economy. In: de Souza Campos LM, Vázquez-Brust DA (eds) The social dimensions of the circular economy, vol 10. Springer, pp 265–282. https://doi.org/10.1007/978-3-031-25436-9_12
55. Vidal U, Obregon M, Ramos ED, Verma R, Coles PS (2024) Sustainable and risk-resilient circular supply chain: a Peruvian paint manufacturing supply chain model. Sustain Fut 7(1–2):100207. https://doi.org/10.1016/j.sftr.2024.100207
56. Sharman S (2025) The rise of the circular economy: how businesses are closing the loop. MYGroup Ltd. https://mygroupltd.com/news/the-rise-of-the-circular-economy-how-businesses-are-closing-the-loop/#:~:text=Resource%20Security%20and%20Efficiency,impact%20is%20a%20growing%20concern. Accessed 14 Jul 2025

57. Iftikhar A, Ali I, Zhan Y, Stevenson M, Tarba SY (2025) Firms' strategic responses to rising uncertainty amid ongoing geopolitical tensions: the synergistic mediating role of network capability and innovation ambidexterity. Transp Res Part E Logist Transp Rev 199:104146. https://doi.org/10.1016/j.tre.2025.104146
58. Tripathi S, Agarwal S, Choudhary R, Choudhury M, Padoin N, van Hullebusch ED et al (2025) Strengthening disaster resilience: emergency response strategies in wastewater management. In: Choudhury M, Majumdar S, Goswami S, Sillanpää M (eds) Smart wastewater systems and climate change: innovations through spatial intelligence, Chapter 10, vol 15. Royal Society of Chemistry, pp 135–154. https://doi.org/10.1039/9781837676880-00135
59. Khan S, Haleem A, Fatma N (2022) Effective adoption of remanufacturing practices: a step towards circular economy. J Remanuf 12:167–185. https://doi.org/10.1007/s13243-021-00109-y
60. Xiao D (2025) Evaluating and prioritizing strategies to reduce carbon emissions in the circular economy for environmental sustainability. J Environ Manag 373:123446. https://doi.org/10.1016/j.jenvman.2024.123446
61. Masterson V (2024) 4 charts to show why adopting a circular economy matters. World Economic Forum. https://www.weforum.org/stories/2024/04/circular-economy-waste-management-unep/
62. Ganguly RK, Chakraborty SK (2024) Plastic waste management during and post Covid-19 pandemic: challenges and strategies towards circular economy. Heliyon 10(4):e25613. https://doi.org/10.1016/j.heliyon.2024.e25613
63. Bredun V, Choudhary R, Kumar A (2024) Regional specifics of using community bins in waste management: a case study of rural communities in Poltava Region (Ukraine). Trends Ecol Indoor Environ Eng 2(4):10–17. https://doi.org/10.62622/TEIEE.024.2.4.10-17
64. Australian Bureau of Statistics (2024) Circular economy. In: Measuring what matters: themes and indicators—sustainable. https://www.abs.gov.au/statistics/measuring-what-matters/measuring-what-matters-themes-and-indicators/sustainable/circular-economy#:~:text=2.95%20tonnes%20of%20waste%20was,2006%2D07%20(50%25). Accessed 14 Jul 2025
65. European Environment Agency (2024) Capturing the climate change mitigation benefits of circular economy and waste sector policies and measures (Briefing No. 25/2023). https://doi.org/10.2800/512574
66. European Environment Agency (2024) Europe's circular economy in facts and figures (EEA Briefing No. 17/2024) [PDF]. Publications Office of the European Union. https://www.eea.europa.eu/en/analysis/publications/europes-circular-economy-in-facts
67. Masterson V (2021) Circularity Gap Report 2021: strategies to address climate breakdown. World Economic Forum. https://www.weforum.org/stories/2021/02/circularity-gap-report-2021-strategies-climate-breakdown/. Accessed 14 Jul 2025
68. Ministry of Housing & Urban Affairs (2022) Circular economy in waste management [PDF]. Government of India. https://mohua.gov.in/pdf/627b8318adf18Circular-Economy-in-waste-management-FINAL.pdf
69. European Environment Agency (2023) The benefits to biodiversity of a strong circular economy (EEA Briefing No. 10/2023) [PDF]. Publications Office of the European Union. https://www.eea.europa.eu/publications/the-benefits-to-biodiversity (Contains ISSN 2467-3196, ISBN 978-92-9480-569-0). https://doi.org/10.2800/489228
70. Earth5R (2025) Earth5R's urban biodiversity revival blueprint through native plantation drives. https://earth5r.org/urban-biodiversity-native-plantation/. Accessed 15 Jul 2025
71. Chowdhury PR, Goswami S, Choudhary R, Choudhury M (2025) Electronic wastes, its impact on wildlife and biodiversity. In: Electronic waste (Chapter 8). CRC Press. https://doi.org/10.1201/9781003582311-8
72. Scheel C, Aguiñaga E, Bello B (2020) Decoupling economic development from the consumption of finite resources using circular economy: a model for developing countries. Sustainability 12(4):1291. https://doi.org/10.3390/su12041291
73. Choudhary R, Kumari S, Kumar A, Kumar P, Choudhury M, Sharma D et al (2025b) Optimizing wastewater management through geospatial analysis. In: Choudhury M, Majumdar S, Goswami

S, Sillanpää M (eds) Smart wastewater systems and climate change: innovations through spatial intelligence, Chapter 6, vol 15. Royal Society of Chemistry, pp 72–87. https://doi.org/10.1039/9781837676880-00072
74. Choudhary R, Kumar A, Narayan S, Mishchenko I, Choudhury M, Bharadwaj LK et al (2025c) Integrating fuzzy logic into life cycle assessment for sustainable municipal solid waste management in metro cities. Ecol Eng Environ Technol 26(7):316–331. https://doi.org/10.12912/27197050/205500
75. Choudhary R, Kumar A, Priyadharsini C, Mude M, Naik M, Moharana C, Khan NA (2025d) Predicting water quality index using stacked ensemble regression and SHAP based explainable artificial intelligence. Sci Rep 15(1):9463. https://doi.org/10.1038/s41598-025-09463-4
76. PRé Sustainability (2022) Life cycle assessment (LCA): basics, meaning & why you need it. https://pre-sustainability.com/articles/life-cycle-assessment-lca-basics/. Accessed 15 Jul 2025
77. Environmental Reports (2023) Sustainable waste management in the circular economy: challenges and opportunities. https://er.researchfloor.org/sustainable-waste-management-in-the-circular-economy-challenges-and-opportunities/. Accessed 15 Jul 2025
78. Strippoli R, Gallucci T, Ingrao C (2025) Life cycle assessment for sustainable business management. Top Ital Sci J 2(3). https://doi.org/10.62684/SBXL2822
79. Centre for Sustainability & Excellence (CSE) (2025) The EU's road to a circular economy by 2025: key steps and challenges. https://cse-net.org/eu-road-to-a-circular-economy-by-2025-key-steps-and-challenges/. Accessed 15 Jul 2025
80. Ellen MacArthur Foundation (2022) The EU's circular economy action plan [case study]. https://www.ellenmacarthurfoundation.org/circular-examples/the-eus-circular-economy-action-plan. Accessed 15 Jul 2025
81. Mathews JA, Tan H (2016) Circular economy: lessons from China. Nature 531:440–442. https://doi.org/10.1038/531440a
82. RLB Europe (2024) Circular economy in construction: a strategic pathway to net zero [Blog post]. https://www.rlb.com/europe/insight/circular-economy-in-construction-a-strategic-pathway-to-net-zero/. Accessed 15 Jul 2025
83. Amend C, Revellio F, Tenner I, Schaltegger S (2022) The potential of modular product design on repair behavior and user experience—evidence from the smartphone industry. J Clean Prod 367:132770. https://doi.org/10.1016/j.jclepro.2022.132770
84. Report Yak (2024) Circular business model: key to sustainable business success. https://reportyak.com/blog/circular-business-model/. Accessed 15 Jul 2025
85. Recovo (2025) How to implement the circular economy in the textile industry. https://recovo.co/en/blog/article/how-to-implement-the-circular-economy-in-the-textile-industry/. Accessed 15 Jul 2025
86. Manutan (2024) Circular economy limitations: how to overcome them [Blog post]. https://www.manutan.com/blog/en/circular-economy/circular-economy-limitations-how-to-overcome-them/. Accessed 15 Jul 2025
87. You Matter (2024) Circular economy: definition, principles, benefits and barriers. https://youmatter.world/en/definition/definitions-circular-economy-meaning-definition-benefits-barriers/. Accessed 15 Jul 2025
88. AlJaber A, Martinez Vazquez P, Baniotopoulos C (2023) Barriers and enablers to the adoption of circular economy concept in the building sector: a systematic literature review. Buildings 13(11):2778. https://doi.org/10.3390/buildings13112778
89. Rizos V, Bryhn J, Alessi M, Righetti E, Fujiwara N, Stroia C (2021) Barriers and enablers for implementing circular economy business models: evidence from the electrical and electronic equipment and agri-food value chains (CEPS Research Report No. RR2021-01) [PDF]. Centre for European Policy Studies. https://cdn.ceps.eu/wp-content/uploads/2021/10/RR2021-01_Barriers-and-enablers-for-implementing-circular-economy-business-models.pdf
90. International Chamber of Commerce & EY (2024) Putting the circular economy into motion: from barriers to opportunities [Report]. International Chamber of Commerce. https://iccwbo.org/wp-content/uploads/sites/3/2024/10/2024_ICC-x-EY-Report_05-1.pdf

91. Roy R (2022) Redesigning traditional linear supply chains into circular supply chains: a study into its challenges. Sustain Prod Consump 31:113–131. https://doi.org/10.1016/j.spc.2022.02.004
92. Fletcher CA, Aureli S, Foschi E, Leal Filho W (2024) Implications of consumer orientation towards environmental sustainability on the uptake of bio-based and biodegradable plastics. Curr Res Environ Sustain 7:100246. https://doi.org/10.1016/j.crsust.2024.100246
93. Gulbrandson T, Bhat R (2024) Prioritizing barriers to circular economy implementation in the toy industry using the Analytical Hierarchy Process. J Sci Ind Res 84(3):9847. https://doi.org/10.56042/jsir.v84i03.9847
94. Halaktionov M, Bredun V, Choudhary R, Goroneskul M, Kumar A, Ouiya F, Sydorenko V, Markina L (2025) AI-enhanced air quality assessment and prediction in industrial cities: a case study of Kryvyi Rih, Ukraine. Ecol Eng Environ Technol 26(6):45–56. https://doi.org/10.12912/27197050/203725

Assessing Traditional Waste Management Practices in Urban Environments and Recent Trends

Dongzi Wu, Wanying Liao, Hanqin Yu, Sushobhan Majumdar, Jiajun Xu, and Moharana Choudhury

Abstract The development of municipal waste management systems reflects the co-evolution of technological advances, regulatory frameworks, and shifting societal norms. Despite notable progress, contemporary systems remain constrained by structural inefficiencies and unresolved sustainability challenges. This chapter critically examines the historical trajectory of waste governance, tracing the insights and trends that have evolved from traditional models. It further explores four core dimensions: the overview of conventional systems, their persistent limitations, emergent global trends, and recent innovations. Drawing on international cases, the analysis contrasts diverse governance approaches, highlighting how low-income contexts often foster adaptive, community-led solutions, while high-income systems exhibit technological sophistication but structural inertia. Meanwhile, transitional economies are increasingly experimenting with hybrid models. Notably, it argues that modern waste governance must transcend mere technological substitution; instead, it requires a paradigm

D. Wu
New Frontiers International Planning & Design (Hubei) Co., Ltd., Wuhan, Hubei, China

W. Liao (✉)
Faculty of Arts and Social Sciences, Universiti Malaya, Kuala Lumpur, Malaysia
e-mail: wanyingliao1998@gmail.com

H. Yu
School of Law, Shanghai University of Finance and Economics, Shanghai, China

S. Majumdar
Department of Geography, Pandit Raghunath Murmu Smriti Mahavidyalaya, Bankura University, Bankura, India

J. Xu (✉)
Institute for Advanced Studies, Universiti Malaya, Kuala Lumpur, Malaysia
e-mail: jiajunxu2000@gmail.com

M. Choudhury
Environmental Research and Management Division, Voice of Environment (VoE), Guwahati, India

J. Xu
Department of Architecture, College of Design and Engineering, National University of Singapore, Singapore, Singapore

S. Majumdar et al. (eds.), *Sustainable Urban Future*, Urban Sustainability,
https://doi.org/10.1007/978-981-95-2976-6_5

shift toward circularity, grounded in ecological balance, systemic integration, and social justice. In synthesizing these trends, it proposes a forward-facing perspective for waste governance—one that is adaptive, inclusive, and resilient, and that reconceives urban waste not as a single problem of disposal, but as an opportunity for systemic socio-environmental transformation.

Keywords Traditional waste management · Historical development · Challenges and limitations · Trends and innovation · Urban sustainability

1 Introduction

Urban governance entails navigating a complex interplay of social, economic, and environmental dimensions [1]. In the context of accelerating urbanization, this complexity has intensified [2]; the rising concentration of populations in metropolitan regions [3, 4], the physical expansion of cities [3, 5], and the mounting strain on aging or insufficient infrastructure [6–8] have increasingly challenged the adaptive capacity of urban systems [9, 10]. Concurrently, cities face escalating environmental degradation [11, 12], rising climate risks [13–15], persistent traffic congestion [16, 17], and growing deficits in energy and natural resources [18]. These pressures highlight the limitations of conventional governance, which are often hindered by fragmented data, slow feedback mechanisms, and rigid institutional structures [19–21].

Within this broader governance landscape, urban waste management has evolved from rudimentary and disjointed practices into a progressively structured domain shaped by modern regulatory and infrastructural developments [22]. Historically dominated by a linear "Use-Dispose" model, traditional systems have contributed to mounting resource depletion, environmental degradation, and recurring economic inefficiencies [23, 24]. As urbanization and industrial activity intensify, the volume and complexity of urban waste have increasingly outpaced the capacity of traditional end-of-pipe treatment approaches [25]. The environmental burdens of such systems are enduring: communities hosting waste facilities often continue to experience ecological and public health risks long after operations cease [26].

Over the past decade, in rapidly expanding urban centers of developing countries, evidence from China suggests that rising levels of social urbanization have suppressed the ecological efficiency of surrounding areas [27]. For instance, despite the massive volumes of waste processed daily in many cities, significant quantities still enter the natural environment through informal channels; this uncontrolled discharge highlights the critical shortcomings of traditional end-of-pipe governance models [22]. Consequently, there is a growing consensus that systemic solutions—particularly the shift from fragmented end-stage interventions to comprehensive material flow management—may hold the key to addressing urban waste challenges more effectively [22].

In response to the above issues, this chapter provides a critical examination of the historical evolution of municipal waste management, revealing how traditional systems—shaped by infrastructure, governance, and material flow—have perpetuated linear, end-of-life approaches rooted in landfilling and incineration [28, 29]. It exposes the structural constraints of these models: inefficient resource recovery, behavioral inertia, and fragmented institutional design. Through comparative analysis of global cases, the chapter delineates distinct governance trajectories—from the improvisational resilience of informal systems in low-income settings to the procedural rigidity of high-income regimes, and the adaptive hybridization emerging in transitional economies. Recent innovations are assessed not merely as technological upgrades, but as signals of more profound systemic shifts: the uptake of digital tools such as Internet of Things (IoT) and big data, the integration of sustainability-driven policy instruments, and the reconfiguration of waste governance around community-based and decentralized frameworks [30]. By situating these developments within a broader socio-ecological and institutional context, this chapter advances a forward-looking understanding of waste governance—one that is environmentally sustainable, socially inclusive, and structurally coherent.

2 Waste Management Practices: Historical Insights and Evolving Trends

The evolution of municipal waste management offers a microcosmic reflection of broader societal transformations, from early urbanization to contemporary sustainability discourses. The earliest documented urban sanitation system, dating back to 800–735 BCE (Before Common Era) in Rome, culminated in a citywide sewage infrastructure by 100 BCE [31]. However, despite its scale, this infrastructure merely channeled untreated effluents into the Tiber River, highlighting an absence of technological capacity for sewage treatment at the time [31]. In parallel, Athens passed what is arguably the earliest known municipal waste law in 320 BCE, prohibiting the disposal of waste on unpaved roads, signaling a nascent recognition of waste as a public concern [32].

Throughout the medieval period, waste management remained rudimentary. Household refuse—often consisting of food scraps and animal waste—was indiscriminately discarded in streets and rivers; in response, emerging urban centers, particularly in the United Kingdom and Italy, introduced ordinances aimed at regulating waste clearance [33, 34]. The outbreak of the Black Death in the fourteenth century further compelled cities to strengthen public sanitation and solid waste management as part of their epidemic response strategies; however, this was not out of environmental foresight [32–34].

The advent of industrialization catalyzed a paradigmatic shift in waste governance. Victorian-era (1837–1901) public health legislation institutionalized urban sanitation as a national legal responsibility, marking a decisive turn toward formalized municipal waste systems in Europe [35]. By the latter half of the nineteenth century, technical advancements in solid waste management began to emerge. Innovations such as waterproof waste containers and purpose-built waste collection vehicles emerged, particularly in the United States of America [32]. In 1874, Nottingham introduced the world's first municipal incinerator, dubbed the "Destructor," designed by Stephen Fryer, laying the foundation for incineration-based urban waste strategies [32].

Mid-twentieth-century waste management saw further technical refinement with the introduction of waste shredders, compactors, and pneumatic collection systems; however, these innovations coincided with growing awareness of the environmental and public health risks associated with open dumping and uncontrolled incineration [32]. In response, sanitary landfills increasingly replaced uncontrolled disposal sites, and incineration was reevaluated and improved. In Brazil, for example, incineration has been promoted as a viable strategy under the national solid waste policy, as it substantially reduces landfill volume while enabling energy recovery through combustion [36]. Nevertheless, public resistance to landfill and incineration projects persists due to their disproportionate impacts on local ecosystems and communities [37].

Crucially, the socio-spatial legacy of traditional waste governance underscores deeper environmental justice concerns [38, 39]. Studies in Bayelsa State, Nigeria, have demonstrated how proximity to former landfill sites correlates with elevated soil heavy metal concentrations—effects that diminish with distance yet reveal long-term environmental degradation [40]. Such cases exemplify how low-income and marginalized populations disproportionately bear the risks of hazardous waste exposure, often as a result of unequal political influence and systemic neglect [37, 41]. Accordingly, "NIMBY (Not in My Backyard)" resistance remains pervasive, particularly in communities that bear disproportionate environmental burdens and lack equitable decision-making authority [42, 43]. Furthermore, on a global scale, the transboundary movement of waste further reveals entrenched asymmetries: affluent nations have long externalized their environmental burdens by exporting waste to developing countries [44]. Thus, while waste may be locally disposed of, its geopolitical and ethical implications often transcend spatial and temporal boundaries.

In contemporary contexts, the escalating environmental, economic, and public health pressures associated with growing volumes of municipal solid waste have prompted governments to align waste management strategies with broader sustainable development goals. In many countries, waste is now classified into hazardous and non-hazardous categories, with separate regulatory frameworks governing their disposal [32]. Integrated solid waste management has emerged as a dominant paradigm, emphasizing the need to tailor treatment strategies to the heterogeneous nature of waste streams [37]. It promotes both centralized and decentralized systems, prioritizing source separation and material-specific processing to enhance efficiency and minimize ecological disruption [37].

Therefore, the evolution of urban waste management systems reflects a condensed history of human civilization. From the rudimentary infrastructure of ancient societies to the institutional and technological advancements of the industrial era, each stage has contributed to progress in waste management. However, contemporary waste management models still face ecological and sustainability challenges. Truly sustainable waste governance must strike a delicate balance between technological advancement and ecological capacity.

3 Overview of Traditional Urban Waste Management Systems

Traditional urban waste management systems are predominantly shaped by a linear, end-of-pipe framework—characterized by a sequential model of collection, transportation, processing, and disposal (Fig. 1). At their core lie two infrastructural pillars: landfills and incineration facilities. Governmental sanitation departments typically collect unsorted waste along fixed routes, transferring it to centralized sites for final disposal—either burial in landfills or incineration for volumetric reduction [25]. This reactive system prioritizes final single-line processing over the circular economy model; by design, it overlooks upstream interventions and fails to incorporate resource circularity, thereby diverting large quantities of potentially recoverable materials into irreversible disposal channels [45]. As a result, significant inefficiencies emerge at the stages of mixed-waste collection and transport, embedding ecological risks deep within the system's operational logic and exacerbating environmental burdens during downstream processing [46].

Traditional waste governance is characterized by a highly centralized, state-led model (Fig. 2), often described as "Vertical Command-and-Control" [47, 48]. Authority and operational responsibility are concentrated within municipal agencies, with environmental protection departments tasked with setting and enforcing regulatory standards, while housing, urban construction, or sanitation departments manage the day-to-day logistics of waste collection, sanitation, and street maintenance. However, this framework is institutionally limited; it systematically excludes producer responsibility and minimizes public engagement, thereby constraining the involvement of market actors and civil society in shaping waste outcomes. The result is a rigid, top-down system that inhibits innovation, undermines shared accountability, and weakens the potential for participatory and decentralized approaches to urban waste governance.

The operation of waste stream management within traditional systems functions within an institutional environment defined by established policy frameworks and infrastructure configurations, and is characterized by the following (Fig. 3). Its policy system demonstrates significant historical continuity; regulatory regimes are frequently designed based on early waste composition and management models, exhibiting adjustment lags in response to accelerating urbanization and evolving consumption patterns [49]. This policy inertia undermines progress toward hierarchical waste treatment, such as source reduction and material separation, by disincentivizing systemic change [50]. Simultaneously, infrastructural limitations perpetuate

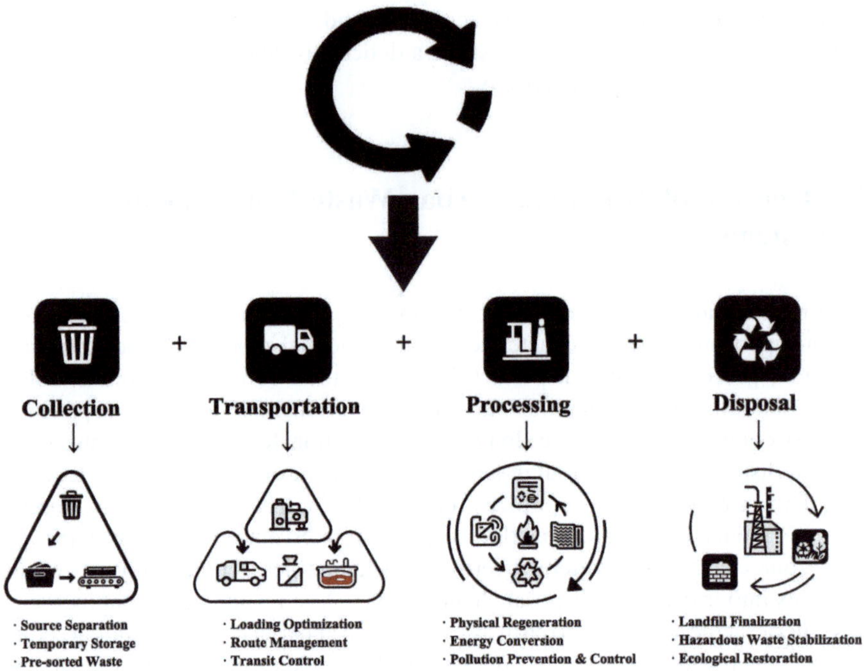

Fig. 1 Core components and processes of traditional waste management systems

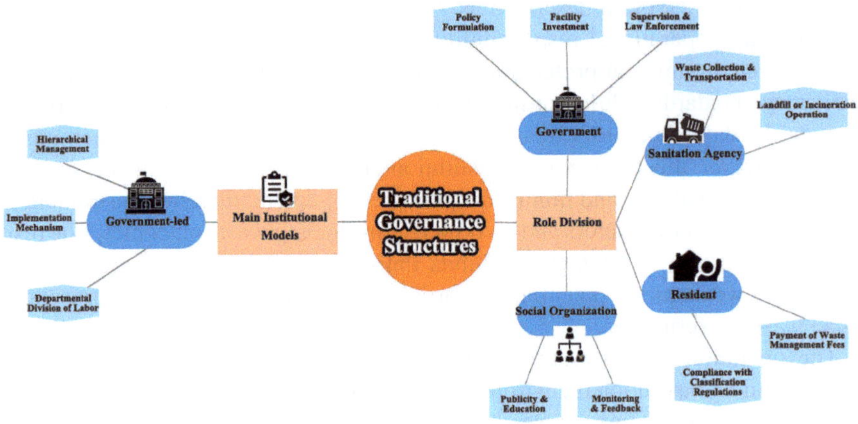

Fig. 2 Traditional management models and division of responsibilities in waste governance

Fig. 3 Main waste sources, management dilemmas, and transformation directions of traditional waste streams

reliance on outdated end-of-pipe solutions, restricting the potential for integrated and forward-looking waste strategies [51].

4 Challenges and Limitations of Traditional Waste Management

4.1 Overreliance on Landfills and Incineration

Traditional disposal methods—landfilling and incineration—have long dominated municipal solid waste management, yet both approaches impose significant ecological and economic burdens [25, 52].

Landfills are a significant source of greenhouse gas emissions, particularly methane generated through the anaerobic decomposition of organic waste, with CO_2 emissions further compounding the problem [53, 54]. Moreover, the spatial and financial demands of landfill construction and maintenance are unsustainable, particularly in densely urbanized areas where land scarcity drives costs upward.

The presumed superiority of incineration as a waste-reducing technology has likewise been steadily dismantled. While it reduces physical volume, incineration emits greenhouse gases and destroys potentially recyclable materials, such as metals, thereby undermining resource recovery goals; the release of toxic by-products, including dioxins, presents acute health and ecological hazards [45]. Furthermore, high operational costs and regulatory complexities only exacerbate these sustainability concerns.

4.2 Resource Recovery Gaps and Public Behavior

Traditional systems remain hindered by structural inefficiencies in material recovery and persistently low public participation [21, 55]. In 2019, only 9% of the 370 million tons of plastics produced globally were recycled, while the majority were both incinerated or discarded into landfills and natural environments—an indictment of systemic inefficiency [56]. Public behavior remains a critical yet often ignored determinant of waste system performance [21, 25]. Entrenched habits, misinformation, and low environmental literacy contribute to improper disposal and sorting, ultimately limiting the effectiveness of even well-designed infrastructure [55, 57]. Thus, previous research has pointed out that addressing these issues requires more than capital investment; it demands sustained behavioral change initiatives and educational outreach to foster public awareness and collective responsibility [58, 59].

4.3 Policy and Infrastructure Barriers

Outdated policy frameworks and policy inertia represent a core bottleneck in modernizing waste management [60, 61]. As urbanization accelerates and consumption patterns evolve, many regulatory regimes remain anchored in obsolete assumptions, lacking both strategic foresight and operational flexibility; this rigidity hampers innovation and disincentivizes systemic reform [62].

Simultaneously, critical infrastructure—ranging from waste collection to processing facilities—remains either non-existent or underdeveloped in many regions [21, 63–65]. In Indonesia, for example, insufficient provision of bins, trucks, and storage containers has severely impaired waste handling in traditional markets, illustrating how infrastructure gaps translate directly into functional failures [66].

Therefore, as a result, transforming entrenched systems requires synergistic progress across technological, institutional, and educational dimensions, such as an integrated waste management (IWM) system.

- Technological upgrades—such as biological treatment processes, artificial intelligence (AI)-assisted sorting, and waste-to-energy conversion—can enhance system efficiency and improve environmental outcomes [58, 67].
- Institutional reform must move beyond fragmented governance. In Pontianak, Indonesia, a SWOT-based framework adapted from Kyoto's best practices facilitated the design of IWM policies that emphasized reduction, recycling, and stakeholder accountability [68].
- Institutional reform must move beyond fragmented governance. In Pontianak, Indonesia, a SWOT-based framework adapted from Kyoto's best practices facilitated the design of IWM policies that emphasized reduction, recycling, and stakeholder accountability [68].
- Behavioral transformation through education offers a crucial lever for change. In the case of Witama National Plus, targeted curriculum interventions significantly improved student awareness, attitudes, and competencies in waste handling, demonstrating that education can serve as both a catalyst and a cornerstone for sustainable waste governance [69].

5 Recent Trends in Global Waste Management

5.1 Grassroots Systems in Low-Income Contexts: Informality and Community Innovation

In many low- and middle-income regions, informal systems remain central to urban waste governance [70–72]. Waste pickers in cities such as Delhi provide essential collecting and recycling services within cities by informally collecting, sorting, and trading recyclable materials, thus contributing significantly to resource recovery despite lacking formal recognition or infrastructural support [73, 74]. In São Paulo, while the "Recycler Rights Law" formally acknowledges grassroots organizations—such as scavenger cooperatives—and guarantees them a share of recyclable materials, these groups remain heavily dependent on municipal authorities; this dependency often results in operational inefficiencies, highlighting the gap between policy frameworks and implementation realities [75]. Meanwhile, experiments with digital payment and waste-tracking platforms in informal settlements, as documented by Guma [76], have shown that technological interventions have significantly improved household livelihoods. However, the collapse of these digital utopian systems reveals the fragility of technological transplants in contexts where power outages are more frequent than meals. These cases highlight both the adaptive capacity and systemic fragility of community-driven models, underscoring the urgent need for integrated policy frameworks that secure their sustainability without undermining local autonomy [77].

5.2 Governance Models in High-Income Countries: Efficiency and Entrenched Limitation

In contrast, high-income countries often employ centralized waste management systems governed by stringent regulatory regimes and advanced technologies [78]. While low-income countries often rely on rudimentary methods such as non-aerated lagoons, natural evaporation, and leachate recirculation for waste and leachate treatment, middle- and high-income countries have adopted more technologically intensive solutions [79]. These include coagulation-flocculation, anaerobic digestion, anammox processes, adsorption via granular activated carbon, reverse osmosis, and advanced oxidation techniques—each selected to comply with increasingly stringent environmental standards [79]. In Europe, the implementation of sophisticated hybrid systems reflects regulatory imperatives and technological capacity, enabling compliance with tight discharge limits; nonetheless, such systems continue to face environmental and logistical pressures, including diminishing landfill capacity, rising operational costs, and stagnant recycling rates [52, 79]. Japan's long-standing reliance on incineration offers a telling example: Constrained by limited land availability, the country had already reached a high incineration rate for combustible municipal waste by the 1990s; although incineration has improved energy efficiency to some extent, the long-term investment in such infrastructure—often bound by extended operational contracts—has hindered technological innovation and system upgrades [80]. Thus, in short, limited land availability led to the early adoption of incineration as the dominant waste treatment method in Japan; however, heavy infrastructural investment has produced a form of technological lock-in, impeding flexibility and innovation in more sustainable alternatives [80]. These examples demonstrate that even technologically advanced systems are not immune to structural inertia and that policy resilience necessitates a continual reassessment of entrenched paradigms.

5.3 Integrative Models in Transitioning Societies: Blending Tradition with Technological Renewal

An emerging category of hybrid and comprehensive systems blends long-standing practices with modern sustainability goals, which has been notably evident in the conceptualization and operationalization of integrated sustainable waste management (ISWM), particularly in developing and transitioning societies [81, 82]. Introduced in the early 1990s, ISWM reconfigures waste governance around three core dimensions: what—the physical system components; who—the diversity of actors involved; and how—the institutional, social, and political mechanisms through which waste is managed [82]. This model has since evolved into a dual-triangle framework, juxtaposing functional goals—public health, environmental protection, and the 3Rs (reduce, reuse, recycle)—against governance strategies that emphasize inclusivity, financial viability, and institutional robustness [82]. An integrated waste system is a waste management model based on the principles of the circular economy; the application of environmental protection technologies, community participation, multi-objective decision-making, and inclusive government policies can improve the effectiveness of waste management [83, 84]. In Sweden, municipal solid waste comprises various fractions, such as food, metals, paper, plastics, and hazardous materials; in response, Sweden exemplifies integrative countermeasures through its highly differentiated municipal solid waste scheme, which emphasizes technology-based rigorous source separation supported by public education, fiscal instruments, and legal mandates [85]. The effective implementation of Sweden's waste management system has resulted in a significant reduction in landfill usage, while simultaneously increasing both material and energy recovery [85]. The system's success lies not only in reducing landfill dependence but also in its adaptability to evolving civic norms and consumer behaviors; moreover, Sweden's openness to international knowledge exchange reinforces its capacity for iterative improvement, offering a scalable model for other transitioning societies. These hybrid systems suggest that transitioning societies are not only integrating technological innovations but are also embedding them within inclusive, context-sensitive governance structures, aligning traditional waste handling practices with the principles of sustainable development.

6 Recent Innovations in Waste Management

6.1 *Digital Technologies: Internet of Things (IoT) and Big Data*

The integration of IoT technologies and big data analytics is redefining waste management paradigms through enhanced operational intelligence [86, 87]. IoT systems, utilizing sensors and RFID (Radio Frequency Identification), enable real-time tracking of waste streams, optimizing collection and transportation logistics while reducing associated costs [88]. These tools also facilitate the dynamic monitoring of processing facilities, enabling early detection of operational inefficiencies or emission breaches and ensuring regulatory compliance [89, 90]. Concurrently, big data analytics extract actionable insights from the vast data generated across waste management processes [91]. For instance, temporal patterns in waste generation can be utilized to inform route planning for collection fleets, thereby improving resource allocation and efficiency. Similarly, emissions data can be continuously analyzed to adjust processing parameters and reduce ecological footprints [91]. The synergistic application of IoT and big data thus fosters a responsive, data-driven system that enhances resource efficiency across the entire waste lifecycle, from waste generation and collection to transportation and processing [92].

6.2 *Emerging Strategies: Integrated Frameworks and Processing Innovations*

Innovative processing frameworks are crucial in transitioning toward sustainable urban waste regimes [93–95]. Singapore's Zero Waste Master Plan exemplifies this shift through a systemic approach combining infrastructural innovation with legislative foresight [93]. Rejecting the literal interpretation of "Zero Waste," the plan advocates for a pragmatic approach that prioritizes waste minimization, resource recovery, and safe disposal [93]. Notably, Singapore's zero waste initiative includes expanding the "Extended Producer Responsibility" scheme, which, following its implementation for electronic waste, will be extended to packaging waste in the future, thereby strengthening upstream accountability [96]. Simultaneously, the Plan supports decentralized food waste treatment, aligning with circular economy principles and enabling adaptive responses to urban density and land scarcity [93]. Singapore's experience underscores how tightly integrated technological, legislative, and infrastructural innovative reforms can shape resilient waste management ecosystems.

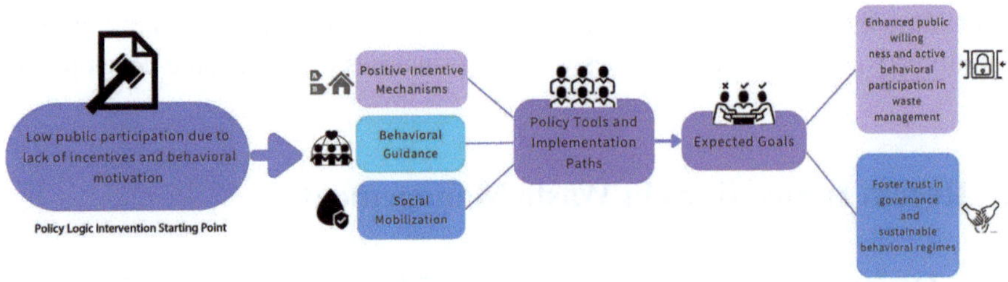

Fig. 4 A framework linking policy instruments to public engagement in waste management

6.3 *Sustainability Initiatives: Policy Reforms and Fiscal Incentives*

Transformative waste governance increasingly hinges on policy innovation and fiscal incentives [97, 98]. Instruments such as carbon taxes internalize environmental costs, compelling firms to adopt emission-reduction technologies and low-carbon production models, thereby advancing waste management reforms [99]. By increasing the cost of carbon emissions, the tax incentivizes companies to engage more actively in resource recovery activities, invest in energy-saving and emissions-reduction technologies, and shift towards low-carbon and sustainable business models; in turn, this promotes resource recycling, drives technological innovation, and facilitates industrial restructuring towards more sustainable practices [99]. In parallel, green subsidies serve as positive incentives that stimulate investment in environmentally friendly technologies and reinforce corporate environmental responsibility. Empirical evidence from China reveals that a positive correlation between green subsidies and companies' clean production and end-of-pipe treatment technologies, with clean production technologies mediating the relationship between subsidies and the company's green image; subsidies not only foster technological upgrades in clean production but also enhance firms' environmental reputations through mediated mechanisms such as absorptive capacity [100]. Moreover, policy frameworks that incorporate public-facing incentives (Fig. 4), e.g., recycling rebates or rewards for waste sorting, can mobilize community engagement and reinforce behavioral shifts toward sustainable practices [101].

6.4 *Community Innovation: Decentralized Systems and Informal Sector Empowerment*

A critical yet underexplored dimension of waste management innovation lies in decentralized, community-led systems that integrate informal sector actors [63, 102, 103]. In many urban contexts, informal waste workers—often excluded from formal governance frameworks—constitute the backbone of material recovery, operating through locally embedded knowledge and adaptive practices [104–107]. Recognizing and integrating these informal actors and sectors within official waste systems not only improves recovery rates but also contributes to social equity [107, 108]. Participatory models that delegate waste governance to local authorities or community cooperatives can foster context-sensitive solutions, increasing responsiveness to local infrastructural heterogeneity [109]. Legal recognition, access to resources, and inclusion in decision-making processes are key enablers of advancing this transition [106]. Decentralization, therefore, is not merely an administrative adjustment but a strategic reconfiguration of governance that foregrounds bottom-up agency, enabling more just and resilient waste management futures [110].

7 Conclusion

The evolution of waste management—from rudimentary disposal to regulated, technology-enhanced systems—offers critical lessons for sustainable urban governance. Early practices such as open dumping and unmanaged landfilling temporarily mitigated waste accumulation but caused long-term environmental degradation, particularly through soil and groundwater contamination. Traditional methods, including household composting, feeding livestock with organic waste, and artisanal repurposing, reflect culturally embedded practices of resourcefulness and ecological sensitivity. Nevertheless, these practices cannot often address the scale and complexity of contemporary waste streams. As awareness of the environmental consequences of waste intensified, many nations introduced legislative frameworks to regulate the generation, transport, and disposal of waste; these legal instruments played a foundational role in institutionalizing waste governance and fostering the transition toward structured systems.

Nowadays, a comparative assessment of traditional and modern systems reveals a broader epistemic shift: from reactive, labor-intensive responses to preventative, data-driven, and circular strategies. Modern waste management must transcend uniform models, adopting differentiated, locally adapted approaches that integrate diverse methods for minimization, resource recovery, and safe disposal. This is evident in the European Union's progressive alignment with circular economy principles, India's Smart City Mission, which targets infrastructure and governance reforms, and the U.S. Environmental Protection Agency's longstanding emphasis on the 3Rs [111]. However, circular economy transitions remain uneven across countries, shaped by socio-economic, political, and cultural contingencies.

In the future, advanced technologies, including IoT, AI, and GIS (Geographic Information Systems), have expanded the scope and precision of waste management operations; however, these innovations bring new challenges—high capital costs, unequal accessibility, regulatory voids, and data governance issues [111, 112]. A pragmatic pathway forward lies in hybridizing traditional practices with technological augmentation: retaining local legitimacy while enhancing efficiency. This integration requires not only technical refinement but also the development of inclusive policy design and coordinated innovation among policymakers, urban planners, and researchers. Life-cycle-oriented regulatory frameworks, anchored in legal reform, fiscal incentives, and transnational cooperation, are essential for systemic transformation. Equally important are iterative mechanisms, such as joint innovation labs, pilot interventions, and adaptive policy platforms, which enable continual feedback between policy and technology. The vision for integrated urban waste management is thus one of convergence: merging innovation with inclusivity, and efficiency with equity. By harnessing technological advances, responsive planning, and participatory governance, cities can not only build resilient and sustainable waste systems but also shape broader paradigms of urban ecological stewardship.

8 Declaration of AI-assisted Technologies in the Writing Process

During the preparation of this work, the authors utilized ChatGPT (GPT-4 Omni) and Grammarly (Pro) to assist with grammar verification, synonym selection, and sentence flow enhancement, thereby improving readability. After using these tools/services, the authors re-viewed and edited the content as needed and take full responsibility for the content of the published chapter.

References

1. Mosbah EB, Alsanad MA, Sassine YN (2022) Assessment of environmental and economic impacts of municipal solid wastes management system: a case study. J Ecol Eng 23(1):137–145. https://doi.org/10.12911/22998993/144074
2. Yu P, Wei Y, Ma L, Wang B, Yung EHK, Chen Y (2024) Urbanization and the urban critical zone. Earth Crit Zone 1(1):100011. https://doi.org/10.1016/j.ecz.2024.100011
3. Angel S (2023) Urban expansion: theory, evidence and practice. Build Cities 4(1):124–138. https://doi.org/10.5334/bc.348
4. Liao W, Wu D, Guo Z, Zhao C, Xu J, Xu Y (2024) Variation analysis in historical demographic distribution within urban agglomeration: experimental evidence from the Wuhan "1+8" City Circle. Sustain Soc Dev 2(4):2778. https://doi.org/10.54517/ssd.v2i4.2778
5. Xu H, Jiao M (2021) City size, industrial structure and urbanization quality—a case study of the Yangtze River Delta urban agglomeration in China. Land Use Policy 111:105735. https://doi.org/10.1016/j.landusepol.2021.105735
6. Xu J, Wu D, Wang H, Fong CS (2023) Introducing sustainable intervention strategies of operation planning and environment design to activate void areas on the urban fringe. E3S Web Conf 437:03002. https://doi.org/10.1051/e3sconf/202343703002
7. Zhang C, Li J, Liao W, Xu J (2024) Sustainable micro renewal and renovation of arts town in China: intermingling design of contemporary art and Chinese classical garden. Eco Cities 5(2):2884. https://doi.org/10.54517/ec2884
8. Zhao J, Abdul Aziz F, Cheng Z, Ujang N, Zhang H, Xu J, Xiao Y, Shi L (2024) Post-occupancy evaluation of the improved old residential neighborhood satisfaction using principal component analysis: the case of Wuxi, China. ISPRS Int J Geo-Information 13(9):318. https://doi.org/10.3390/ijgi13090318

9. Choudhury M, Suhag M, Jha P, Harne KR, Samanta P, Choudhury M, Borgohain RB, Majumdar S, Kravchenko Y, Choudhary R (2025a) Urban waste management and challenges in emerging smart city systems governance and circular economy perspectives. Ecol Eng Environ Technol 26(10):243–266. https://doi.org/10.12912/27197050/211014
10. Choudhury M, Choudhary R, Sharma D, Bhattacharjee B, Majumdar S, Chegini S, Kumar A, Kumari S, van Hullebusch ED (2025b) Building Resilience in Wastewater Management: Disaster Preparedness and Emergency Response Strategies. In Choudhury M, Majumdar S, Goswami S, Sillanpää M (eds) Smart wastewater systems and climate change: innovations through spatial intelligence. pp. 155–169. https://doi.org/10.1039/9781837676880-00155
11. Ogwu MC, Kosoe EA, Alaka AC, Xu J (2025) Green infrastructure and environmental health security: the role of nature-based solutions in improving urban health. In: Chibueze Izah S, Chidozie Ogwu M (eds) Innovative approaches in environmental health management: processes, technologies, and strategies for a sustainable future. Springer Nature, pp 225–255. https://doi.org/10.1007/978-3-031-81966-7_9
12. Thompson OP, Kosoe EA, Xu J (2024) Green infrastructure and urban planning for sustainable clean air. In: Ogwu MC, Izah SC (eds) Sustainable strategies for air pollution mitigation: development, economics, and technologies. Springer Nature, pp 343–375. https://doi.org/10.1007/698_2024_1122
13. Choudhury M, Adithya A, Xu J, Liao Y, Herlekar M, Srivastava V (2025) In: Choudhury M, Majumdar S, Goswami S, Sillanpää M (eds) Smart wastewater systems and climate change innovations through spatial intelligence, Chapter 17, vol 15. Royal Society of Chemistry, pp 246–256. https://doi.org/10.1039/9781837676880-00246
14. Liao Y, Liao W, Wu D, Zhang C, Xu J, Majumdar S (2024) Inspired applications of artificial intelligence in disaster-resilient regional construction for climate risk. In: Ghosh S, Majumdar S, Cheshmehzangi A (eds) Cities of tomorrow: urban resilience and climate change preparedness. Springer Nature, pp 305–315. https://doi.org/10.1007/978-981-97-9658-8_16
15. Wang H, Xu J, Lim NHH, Liao W, Fong CS (2024) The impact of climate change on China's central region grain production: evidence from spatiotemporal pattern evolution. AIMS Geosci 10(3):460–483. https://doi.org/10.3934/geosci.2024024
16. Liao W, Wang H, Xu J (2023) The spatial structure characteristic and road traffic accessibility evaluation of A-level tourist attractions within Wuhan Urban Agglomeration in China. 3C Tecnol 12(2):388–409. https://doi.org/10.17993/3ctecno.2023.v12n3e45.388-409
17. Xu J, Liao W, Fong CS (2023) Identification and simulation of traffic crime risk posture within the Central City of Wuhan in China. Proc SPIE 12797:1279712. https://doi.org/10.1117/12.3007821
18. Zhong M, Wang C, Xu J, Cai Y, Xiao B, Xu T, Wang X (2024) Composite $xNiFe_2O_4$/$(1-x)SrFe_{12}O_{19}$ oxygen carriers for chemical looping reforming of bioethanol coupled with water splitting to coproduce syngas and hydrogen. J Energy Inst 117:101780. https://doi.org/10.1016/j.joei.2024.101780
19. Newman J, Collinson S, Driffield N, Gilbert N, Hoole C (2023) Mechanisms of metagovernance as structural challenges to levelling up in England. Regional Stud 58(4):876–892. https://doi.org/10.1080/00343404.2023.2217215
20. Taşkaya S, Wu D, Kurt M, Liao Y, Xu J, Liao W (2024) Exploring the application of Building Information Modeling (BIM) in town planning: key roles in the relationship between buildings and parcels. Int J Comput Exp Sci Eng 10(4):701–717. https://doi.org/10.22399/ijcesen.459
21. Xu J, Liao W, Lim NHH, Fong CS, Wang H, Xu Y (2024) Operational and environmental assessment for night markets under the informal economy impacts: insights from the case study of Selangor, Malaysia. 3C TIC 13(1):189–219. https://doi.org/10.17993/3ctic.2024.131.189-219
22. Wilson DC (2023) Learning from the past to plan for the future: an historical review of the evolution of waste and resource management 1970–2020 and reflections on priorities 2020–2030—the perspective of an involved witness. Waste Manag Res 41(12):1754–1813. https://doi.org/10.1177/0734242X231178025

23. Becerra L, Carenzo S, Juarez P (2020) When circular economy meets inclusive development. Insights from urban recycling and rural water access in Argentina. Sustainability 12(23):9809. https://doi.org/10.3390/su12239809
24. Siahaan G, Junianto P, Pada AT, Sembiring CF, Regina D (2023) Zero waste business model: building a regenerative business model through innovation and collaboration. Int J Sci Soc 5(4):404–415. https://doi.org/10.54783/ijsoc.v5i4.795
25. Zhang Z, Chen Z, Zhang J, Liu Y, Chen L, Yang M, Osman AI, Farghali M, Liu E, Hassan D, Ihara I, Lu K, Rooney DW, Yap P (2024) Municipal solid waste management challenges in developing regions: a comprehensive review and future perspectives for Asia and Africa. Sci Total Envir 930:172794. https://doi.org/10.1016/j.scitotenv.2024.172794
26. Tao X, Che Y (2018) Discussions on disposal and management of urban solid waste. Xiandai Hortic 15:181–182. https://doi.org/10.14051/j.cnki.xdyy.2018.15.102
27. Yao J, Xu P, Huang Z (2021) Impact of urbanization on ecological efficiency in China: an empirical analysis based on provincial panel data. Ecol Indic 129:107827. https://doi.org/10.1016/j.ecolind.2021.107827
28. Mondal T, Choudhury M, Kundu D, Dutta D, Samanta P (2023) Landfill: an eclectic review on structure, reactions and remediation approach. Waste Manag 164:127–142. https://doi.org/10.1016/j.wasman.2023.03.034
29. Paul S, Choudhury M, Deb U, Pegu R, Das S, Bhattacharya SS (2019) Assessing the ecological impacts of ageing on hazard potential of solid waste landfills: a green approach through vermitechnology. J Clean Prod 236:117643. https://doi.org/10.1016/j.jclepro.2019.117643
30. Liao W, Tan X, Xu J, Kumar P, Choudhury M (2025) In: Choudhury M, Majumdar S, Goswami S, Sillanpää M (eds) Smart wastewater systems and climate change innovations through spatial intelligence, Chapter 5, vol 15. Royal Society of Chemistry, pp 58–71. https://doi.org/10.1039/9781837676880-00058
31. Liu L (2014) The construction of sewers in ancient Rome and their effect on urban development. Soc Sci 9:145–149. [In Chinese] 刘琳琳. (2014). 古罗马城下水道的修建及对城市发展的作用. 社会科学家, 9, 145–149
32. Nathanson JA (2025) Solid-waste management. Encyclopedia Britannica. https://www.britannica.com/technology/solid-waste-management. Accessed 31 Mar 2025
33. News Peach Uptide (2020) Fudan general knowledge·scholars' thoughts on epidemic|Xiangrong: the black death and the fight against epidemics in Italy and the UK. https://www.sohu.com/a/402925419_260616. Accessed 31 Mar 2025
34. Qulishi (2024) Urban sanitation in medieval Europe: the truth and enlightenment of feces and urine everywhere. https://www.qulishi.com/news/v622122.html. Accessed 31 Mar 2025
35. Middleton J (2018) Public health at 170. BMJ 362:k3653. https://doi.org/10.1136/bmj.k3653
36. Silva LJDV, Santos IFSD, Mensah JHR, Gonçalves ATT, Barros RM (2020) Incineration of municipal solid waste in Brazil: an analysis of the economically viable energy potential. Renew Energy 149:1386–1394. https://doi.org/10.1016/j.renene.2019.10.134
37. Menon VJ, Palackal A (2022) Centralized and decentralized approaches to solid waste management—a case study. Holistic Approach Environ 12(4):155–164. https://doi.org/10.33765/thate.12.4.3
38. Choudhury M, Jyethi DS, Dutta J, Purkayastha SP, Deb D, Das R et al (2022) Investigation of groundwater and soil quality near to a municipal waste disposal site in Silchar, Assam, India. Int J Energy Water Resour 6(1):37–47. https://doi.org/10.1007/s42108-021-00117-5
39. Kamboj N, Choudhury M (2013) Impact of solid waste disposal on ground water quality near Gazipur dumping site, Delhi, India. J Appl Nat Sci 5(2):306
40. Uzoekwe AS, Richard G (2020) Level and ecological risk assessment of heavy metals in old landfill in Bayelsa state, Nigeria. J Environ Chem Ecotoxicol 12(1):32–44. https://doi.org/10.5897/JECE2020.0461
41. Rentschler J, Leonova N (2023) Global air pollution exposure and poverty. Nat Commun 4(1):4432. https://doi.org/10.1038/s41467-023-39797-4
42. Xu M, Lin B (2020) Exploring the "not in my backyard" effect in the construction of waste incineration power plants—based on a survey in metropolises of China. Environ Impact Assess Rev 82:106377. https://doi.org/10.1016/j.eiar.2020.106377

43. Xu M, Liu Y, Cui C, Xia B, Ke Y, Skitmore M (2023) Social acceptance of NIMBY facilities: A comparative study between public acceptance and the social license to operate analytical frameworks. Land Use Policy 124:106453. https://doi.org/10.1016/j.landusepol.2022.106453
44. Liu W, Jiang Q (2013) Strict controls of illegal foreign garbage transfer to guard against imported waste environmental risk. China WTO Tribune 6:75–77. https://doi.org/10.3969/j.issn.1672-1160.2013.06.022
45. Yong ZJ, Bashir MJK, Ng CA, Sethupathi S, Lim JW, Show PL (2019) Sustainable waste-to-energy development in Malaysia: appraisal of environmental, financial, and public issues related with energy recovery from municipal solid waste. Processes 7(10):676. https://doi.org/10.3390/pr7100676
46. Shahbaz M, Rashid N, Saleem J, Mackey H, McKay G, Al-Ansari T (2023) A review of waste management approaches to maximise sustainable value of waste from the oil and gas industry and potential for the State of Qatar. Fuel 332:126220. https://doi.org/10.1016/j.fuel.2022.126220
47. Huang C, Yi H, Chen T, Xu X, Chen S (2020) Networked environmental governance: formal and informal collaborative networks in local China. Policy Stud 43(3):403–421. https://doi.org/10.1080/01442872.2020.1758306
48. Xu M, Wu J (2020) Can Chinese-style environmental collaboration improve the air quality? A quasi-natural experimental study across Chinese cities. Environ Impact Assess Rev 85:106466. https://doi.org/10.1016/j.eiar.2020.106466
49. Hezri AA, Nordin Hasan M (2006) Towards sustainable development? The evolution of environmental policy in Malaysia. Nat Resour Forum 30(1):37–50. https://doi.org/10.1111/j.1477-8947.2006.00156.x
50. Van Ewijk S, Stegemann JA (2016) Limitations of the waste hierarchy for achieving absolute reductions in material throughput. J Clean Prod 132:122–128. https://doi.org/10.1016/j.jclepro.2014.11.051
51. Mngomezulu S, Mbanga S, Adeniran A (2024) The factors influencing waste management for economic development—the perspective of Nelson Mandela bay municipality residents. Front Sustain 5:1469207. https://doi.org/10.3389/frsus.2024.1469207
52. Khan WS, Asmatulu E, Uddin MN, Asmatulu R (2022) Waste landfill reclamation. In: Khan WS, Asmatulu E, Uddin MN, Asmatulu R (eds) Recycling and reusing of engineering materials. Elsevier, pp 311–319. https://doi.org/10.1016/B978-0-12-822461-8.00016-4
53. EPA (2024) Basic information about landfill gas. https://www.epa.gov/lmop/basic-information-about-landfill-gas. Accessed 4 Apr 2025
54. Toha M, Ariyan TN, Alam M (2025) Methane hotspots in landfills: a systematic review on transboundary dispersion and impacts on heatwaves in South Asia. Clean Waste Syst 10:100235. https://doi.org/10.1016/j.clwas.2025.100235
55. Hasan SE (2004) Public awareness is key to successful waste management. J Environ Sci Health A 39(2):483–492. https://doi.org/10.1081/ESE-120027539
56. Kumar R, Verma A, Shome A, Sinha R, Sinha S, Jha PK, Kumar R, Kumar P, Shubham, Das S, Sharma P, Vara Prasad PV (2021) Impacts of plastic pollution on ecosystem services, sustainable development goals, and need to focus on circular economy and policy interventions. Sustainability 13(17):9963. https://doi.org/10.3390/su13179963
57. Adekola PO, Iyalomhe FO, Paczoski A, Abebe ST, Pawłowska B, Bąk M, Cirella GT (2021) Public perception and awareness of waste management from Benin City. Sci Rep 11(1):306. https://doi.org/10.1038/s41598-020-79688-y
58. Moharir RV, Kumar S (2021) Traditional solid waste treatment technologies. In: Kumar S, Kumar R, Pandey A (eds) Current developments in biotechnology and bioengineering. Elsevier, pp 23–45. https://doi.org/10.1016/B978-0-12-821009-3.00016-6
59. Wang H, Liu X, Wang N, Zhang K, Wang F, Zhang S, Wang R, Zheng P, Matsushita M (2020) Key factors influencing public awareness of household solid waste recycling in urban areas of China: a case study. Resour Conserv Recycl 158:104813. https://doi.org/10.1016/j.resconrec.2020.104813

60. Dotsenko E, Galoyan D, Burdenko E, Mudrova S (2020) Structural inertia as a constraint for sustainable development and interregional differences in quality of life reducing. E3S Web Conf 174:04041. https://doi.org/10.1051/e3sconf/202017404041
61. Kotyal K (2024) Sustainable waste management in the circular economy: challenges and opportunities. Environ Rep. https://doi.org/10.51470/ER.2023.5.2.01
62. Farrelly T, Chitaka TY (2023) Policy implications for gaps in traditional plastic waste material flow analysis: Palmerston North, New Zealand. Front Sustain 4:982357. https://doi.org/10.3389/frsus.2023.982357
63. Mihai FC, Gündoğdu S, Markley LA, Olivelli A, Khan FR, Gwinnett C, Gutberlet J, Reyna-Bensusan N, Llanquileo-Melgarejo P, Meidiana C, Elagroudy S, Ishchenko V, Penney S, Lenkiewicz Z, Molinos-Senante M (2022) Plastic pollution, waste management issues, and circular economy opportunities in rural communities. Sustainability 14(1):20. https://doi.org/10.3390/su14010020
64. Sesay REV, Fang P (2025) Circular economy in municipal solid waste management: innovations and challenges for urban sustainability. J Environ Protect 16(2):35–65. https://doi.org/10.4236/jep.2025.162003
65. Unegbu HCO, Yawas DS (2024) Optimizing construction and demolition waste management in Nigeria: challenges, regulatory frameworks, and policy solutions. Discov Civil Eng 1(1):141. https://doi.org/10.1007/s44290-024-00142-3
66. Mangarengi NAP, Zubair A, Abdurrahman MA (2020) Analysis of infrastructure needs and operational systems for traditional market solid waste management (A case study on Makassar—Niaga Daya traditional market). IOP Conf Ser Earth Environ Sci 419(1):12156. https://doi.org/10.1088/1755-1315/419/1/012156
67. Choudhary R, Kumar A, Narayan S, Mishchenko I, Choudhury M, Bhardwaj LK et al (2025) Integrating fuzzy logic into life cycle assessment for sustainable municipal solid waste management in metro cities. Ecol Eng Environ Technol 26(7):316–331. https://doi.org/10.12912/27197050/205500
68. Rahman S, Supriyono B, Hariswanto A, Koyama M (2020) Policy reform towards waste management in Pontianak Municipality, West Kalimantan Indonesia (a SWOT analysis and lessons learned from Kyoto City). Jurnal Ilmiah Administrasi Publik 6(1):147–155. https://doi.org/10.21776/ub.jiap.2020.006.01.17
69. El Amady MR (2023) Impact of waste management education at Smu Witama Pekanbaru Students. Int J Multidiscip Res Anal 6(6):2431–2437. https://doi.org/10.47191/ijmra/v6-i6-38
70. Aleluia J, Ferrão P (2016) Characterization of urban waste management practices in developing Asian countries: a new analytical framework based on waste characteristics and urban dimension. Waste Manag 58:415–429. https://doi.org/10.1016/j.wasman.2016.05.008
71. Aparcana S (2017) Approaches to formalization of the informal waste sector into municipal solid waste management systems in low- and middle-income countries: review of barriers and success factors. Waste Manag 61:593–607. https://doi.org/10.1016/j.wasman.2016.12.028
72. Scheinberg A, Spies S, Simpson MH, Mol APJ (2011) Assessing urban recycling in low- and middle-income countries: building on modernised mixtures. Habitat Int 35(2):188–198. https://doi.org/10.1016/j.habitatint.2010.08.004
73. Agarwal A, Singhmar A, Kulshrestha M, Mittal AK (2005) Municipal solid waste recycling and associated markets in Delhi, India. Resour Conserv Recycl 44(1):73–90. https://doi.org/10.1016/j.resconrec.2004.09.007
74. Bercegol Rd (2020) From Trash to Cash, recovering practices, wholesale markets and industrial recycling in Delhi. Sci J Arch Construct 38:21–27. https://shs.hal.science/halshs-03491909v1
75. Ribeiro H, Besen GR, Günther WR, Jacobi P, Demajorovic J (2006) Recycling programs in partnership with scavenger associations as a sustainability factor in Metropolitan São Paulo, Brazil. WIT Trans Ecol Environ 93:409–418. https://doi.org/10.2495/SC060391
76. Guma PK (2021) Rethinking Smart Urbanism: city-making and the spread of digital infrastructures in Nairobi. Doctoral Dissertation, Utrecht University. https://doi.org/10.33540/497

77. Agya BA, Rückert A, Dornack C (2024) Effectiveness of traditional solid waste management system of rural communities: a case study in the Kwahu East District, Ghana. Environ Chall 15:100869. https://doi.org/10.1016/j.envc.2024.100869
78. Hoareau CE, Ahmad N, Nuid M, Rubiyatno, Khoi DN, Kristanti RA (2021) Sustainable technology in developed countries: waste municipal management. Ind Domest Waste Manag 1(1):48–55. https://doi.org/10.53623/idwm.v1i1.49
79. Clemente E, Domingues E, Quinta-Ferreira RM, Leitão A, Martins RC (2024) European and African landfilling practices: an overview on MSW management, leachate characterization and treatment technologies. J Water Process Eng 66:105931. https://doi.org/10.1016/j.jwpe.2024.105931
80. Higuchi S (2021) Municipal solid waste landfill technology in Japan. Springer Nature. https://doi.org/10.1007/978-981-16-2734-7
81. Seadon JK (2010) Sustainable waste management systems. J Clean Prod 18(16):1639–1651. https://doi.org/10.1016/j.jclepro.2010.07.009
82. Wilson DC, Velis CA, Rodic L (2013) Integrated sustainable waste management in developing countries. Proc Inst Civil Eng Waste Resour Manag 166(2):52–68. https://doi.org/10.1680/warm.12.00005
83. Tirkolaee EB, Simic V, Ghobakhloo M, Foroughi B, Asadi S, Iranmanesh M (2024) Integrated design of a sustainable waste management system with co-modal transportation network: a robust bi-level decision support system. J Clean Prod 449:141760. https://doi.org/10.1016/j.jclepro.2024.141760
84. Yusri M, Kholik K, Hajar S, Warjio W (2025) Healthy and waste-free area: integrated waste management, regional resilience through sustainable development in Deli Serdang Regency, Indonesia. J Posthuman 5(3):980–1000. https://doi.org/10.63332/joph.v5i3.832
85. Bolton K, Rousta K (2019) Solid waste management toward zero landfill: a Swedish model. In: Taherzadeh MJ, Bolton K, Wong J, Pandey A (eds) Sustainable resource recovery and zero waste approaches. Elsevier, pp 53–63. https://doi.org/10.1016/B978-0-444-64200-4.00004-9
86. Idoko DO, Imarenakhue WU, Olade AD, Oppong RA, Bah MB, Elue HC, Fadeke AA (2024) Development of smart waste management technologies using IoT solutions for environmental sustainability in urban infrastructure planning. Int J Innov Sci Res Technol 9(12):1–8. https://doi.org/10.5281/zenodo.14353675
87. Kalla D, Smith N (2024) Integrating Iot, AI, and Big Data for enhanced operational efficiency in smart factories. Educ Admin Theory Pract 30(5):14235–14245. https://doi.org/10.53555/kuey.v30i5.6492
88. Chaturvedi V, Yadav M, Tiwari N (2021) Smart waste management system using Internet of Things and RFID Technology. New Front Commun Intell Syst:63–68. https://doi.org/10.52458/978-81-95502-00-4-8
89. Chavhan N, Bhattad R, Khot S, Patil S, Pawar A, Pawar T, Gawli P (2025) APAH: an autonomous IoT driven real-time monitoring system for Industrial wastewater. Digit Chem Eng 14:100217. https://doi.org/10.1016/j.dche.2025.100217
90. Lakhouit A (2025) Revolutionizing urban solid waste management with AI and IoT: a review of smart solutions for waste collection, sorting, and recycling. Results Eng 25:104018. https://doi.org/10.1016/j.rineng.2025.104018
91. Ferrão CC, Moraes JAR, Fava LP, Furtado JC, Machado E, Rodrigues A, Sellitto MA (2024) Optimizing routes of municipal waste collection: an application algorithm. Manag Environ Qual 35(5):965–985. https://doi.org/10.1108/MEQ-08-2023-0267
92. Bibri SE, Huang J, Jagatheesaperumal SK, Krogstie J (2024) The synergistic interplay of artificial intelligence and digital twin in environmentally planning sustainable smart cities: a comprehensive systematic review. Environ Sci Ecotechnol 20:100433. https://doi.org/10.1016/j.ese.2024.100433
93. Heng K, Tan K, Chan A, Lee CCC (2024) Food waste valorization: leveraging Singapore's Zero Waste Master Plan and 30-by-30 goal. Sustainability 16(17):7321. https://doi.org/10.3390/su16177321

94. Li Y, Beeton RJS, Zhao X, Fan Y, Yang Q, Li J, Ding L (2024) Advancing urban sustainability transitions: a framework for understanding urban complexity and enhancing integrative transformations. Human Social Sci Commun 11(1):1064. https://doi.org/10.1057/s41599-024-03598-x
95. Ma W, de Jong M, Zisopoulos F, Hoppe T (2023) Introducing a classification framework to urban waste policy: analysis of sixteen zero-waste cities in China. Waste Manag 165:94–107. https://doi.org/10.1016/j.wasman.2023.04.012
96. Bea E, Low M (2019) Extended producer responsibility in Singapore's Resource Sustainability Act. https://law.nus.edu.sg/wp-content/uploads/2020/04/024_2019_Eric-Bea.pdf
97. Asare W, Oduro Kwarteng S, Donkor EA, Rockson MAD (2022) Cost-effectiveness of incentive schemes for waste material resource recovery. Clean Waste Syst 2:100019. https://doi.org/10.1016/j.clwas.2022.100019
98. Laureti L, Costantiello A, Anobile F, Leogrande A, Magazzino C (2024) Waste management and innovation: insights from Europe. Recycling 9(5):82. https://doi.org/10.3390/recycling9050082
99. Hartono D, Indriyani W, Iryani BS, Komarulzaman A, Nugroho A, Kurniawan R (2023) Carbon tax, energy policy, and sustainable development in Indonesia. Sustain Dev 31(4):2332–2346. https://doi.org/10.1002/sd.2511
100. Xie X, Zhu Q, Wang R (2019) Turning green subsidies into sustainability: how green process innovation improves firms' green image. Bus Strat Environ 28(7):1416–1433. https://doi.org/10.1002/bse.2323
101. Gibovic D, Bikfalvi A (2021) Incentives for plastic recycling: how to engage citizens in active collection. Empirical evidence from Spain. Recycling 6(2):29. https://doi.org/10.3390/recycling6020029
102. Ambastha R, Aich S (2020) Decentralized community-led solid waste management. In: Ghosh S (ed) Sustainable waste management: policies and case studies. Springer, Singapore. https://doi.org/10.1007/978-981-13-7071-7_7
103. Wang S, Bai X, van der Heijden J, Tong X (2024) The evolving roles of actors in sustainability experiments: evidence from community waste management in a Chinese city. Technol Forecast Soc Change 205:123469. https://doi.org/10.1016/j.techfore.2024.123469
104. Nawaz M, Yousafzai MT, Khan S, Ahmad W, Salman M, Han H, Ariza-Montes A, Vega-Muñoz A (2021) Assessing the formal and informal waste recycling business processes through a stakeholders lens in Pakistan. Sustainability 13(21):11717. https://doi.org/10.3390/su132111717
105. Nyakabawu S, Schenck CJ (2025) Exploring waste management practices among backyard dwellers in Fisantekraal, South Africa. Urban Forum. https://doi.org/10.1007/s12132-025-09535-8
106. Rizk M, Massoud MA, Chalak A, Abiad MG (2025) Inside the divide: a stakeholder's perspective on formal vs. informal waste practices. Integr Environ Assess Manag f47. https://doi.org/10.1093/inteam/vjaf047
107. Zisopoulos FK, Steuer B, Abussafy R, Toboso-Chavero S, Liu Z, Tong X, Schraven D (2023) Informal recyclers as stakeholders in a circular economy. J Clean Prod 415:137894. https://doi.org/10.1016/j.jclepro.2023.137894
108. Marello M, Helwege A (2017) Solid waste management and social inclusion of wastepickers: opportunities and challenges. Latin Am Perspect 45(1):108–129. https://doi.org/10.1177/0094582X17726083
109. Utami, Istiqomah T, Corsita L (2024) Implementation of community-based waste management to improve environmental health in villages. Sustain Appl Modif Evid Commun 1(2):1–8. https://doi.org/10.69855/samec.v1i2.90
110. Ansari A, Dutt D, Kumar V (2024) Catalyzing paradigm shifts in global waste management: a case study of Saharanpur Smart city. Waste Manag Bull 2(1):29–38. https://doi.org/10.1016/j.wmb.2023.12.003
111. Pal MS, Bhatia M (2023) Smart solid waste management system using IoT technology: comparative analysis, gaps, and challenges. In: Hemanth J, Pelusi D, Chen JIZ (eds) Intelligent

cyber physical systems and Internet of Things. ICoICI 2022. Engineering Cyber-Physical Systems and Critical Infrastructures, vol 3. Springer, Cham. https://doi.org/10.1007/978-3-031-18497-0_58

112. Choudhury M, Adithya A, Xu J, Liao Y, Herlekar M, Srivastava V (2025c) Transforming wastewater management: machine learning and artificial intelligence in decision-making processes. In Choudhury M, Majumdar S, Goswami S, Sillanpää M (eds) Smart wastewater systems and climate change: innovations through spatial intelligence. pp. 246–256. https://doi.org/10.1039/9781837676880-00246

Challenges in Linear Waste Systems: A Critical Look at Conventional Models and Their Environmental Effects

Shivam Attri, Avnish Chauhan, Aanchal Saklani, Gaurav Pandey, Abhishek Chauhan, F. F. Amirdzhanov, Samuele Barrili, and Kaiss Redouane

Abstract Nature's bounty is inherently limited, and current rates of human consumption will inevitably lead to the exhaustion of such bounty. What then are the appropriate rules to govern our collective use of environmental resources the high urbanization rate and increasing population have resulted in the continuous release of large quantities of waste into the environment. Furthermore, from cradle to grave, linear waste handling systems face significant difficulties in meeting environmental and economic sustainability objectives due to the take–make–dispose paradigm. In this chapter, several key problems associated with such systems are discussed, including high energy costs, significant environmental damage, and resource depletion. It also highlights the economic burden resulting from waste management, particularly due to low recycling rates and the ultimate loss of value from these materials. Additionally, the chapter examines the broader social and ecological implications of linear systems, including heightened toxicity, pressure on land use, threats to public health, and increased social disparities.

S. Attri · A. Chauhan (✉) · A. Saklani
Department of Environmental Sciences, Graphic Era Hill University, Dehradun, Uttarakhand, India

G. Pandey
Chemistry and Bioprospecting Division, Forest Research Institute, Dehradun, Uttarakhand, India

A. Chauhan
Amity Institute of Environmental Toxicology, Safety and Management, Amity University, Noida, Uttar Pradesh, India

F. F. Amirdzhanov
Academy of Biology and Biotechnology, Southern Federal University, Rostov-on-Don, Russia

S. Barrili
M4W—Marketing for Waste, the First Marketing Agency for Small Waste Companies and Junk Removals, Rome, Italy

K. Redouane
Research Laboratory in Economics, Management, and Business Administration, Faculty of Economics and Management, Hassan 1st University, Settat, Morocco

S. Majumdar et al. (eds.), *Sustainable Urban Future*, Urban Sustainability,
https://doi.org/10.1007/978-981-95-2976-6_6

Keywords Recycling · Toxicity · Linear waste systems · Environmental sustainability · Urbanization · Recycling efficiency

1 Introduction

Excessive waste generation has become a global concern today due to its high handling costs and adverse environmental impacts [1, 2]. Waste is a significant threat to sustainability, and it is undeniable that both population growth and the global GDP increase will also lead to higher waste levels. Global waste generation is expected to increase by around 70% by 2050 from 2.01 billion tons in 2016 to 3.40 billion tons, according to the World Bank. The "waste disposal industry," which ranks fourth globally in terms of greenhouse gas emissions, is one of the primary sources of these emissions caused by human activities. Additionally, the primary GHG (Greenhouse gas) evaluations rely on generic information due to a lack of data [3]. As a result, managing solid waste has become crucial to a country's ability to meet the SDGs and reach relative zero emissions. The primary challenges posed by trash include environmental damage, health issues, and climate change.

The concepts of the linear economy are currently widely applied in all disciplines. In linear waste systems, the model is based on extraction, production, consumption, and disposal of waste. Until the twentieth century, the linear system model was more effective in producing goods in industries. In the current situation and considering the future prognosis, it has since reached a higher level of failure. Linear systems can lead to the depletion of natural resources, the generation of excessive waste, and environmental degradation, including pollution and greenhouse gas emissions [4, 5]. Most waste materials are incinerated or disposed of in landfills instead of being recovered or repurposed. By doing this, many valuable resources have been lost. The concept of a circular economy was introduced in the 1960s as global concerns over resource depletion and environmental degradation began to emerge [6]. The fundamental principles of this system include recycling, reuse, and resource recovery, with businesses and industries playing an active role in optimizing resource use and regenerating materials [7]. Unlike the linear economy, which results in resource depletion and waste accumulation, the circular model ensures that products and materials are continuously reused, repaired, and recycled to minimize waste and maximize economic value [8].

Additionally, the circular economy redefines production and consumption patterns by incorporating eco-design, repair, and waste-to-resource strategies to extend the lifespan of materials [9]. The 6R framework (Reduce, Reuse, Recycle, Recover, Redesign, and Remanufacture) serves as a key implementation strategy for sustainable waste management. Given the increasing pressure on natural resources and environmental concerns, adopting a circular economy is no longer an option but a necessity for achieving a balance between economic prosperity, ecological sustainability, and social well-being [10, 11]. Transitioning to a circular model is essential for addressing the challenges of linear waste systems and ensuring a sustainable urban future. This transformation requires policy reforms, technological advancements, and stakeholder engagement to integrate circular practices into waste management systems successfully.

2 Bibliography Study

This chapter offers a comprehensive review of Review articles, Book chapters and Research articles from the "Scopus" database, focusing on the period between 1986 and 2025 in English language with the search terms 'challenges,' 'linear,' 'waste,' and 'systems' also the operator (IN) were used to combine the search terms. The analysis prioritizes recent publications to showcase advancements in the sensitive areas of waste systems economy, linear waste systems, and circular economy practices worldwide. A total of 605 distinct publications from Scopus-indexed journals and books were used for this analysis.

The data were collected from Scopus database using keywords for bibliometric analysis. The datasets were processed using VOSviewer 1.6.18 software and Co-occurrence analysis were done concerning all keywords unit and complete counting method and for threshold minimum number of occurrences of a keyword is 15 using this 61 keywords meet the threshold in 5 clusters having 18 keywords in first cluster, 14 keywords in second cluster, 12 keywords in third cluster, 9 keywords in fourth cluster and 8 keywords in fifth cluster. Figure 1a illustrates a network visualization, where larger frame sizes represent greater prominence, and a greater number of connecting lines indicates a more substantial networking potential. Keywords that appeared more frequently were visualized with larger frames, reflecting their co-occurrence in the analyzed publications. Figure 1 presents a network and overlay visualization map displaying the most cited terms in the 605 papers published between 1986 and 2025.

3 Linear Waste Systems: Structure and Properties

The linear waste model is typical of the so-called "take–make–dispose" logic in resource use. It is intrinsically unsustainable because virgin materials are used, waste is generated, and there are no components for resource recovery or circularity. This system is deeply ingrained in the industrial system's mode of production and consumption, significantly contributing to environmental degradation, resource depletion, and economic irrationalities. The main steps in this system—extraction of resources, industrial production, consumption, and end-of-life processing—are described [12].

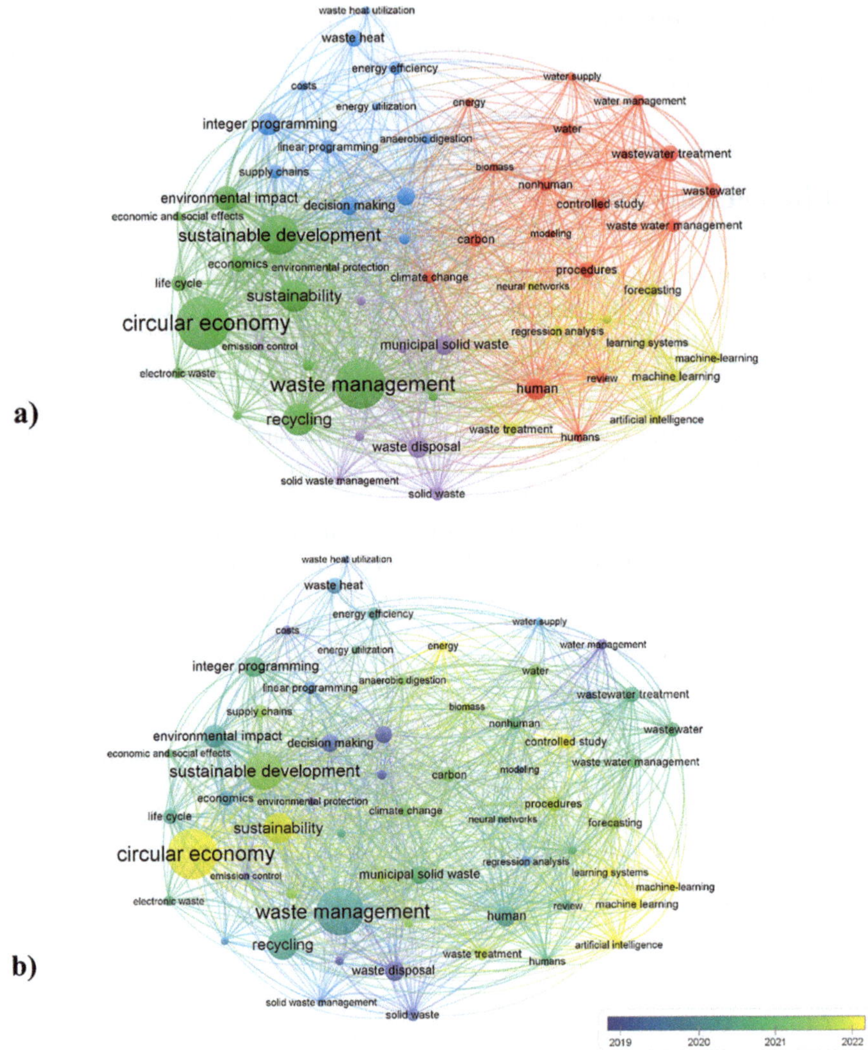

Fig. 1 (**a**, **b**) Bibliometric analysis centered on network clustering of 605 published articles from 1986 to 2025 which have high matches of selected keywords and their citations (**a**) Network Visualization (**b**) Overlay visualization

3.1 Resource Extraction and Processing

Linear systems originate from the extraction of natural resources, followed by the conversion of raw materials into goods with significant energy inputs. This is the age of exploiting non-renewable resources, including fossil fuels, minerals, and metals. Not only do virgin inputs deplete resources and require additional materials, but the manufacturing processes also emit excessive amounts of greenhouse gases, disrupt the ecosystem, and even damage the Earth's surface. Additionally, there is a lack of resource efficiency, and excessive production results in an abundance of products that, in many cases, are not durable and recyclable. While industries value cost-effectiveness and scale, environmental concerns are often secondary and contribute to the overall environmental impact of production.

3.2 Consumption and Short Lives of Products

The consumption circuit in linear systems is enabled by market mechanisms that encourage quick turnover of products and materials (termed planned obsolescence). Items are meant to be disposable, inexpensive, and have broad appeal, rather than being durable or reusable. This means that the life of many products and consumer goods is artificially shortened, and waste increases. The advancement of consumerism along with aggressive marketing strategies and lifestyle changes adds to the pace of disposability. This stage is characterized by low consumer incentives for repairing, returning, or reusing used products, contributing to the throwaway culture that underpins linear economic models.

3.3 End-of-Life Treatment

Products within a linear system are traditionally disposed of in landfills, through open dumping, or by incineration. These processes are environmentally and health-damaging. Landfills occupy valuable land, emit methane, a potent greenhouse gas, and pose a threat to contaminate soil and waterways with leachate. The incineration process, although capable of reducing waste volume, emits toxic gases and does not provide a solution to the problem of material value loss. In addition, the lack of effective source separation and recycling or composting infrastructure results in the loss of resources, such as metals, plastics, and organics, through surface disposal. This inefficient end-of-life treatment highlights the non-sustainability of the linear model and its incompatibility with long-term environmental targets [13].

4 Environmental Impacts of End-to-End Waste Systems

By design, linear waste systems are heavy on the environment. Predicated on a linear trajectory of materials (from extraction of resources to disposal), these systems result in pervasive ecological devastation, climate inconsistency, and irreversible damage to resources. The overall environmental effects of such systems are multiple and profound, encompassing air, water, soil, and biodiversity. The significant environmental effects are discussed below.

4.1 Pollution and Destruction of the Ecosystem

The most direct and obvious consequences of linear waste systems are environmental contamination. Indiscriminate disposal, such as open dumping and uncontrolled landfills, leads to soil and groundwater contamination due to the leaching of leachate. Improper waste sorting causes hazardous, Bio waste, emerging contaminants and non-hazardous waste to mix, thus increasing poisoning levels within the ecodiversity [14, 15]. Moreover, pollution emissions from incineration and e-waste treatment, including dioxins, heavy metals, and persistent organic pollutants (POPs), also have a direct impact on both terrestrial and aquatic ecosystems. One of the ecosystems most severely impacted by plastic pollution is the marine ecosystem, which has disrupted food chains, destroyed habitats, and introduced microplastics into the food web. These losses of resilience in ecosystems deplete biodiversity and undermine the provision of ecosystem services to human communities [16].

4.2 Greenhouse Gas Emissions

Linear waste infrastructures are significant contributors to greenhouse gas (GHG) emissions due to their high carbon intensity.

In several processes, including resource extraction, production, transportation, unloading, and waste emission, greenhouse gases (GHGs) and pollutants are emitted in large quantities, such as carbon dioxide (CO_2), methane (CH_4), and nitrous oxide (N_2O). At the same time, landfills are leading source of methane, a gas with a warming potential 28–36 times greater than that of CO_2 over 100 years. Incinerating waste, especially plastics and organic materials, also generates CO_2 and other harmful emissions [17]. In addition, materials are not recycled or reused, so new products are made from scratch, which perpetuates the reliance on fossil fuels and increases GHG emissions. Thus, linear waste systems pose significant obstacles to global responses to climate change and the goals established under agreements such as the Paris Agreement and the United Nations Sustainable Development Goals (SDGs).

4.3 Resource Depletion

In a finite world, the linear model of "take–make–dispose" is unsustainable. The unrelenting extraction of natural resources is placing unsustainable pressure on the Earth's natural capital, including forests, mineral resources, fossil fuels, and freshwater. Such extraction not only exhausts these resources but also results in land degradation, a water shortage, and a loss of habitat. Moreover, the lack of recovery and recycling systems leads to the definitive waste of materials that could be reduced through a redeveloped production cycle. As examples, precious metals in electronic waste are often discarded, and organic waste is rarely composted for nutrient reclamation. Utilizing resources without ceasing intensifies tensions about access to resources among nations and weakens the future resilience of natural systems and economies.

4.4 Biodiversity Loss

Linear waste systems are associated with the loss of biodiversity, which is a crucial, albeit underexplored, consequence. The rampant environmental destruction resulting from unmitigated extraction, pollution, and unverifiable waste disposal methods poses a direct threat to ecosystems and the species that depend on them. Deforestation, mining, and landfills, along with their accompanying habitat loss or destruction, alter or destroy the natural environment, thereby reducing food, shelter, and breeding space for wildlife. Aquatic ecosystems, in particular, are significantly impacted by the influx of raw sewage, plastic waste, and chemical pollution, resulting in oxygen deprivation, the bioaccumulation of toxins, and reproductive disturbances in aquatic fauna. Toxicants in e-waste, including heavy metals and persistent organic pollutants, enter the food chain, disturbing species at various trophic levels. Once taken up by smaller marine creatures, microplastics can move up the food chain, affecting fish, birds—and even, potentially, humans. Invasion by such species, which are typically introduced with waste and often become cache-sensitive due to the removal of natural environmental control features, poses an additional threat to native biota by displacing or predating on local species.

Furthermore, with linear monocultural production systems in agriculture, as well as in forestry, genetic diversity is eroding, vulnerability to pests and diseases is increasing, and ecosystem services, such as pollination, nutrient cycling, and carbon sequestration, are declining. The resulting loss of diversity reduces the resilience of ecosystems, making them less likely to recover from environmental insults and thus rendering them more susceptible to ecological collapse [18].

5 Case Studies of Linear Waste System Failures

5.1 The Growing Municipal Solid Waste Crisis

Solid waste management is a technological, financial, and environmental concern, particularly for developing nations. Most of the garbage is still thrown in the open, despite urban local bodies in these nations rising in the waste management hierarchy. The failure to take into consideration (a) the societal costs of open dumping (OD) and (b) the direct and indirect advantages of other choices is a significant factor in this decision [19].

For example, the Ghazipur dumping site in Delhi is one of the largest garbage dumping sites in India and has exceeded its capacity. This waste dumping site frequently catches fire due to methane gas emissions from decomposing municipal organic waste [20]. Similarly, in Lagos, only 40% of the waste has been collected. The remaining 40% of waste has been illegally dumped in rivers, drains, and streets, leading to floods and disease outbreaks [21].

5.2 E-Waste Management Challenges

E-waste, or waste generated by electrical and electronic devices, remains a significant environmental, economic, and societal issue. With a projected 53.6 million metric tons produced in 2019 alone, electronic waste, also known as "e-waste," is one of the waste streams with the most significant growth rate globally. The prevention of electronic trash from ending up in landfills, the recovery of valuable materials via recycling, and the reduction of greenhouse gas emissions are just a few advantages of sustainable e-waste management [22, 23]. Hazardous effects on the environment and human health have been caused by the linear approach to electronics disposal, especially in underdeveloped countries where e-waste is disposed of or processed informally [24].

For example, one of the most notorious slums in the world is Agbogbloshie, situated in the heart of Accra, Ghana. The Agbogbloshie e-waste dump was ranked as "among the top ten most toxic sites in the world" by the Blacksmith Institute, which changed its name to Pure Earth in 2015. Large quantities of abandoned electronics are shipped to Africa from developed nations, particularly from Europe and North America, where they are often burned in open fields to extract precious metals such as copper and gold. Workers and the surrounding populations are negatively impacted by the poisonous fumes released by this process, which include lead, mercury, and dioxins. Children who work in these unregulated e-waste industries risk long-term health issues, respiratory disorders, and neurological damage [25].

5.3 *Plastic Waste and Marine Pollution*

A significant global issue affecting marine and coastal ecosystems is plastic pollution. Plastic pollution is another major failure of the linear waste system, with over 400 million tons of plastic produced annually, yet only 9% is recycled globally. It disrupts the structure, functionality, and services of ecosystems, primarily caused by land-based and marine sources. Megaplastic, macroplastic, mesoplastic, and microplastic are among the several kinds of pollution. With concentrations ranging from 0.001 to 140 particles/m^3 in water and 0.2 to 8766 particles/m^3 in sediments, microplastics are commonly found in marine and coastal environments. Invasive species introduction, toxicological impacts, habitat supply, and entanglement are some of the ecological and socioeconomic repercussions of pollution. Prevention, the 3Rs, awareness, capacity building, and producer responsibility are all effective strategies to address this problem.

One of the most notable examples is the Great Pacific Garbage Patch, which is the world's greatest concentration of ocean plastic and is situated between Hawaii and California. Over 1.8 trillion plastic particles are thought to be present, spanning an area of nearly 1.6 million square kilometers, which is almost three times the size of France. Microplastics, single-use plastics, and abandoned fishing gear comprise the majority of this waste, which traps marine life, contaminates the food chain, and contributes to ocean acidification [26].

6 Barriers to Transitioning from Linear to Circular Waste Systems

6.1 *Inconsistency Between Intention and Execution*

The "intention-action gap" among customers has proven to be a significant barrier to circularity. According to a study, consumers play a larger part in society's transition to sustainability. People naturally wish to live better, more moral lives and to preserve the environment. Consequently, people are more likely to select sustainable options over mass-market ones if they are more conscious, interested, and worried about environmental issues [27]. For example, regarding recycled products, consumers view them as used goods and are worried about their quality and safety. These worries are heightened for products that are associated with strength. Even if a product's performance is unaffected by its green characteristics, consumers frequently believe that green items are less effective. Truck drivers, for example, were hesitant to use recycled tires because they thought they would not perform as well or last as long as non-recycled tires [28].

6.2 Risk Tendencies

Another notable sociocultural obstacle is the perception of risk. It is found that the majority of small and mid-sized business leaders are risk-averse when it comes to making changes at the organizational level. Due to these risk-averse tendencies, managers unintentionally create unnecessary inertia, preventing them from adapting their companies to participate in the circular economy. Their choices and actions thus have a ripple effect on the entire business, hindering its ability to innovate and move closer to circularity [29]. Risk-averse company executives also frequently encounter conflicting reactions when attempting to collaborate with stakeholders in their value chain [30].

6.3 Policy Initiatives

A policy is a legislative, regulatory, procedural, administrative, or voluntary action implemented by governments and other organizations. They can be both motivating and punitive. Punitive actions encourage responsible production and consumption while discouraging undesired conduct. A thorough and effectively applied punitive action is considered a potent regulatory motivator. For instance, Companies operating in China and the United States may face financial fines if they decide to discard unsought but useful products rather than recycle or resell them [31]. However, to promote the circular economy, there are currently few comprehensive punitive measures in place. Nowadays, most policies primarily address waste and pollution. For instance, in most jurisdictions, organizations and their directors who are found to be harming the environment face substantial fines, penalties, or even imprisonment [32]. As for supportive measures, the most common forms of legislation and policies are financial help, grants, awards, and subsidies. Due to various sociocultural barriers, corporations often prefer to extend the usable life of their current linear-centric systems until an unavoidable event occurs or a more promising alternative emerges. Therefore, firms would be encouraged to transition if there were strong supportive measures that reduce their financial burden and increase efforts to align stakeholders. As a result, federal support is a crucial regulatory factor. Unfortunately, a study conducted by Hegab et al. [28] found that current government assistance is severely insufficient [33].

6.4 Technical Resources, Intelligence, and Infrastructure

Strong legislative measures can promote the development of vital ecosystem infrastructures, information exchange, and technology breakthroughs.

These components are necessary but now woefully insufficient. One notable obstacle is the absence of knowledge about the circular economy. In-depth subject-matter expertise, technical proficiency in transitioning from linear to circular product life cycles (e.g., establishing a circular company strategy, product design, and material flow), and other relevant knowledge are provided [34].

Businesses must also pay for purchasing and ongoing maintenance of their technology and knowledge. Even though some businesses might not have the funds or the desire to buy it, De Jesus, Mendonça, along with Hart et al., made the case that they must do so eventually. Maintaining the alignment of key stakeholders enables them to share their knowledge both internally and throughout the value chain, and to identify potential opportunities. According to several studies, companies that possess technological expertise and domain knowledge are often reluctant to share, as they view it as a competitive advantage. By keeping it all to themselves, they may extract as much profit as possible. Therefore, if such a Homo economicus mindset and actions continue, companies and their executives will essentially prioritize profit-maximizing objectives over all other objectives. This causes the competence gap between enterprises to gradually increase even if everyone is working toward the same goal of circularity. Such actions will therefore take place and have an impact on the development of vital ecosystem infrastructures that are necessary for a smooth shift.

Therefore, to promote a circular flow of materials, businesses and other key stakeholders, including governments, must, at the very least, collaborate to support one another in developing and implementing these crucial components.

7 Strategies for Overcoming Linear Waste System Challenges

Over time, numerous concepts, models, and approaches have been established to encourage holistic thinking and long-term sustainability. The cradle-to-cradle philosophy stands out as one of the significant theoretical influences shaping resource recovery systems which concentrates on creating products that reduce waste to zero; the performance economy which stresses prolonging product lifespans through reuse and remanufacturing; and industrial ecology and industrial symbiosis, which encourage inter-industry cooperation to minimize waste and optimize resource utilization. A more comprehensive explanation of these ideas can be found in earlier studies [35]. Transition management bridges the divide between theoretical concepts and practical application by acknowledging the importance of societal and technological factors in achieving sustainable transformations. This approach suggests that achieving sustainability necessitates radical socio-technical transitions that involve governments, individuals, communities, and businesses. It provides a structured approach to developing and implementing practical strategies for managing these transitions.

Technological Innovation Systems (TIS) is a new framework in sustainability transitions that examines how specific technologies contribute to broader socio-technical shifts. This approach identifies essential processes, referred to as functions, which need to be evaluated in terms of their catalysts and obstacles [36]. These may include ineffective networks, institutional impediments, or gaps in infrastructure [37]. This structure supports groundbreaking, long-lasting innovations from their inception, facilitating their successful creation and adoption [38]. Systemic sustainability issues can be addressed in an organized manner with the help of the Framework for Strategic Sustainable Development (FSSD). Understanding the system, establishing sustainability goals using guiding principles, planning various strategies, creating a strategic action plan, and utilizing technologies to track and assist decision-making are its five stages [39].

There are various levels at which circular economy (CE) strategies function. The micro-scale tackles sustainability at the level of specific goods, enterprises, and customers; the meso-scale encompasses eco-industrial parks and corporate partnerships; and the macro-scale concentrates on national, regional, or international regulations [40].

Innovation, infrastructure development, and technological breakthroughs are essential to CE and resource recovery. However, after entering the technosphere, some materials deteriorate, making recovery and reuse more difficult [41, 42]. To overcome these obstacles and ensure long-term sustainability, regulations, business models, and technology must be continually improved.

7.1 Strategy 1: Maintain Resource Value Through Conservation

Maximizing the value of resources through efficient use to generate valued goods or services is the basic principle of production systems. Careful planning is the first step in this process, which also involves making effective use of scarce resources. Although attempts have been made to minimize resource use, more methods and approaches are required to maximize their use while also preserving them through reuse or recycling. Numerous production techniques and practical strategies have been developed to achieve this goal. Key goals and strategies for maintaining and increasing the usability of resources in production systems are outlined in the Preserve Resource Wheel (Fig. 2).

Fig. 2 Strategies maintaining resource value through conservation. (Prepared by authors)

7.2 Strategy 2: Sustaining Resources Through Continuous Use

By ensuring that resources and goods remain functional for as long as possible, here are a few methods to accomplish this:

1. Improved product design: Creating items that are simple to repair, upgrade, or reuse.
2. Using superior materials: Replace items that are difficult to recycle with environmentally friendly alternatives.
3. Holding manufacturers accountable: Businesses should not only sell their products but also be responsible for managing the waste generated by their products.
4. Reducing waste: from the outset by encouraging online work and using less packaging.
5. Recycling and reusing, figuring out new uses for outdated goods rather than discarding them.

7.3 Strategy 3: Enhance Leadership's Commitment to Sustainability

Management needs to understand the significance of sustainability in order to make informed decisions that work. Motivated managers who are prepared to incorporate sustainability principles into production systems are essential. Organizations can adopt and maintain eco-friendly practices by increasing awareness at the highest levels of management. Four crucial steps are highlighted in (Fig. 3) to support management's emphasis on sustainability: increasing understanding about sustainability, motivating people to take part in sustainable projects, incorporating sustainability into business plans, and coordinating production systems with sustainability objectives. These actions help manage risks, meet customer requests, enhance stakeholder relationships, and minimize environmental impact. Consequently, businesses can increase productivity, expand the market for eco-friendly products, foster stakeholder collaboration, and promote ethical corporate behavior [43].

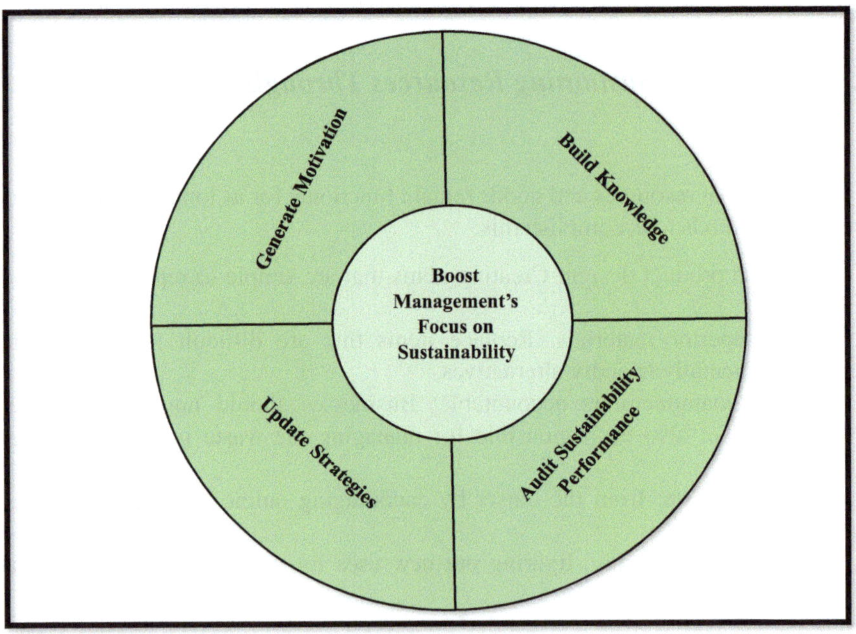

Fig. 3 Strategies of Enhancing leadership's commitment to sustainability. (Prepared by authors)

8 Future Prospects

The traditional linear waste management model, characterized by a "take, make, dispose" approach, poses significant environmental and resource-related challenges. The integration of Artificial Intelligence (AI) and Industry 5.0 offers promising solutions to overcome these limitations by shifting towards a circular economy. The following prospects highlight how AI, IoT, and Machine Learning (ML) can address key challenges in linear waste management:

Linear waste systems often struggle with inefficient waste collection, leading to overflowing landfills and environmental hazards. AI-powered waste monitoring and IoT-enabled smart bins can optimize collection routes, minimize overflow, and reduce operational costs through predictive waste generation analysis.

A significant limitation of linear waste management is the low recovery rate of recyclable materials. Advanced ML and deep learning (DL) models, such as Convolutional Neural Networks (CNNs), can improve automated waste sorting, accurately distinguishing between materials and boosting recycling efficiency. This will prevent valuable resources from being sent to landfills.

Poor waste segregation and disposal contribute to pollution and greenhouse gas emissions. AI-driven tracking systems can monitor waste streams, ensuring hazardous and non-biodegradable waste are processed correctly, thereby reducing soil and water contamination.

Linear waste management lacks real-time data analytics for effective decision-making. AI-powered platforms can enhance data sharing among stakeholders, including municipalities, recycling companies, and policymakers, allowing for informed strategies that improve waste diversion rates and sustainability efforts.

Industry 5.0 will promote human-machine collaboration, requiring investment in automation, robotics, and workforce upskilling. This shift will address labor shortages in waste management while increasing efficiency and safety in waste handling and processing.

9 Conclusion

The linear "take–make–dispose" waste management model is no longer viable for addressing the growing environmental and economic challenges of waste generation. The unsustainable exploitation of resources and excessive waste production have demonstrated the limitations of the linear economy, as both natural resource depletion and increasing demand put immense pressure on the environment and economic systems. A shift towards Circular Integrated Waste Management Systems is essential to ensure long-term sustainability. One of the key proposed improvements is expanding the boundaries of traditional waste management systems to include upstream subsystems, thereby linking raw material transformation to waste treatment and management. This approach enables better waste prevention, resource recovery, and quantification of biogenic carbon. A well-designed Circular Integrated Waste Management System can provide multiple benefits, including material recovery, energy generation, and nutrient extraction from waste. However, due to the complexity of waste composition and the variety of available treatment technologies, mathematical programming, Material Flow Analysis, and system assessment tools like Life Cycle Costing and Life Cycle Assessment are necessary for practical design. Despite the clear environmental benefits of Circular Integrated Waste Management Systems, their economic viability requires in-depth analysis to demonstrate cost-effectiveness. One of the significant challenges in assessing these systems is the diversity of materials managed, which complicates the selection of appropriate functional units in Life Cycle Assessment studies.

Additionally, while integrating multiple methodologies enhances system performance, the inclusion of upstream subsystems adds complexity to modeling efforts. Ultimately, transitioning to circular waste management models is not just an option but a necessity. Without this shift, resource scarcity, environmental degradation, and economic constraints will continue to escalate. A comprehensive and systemic approach to waste management, incorporating circular economic principles and technological advancements, is required to ensure a sustainable urban future.

References

1. Chauhan A, Chauhan M, Sethi M, Bodhe A, Tomar A, Shikha, Singh N (2024) Application of flower wastes to produce valuable products. In: Srivastav AL, Bhardwaj AK, Kumar M (eds) Valorization of biomass wastes for environmental sustainability: green practices for the rural circular economy. Springer, pp 251–268. https://doi.org/10.1007/978-3-031-52485-1_14
2. Chauhan A, Rajput V, Stanikzai K, Kumar S (2025) Understanding the global status of electronic wastes. In: Goswami S, Choudhury M, Agarwal S (eds) Electronic waste. CRC Press, pp 31–39. https://doi.org/10.1201/9781003582311-4
3. Oo PZ, Prapaspongsa T, Strezov V, Huda N, Oshita K, Takaoka M et al (2024) The role of global waste management and circular economy towards carbon neutrality. Sustain Prod Consump 52:498–510
4. Mondal T, Choudhury M, Kundu D, Dutta D, Samanta P (2023) Landfill: an eclectic review on structure, reactions and remediation approach. Waste Manag 164:127–142. https://doi.org/10.1016/j.wasman.2023.03.034
5. Rajpal A, Choudhury M, Goswami S, Chakravorty A, Raghavan V (eds) (2024) Waste management and treatment: advances and innovations, 1st edn. CRC Press. https://doi.org/10.1201/9781003258377
6. MacArthur E (2013) Towards the circular economy. J Ind Ecol 2(1):23–44
7. Ulian G, Cojocaru M, Rusu E, Ulian E (2020) The role of sustainable public procurements in the transition process from the linear economy model to the circular economy model. J Public Adm Finan Law:100–110
8. Morseletto P (2023) Sometimes linear, sometimes circular: states of the economy and transitions to the future. J Clean Prod 390:136138
9. Vence X, Pereira A, Laperche B (2022) Overcoming the circular economy paradox through innovation: pitfalls in the transition pathways. J Innov Econ Manag 39(3):1–13
10. Chowdhury R, Barman S, Choudhury M, Kim KH, Bhattacharya SS (2024) Earthworm modifies microbial community and functional genes for lignocellulosic waste valorization:

isolating plant-growth-promoting bacteria via next generation sequencing. Int Biodeterior Biodegradation 193:105854. https://doi.org/10.1016/j.ibiod.2024.105854
11. Paul S, Choudhury M, Deb U, Pegu R, Das S, Bhattacharya SS (2019) Assessing the ecological impacts of ageing on hazard potential of solid waste landfills: a green approach through vermitechnology. J Clean Prod 236:117643. https://doi.org/10.1016/j.jclepro.2019.117643
12. Self JL, Zervoudakis AJ, Peng X, Lenart WR, Macosko CW, Ellison CJ (2022) Linear, graft, and beyond: multiblock copolymers as next-generation compatibilizers. JACs Au 2(2):310–321. https://doi.org/10.1021/jacsau.1c00500
13. Choudhury M, Roy P (2025) Challenges with microplastic pollution in the regime of UN sustainable development goals. World Dev Sustain 6:100216. https://doi.org/10.1016/j.wds.2025.100216
14. Choudhury M, Adhikari MD, Agarwal S, Samanta P, Sharma A, Kundu D, Kumar S (2025) Pharmaceuticals and personal care products in soil: sources, impacts and myco-remediation strategies. Emerg Contam 2025:100488. https://doi.org/10.1016/j.emcon.2025.100488
15. Choudhury M, Rajpal A, Goswami S, Chakravorty A, Raghavan V (eds) (2024) Analytical case studies on municipal and biomedical waste management: perspectives on sustainable development goals. CRC Press. https://doi.org/10.1201/9781003499695
16. Choudhury M, Sharma A, Pervez A, Upadhyay P, Dutta J (2021) Growing menace of microplastics in and around the coastal ecosystem. In: Coastal ecosystems: environmental importance, current challenges and conservation measures. Springer International Publishing, Cham, pp 117–137. https://doi.org/10.1007/978-3-030-84255-0_6
17. Kurniawan TA, Hassan GK, Al-Hazmi HE, Othman MHD, Goh HH, Aziz F et al (2024) Landfill mining: a step forward to reducing CH_4 emissions and enhancing CO_2 sequestration from landfill. J Hazard Mater Adv 2024:100512. https://doi.org/10.1016/j.hazadv.2024.100512
18. Mayes WM, Hull SL, Gomes HI (2022) From linear economy legacies to circular economy resources: maximising the multifaceted values of legacy mineral wastes. In: Circular economy and sustainability. Elsevier, pp 409–431. https://doi.org/10.1016/B978-0-12-819817-9.00009-0
19. Jardosh N, Kathuria V (2025) Social cost–benefit analysis of solid waste management options with application to Mumbai, India. Waste Manag Res 43(1):39–49
20. Chakraborty M, Singh R, Awasthi A, Kumar V, Kumar A, Kaur S (2025) Assessment of greenhouse gas emissions in Delhi landfills: analyzing factors affecting emissions and environmental implications. J Mater Cycles Waste Manag 27(1):330–342
21. Akpan SN, van Hooft P, Happi AN, Buij R, van Langevelde F, Cook EA et al (2025) Structure, conservation and health implications of urban wild meat value chains: a case study of Lagos, Nigeria. One Health 20:100992
22. Agarwal S, Darbar S, Saha S, Choudhury M, Singh RP (2023) E-waste management using different cost-effective, eco-friendly biological techniques: an overview. In: Waste management and resource recycling in the developing world. Elsevier, pp 205–235. https://doi.org/10.1016/B978-0-323-90463-6.00026-9
23. Goswami S, Choudhury M, Agarwal S (eds) (2025) Electronic waste: impact on health, animals, and the environment, 1st edn. CRC Press. https://doi.org/10.1201/978100358231
24. Bhardwaj LK, Rath P, Jain H, Choudhury M (2025) Exploring the effects of e-waste on soil, water quality and human health. Discov Civil Eng 2(1):1–16. https://doi.org/10.1007/s44290-025-00167-2
25. Daum K, Stoler J, Grant RJ (2017) Toward a more sustainable trajectory for e-waste policy: a review of a decade of e-waste research in Accra, Ghana. Int J Environ Res Public Health 14(2):135
26. Lebreton L, Slat B, Ferrari F, Sainte-Rose B, Aitken J, Marthouse R et al (2018) Evidence that the Great Pacific Garbage Patch is rapidly accumulating plastic. Sci Rep 8(1):1–15
27. B.-B.I. Consortium (2021) The bio-based industries a public-private partnership on bio-based industries vision: accelerating innovation and market uptake of bio-based products
28. Hegab H, Shaban I, Jamil M, Khanna N (2023) Toward sustainable future: strategies, indicators, and challenges for implementing sustainable production systems. Sustain Mater Technol 36:e00617. https://doi.org/10.1016/j.ibiod.2024.105854

29. Kanellou E, Alexakis K, Kapsalis P, Kokkinakos P, Askounis DJPM (2021) The DigiPrime KPIs' framework for a circular economy transition in the automotive industry. Procedia Manuf 54:302–307
30. Kinnunen J, Saunila M, Ukko J, Rantanen H (2022) Strategic sustainability in the construction industry: impacts on sustainability performance and brand. J Clean Prod 368:133063
31. De Giovanni P, Folgiero P (2023) Strategies for the circular economy: circular districts and networks. Taylor & Francis
32. Bataille C, Waisman H, Briand Y, Svensson J, Vogt-Schilb A, Jaramillo M, Delgado R, Arguello R, Clarke L, Wild TJESR (2020) Net-zero deep decarbonization pathways in Latin America: challenges and opportunities. Energ Strat Rev 30:100510
33. Ben Amor L, Hammami SJR (2022) An empirical study on the main determinants of recycling plastic waste in Tunisia. Recycling 7(1):1
34. Kuei C-H, Lu MH (2013) Integrating quality management principles into sustainability management. Total Qual Manag Bus Excell 24(1–2):62–78
35. Lacovidou E, Millward-Hopkins J, Busch J., Purnell P, Velis CA, Hahladakis JN, Zwirner O, Brown A (2017) A pathway to circular economy: developing a conceptual framework for complex value assessment of resources recovered from waste. J Clean Prod 168:1279–1288. https://doi.org/10.1016/j.jclepro.2017.09.002
36. Choudhary R, Kumar A, Narayan S, Mishchenko I, Choudhury M, Bhardwaj LK et al (2025) Integrating fuzzy logic into life cycle assessment for sustainable municipal solid waste management in metro cities. Ecol Eng Environ Technol 26(7):316–331. https://doi.org/10.12912/27197050/205500
37. Bergek A, Jacobsson S, Sandén B (2008) 'Legitimation' and 'development of positive externalities': two key processes in the formation phase of technological innovation systems. Tech Anal Strateg Manag 20(5):575–592. https://doi.org/10.1080/09537320802292768
38. Markard J, Raven R, Truffer B (2012) Sustainability transitions: an emerging field of research and its prospects. Res Policy 41(6):955–967. https://doi.org/10.1016/j.respol.2012.02.013
39. Robèrt KH (2000) Tools and concepts for sustainable development: how do they relate to a general framework for sustainable development, and to each other? J Clean Prod 8(3):243–254. https://doi.org/10.1016/S0959-6526(00)00011-1
40. Moraga G, Huysveld S, Mathieux F, Blengini GA, Alaerts L, van Acker K, Meester S de, Dewulf J (2019) Circular economy indicators: what do they measure? Res Conserv Recycl 146:452–461. https://doi.org/10.1016/j.resconrec.2019.03.045
41. Raig, W. (2001). The role of materials in the circular economy. Environ Sci Technol 35(1):10–15. https://doi.org/10.1021/es0101234
42. Korhonen, J. (2001). Industrial ecosystem renewal towards circularity to achieve sustainability. J Clean Prod 9(3):253–259. https://doi.org/10.1016/S0959-6526(01)00017-3
43. Roy P, Choudhury M (2022) Bioeconomy and green plastic production. In: Baskar C, Dharmalingam S, Christuraj R (eds) Encyclopedia of green materials. Springer Nature, Singapore, pp 1–7. https://doi.org/10.1007/978-981-16-4921-9_189-1

Barriers to Circular Economy Implementation: Examining Regulatory and Policy Obstacles

Omid Reza Baghchesaraei, Alireza Baghchesaraei, and Amirali Khademi

Abstract The circular economy is an economic model based on waste reduction and resource recycling. This model presents both sustainable economic solutions and social challenges. In this context, artificial intelligence demonstrates the potential to optimize waste management processes and reduce associated costs through tools such as machine learning, computer vision, and data analysis. Many challenges to implementing sustainable strategies remain unresolved. The advantages of a circular economy include reducing costs associated with waste management, improving resource efficiency, creating employment opportunities in recycling and reuse sectors, and mitigating adverse environmental impacts. These benefits align with the United Nations' Sustainable Development Goals (SDGs). To adopt the circular economy as the economic model of the future, significant efforts are required to reform existing practices and formulate more comprehensive policies. As demonstrated by international experiences, economic instruments remain among the most effective means of environmental protection and resource conservation. The analysis concludes with a discussion of CE's current state and policy recommendations for its future growth. Additionally, research and development investments at the corporate and university levels can enhance domestic technological innovation capacity.

Keywords Circular economy · Policy obstacles · Waste management · Artificial intelligence

O. R. Baghchesaraei (✉)
Centre for Infrastructure Engineering, Sydney, NSW, Australia
e-mail: omidrezabaghchesaraei@gmail.com

A. Baghchesaraei
Infrastructure Management (Engineering), Construction Industry, Sydney, NSW, Australia
e-mail: baghchesaraeialireza@gmail.com

A. Khademi
Department of Civil Engineering, Sharif University of Technology, Tehran, Iran
e-mail: amirali.khademi81@sharif.edu

S. Majumdar et al. (eds.), *Sustainable Urban Future*, Urban Sustainability,
https://doi.org/10.1007/978-981-95-2976-6_7

1 Introduction

Waste management poses a significant challenge globally. Population growth, urbanization, and evolving lifestyles have escalated waste generation across various sectors as household, industrial, commercial, and medical [1, 2]. This surge in waste not only increases management costs but also causes severe environmental and social problems, including pollution. In today's complex market, characterized by product diversification, fluctuating political and economic landscapes, climate change, resource scarcity, and pollution, economics has moved beyond the traditional linear model (production-distribution-consumption) to embrace a circular economy (production-consumption-recycling) [3].

The circular economy significantly improves waste management and mitigates environmental harm [4]. By minimizing waste generation and reducing primary resource consumption, it addresses the increasing demands for resources and the limited availability of natural reserves, making sustainable resource and waste management models crucial [5]. Several nations, notably in Europe and East Asia, have successfully implemented this model through advanced technologies and infrastructure development, achieving substantial gains in waste management and environmental conservation [6]. Based on recent articles published about political barriers to reaching the circular economy (Fig. 1), we face some important problems [7–10]:

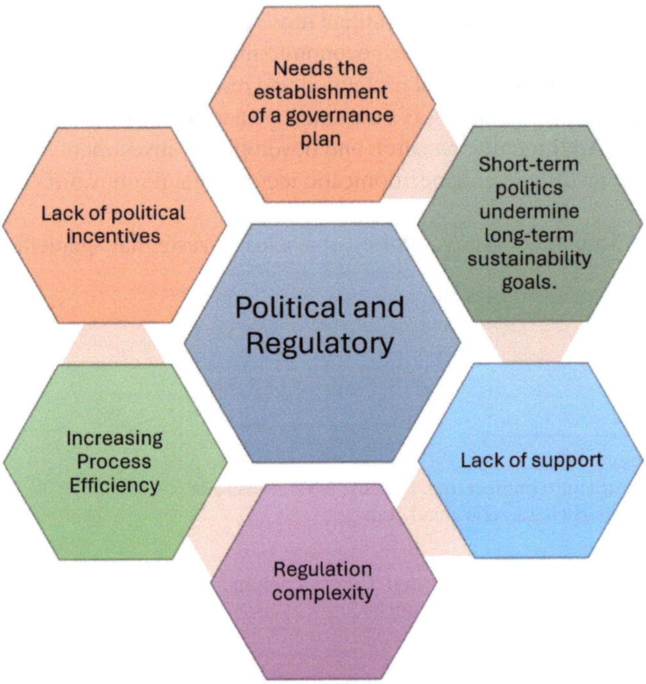

Fig. 1 Political main barriers in the circular economy

The circular economy prioritizes waste reduction through process redesign, efficient consumption, recycling, and material recovery. Conversely, the long-dominant linear model (extraction-production-consumption-disposal) strains natural resources and the environment [11]. The circular economy minimizes environmental impact by preserving resources, extending product lifespans, reducing waste, and preventing pollution [12]. It seeks to retain and recycle materials, products, and energy to avoid excessive resource extraction and waste disposal. This model is built upon principles such as sustainable design, reuse and recycling, and material recovery. One of its key components is design for durability, which emphasizes the production of long-lasting, repairable, upgradable, and recyclable products [13]. In this framework, raw materials and products are designed to remain useful even after their initial use, ensuring that they are not discarded but instead re-enter the production cycle. This approach reduces the need for raw material extraction and mitigates the environmental impacts associated with both resource extraction and waste disposal (Table 1).

Industrial advancements have led to an increase in the production of hazardous waste. New technologies and expanded industrial activity have increased the volumes of chemical waste and hazardous materials, including toxic chemicals, heavy metals, and non-biodegradable plastics [14, 15]. These persistent pollutants have a severe impact on both ecosystems and human health. Therefore, proper waste management and principles of the circular economy are crucial for reducing waste and enhancing resource recycling. By emphasizing reduction, reuse, and recycling, the circular economy mitigates pollution, conserves resources, and contributes to environmental sustainability and improved quality of life [16].

As previously mentioned, the rapid increase in population and industrial expansion has led to a significant rise in municipal and industrial waste generation. This trend is driven not only by increased consumption of materials and products but also by shifts in consumption patterns, resulting in more complex and hazardous waste compositions. Industrial and municipal waste often contains hazardous chemicals, heavy metals, and non-biodegradable plastics, which pose serious risks to both the environment and human health. Consequently, effective waste management has become a necessity to prevent environmental pollution and mitigate health risks (Fig. 2).

2 Key Regulatory and Policy Barriers

Sustainable development is a process aimed at achieving societal stability in resource utilization, ensuring the replenishment of resources. It encompasses economic growth and human development, seeking continuous progress beyond basic economic gains [17]. It prioritizes the sustainability of non-renewable resources for future generations, creating a positive future by integrating economic, social, and environmental factors.

Table 1 Elements comparison of linear and circular economy

Feature/indicator	Linear economy	Circular economy
Material Flow	Linear (Extraction—Production—Consumption—Disposal)	Circular (Reduce—Reuse—Recycle—Remanufacture—Regeneration)
Objective	Economic growth through increased production	Sustainable economic growth by reducing waste and enhancing resource efficiency
Focus	Product manufacturing	Reuse, repair, and recycling
Value Consideration	Only the final product	Entire product lifecycle
Waste Generation	High waste levels with no reuse of products	Waste reduction through recycling and reuse
Resource Efficiency	Low efficiency with continuous demand for new resources	High efficiency through multiple use of materials and existing energy
Environmental Impact	Increased pollution and environmental degradation	Reduced pollution, biodiversity conservation, and improved sustainability
Product Lifespan	Designed for short-term use	Designed for longevity and recyclability
Economic Costs	High costs due to resource dependency and waste disposal	Cost reduction through remanufacturing and lower raw material demand

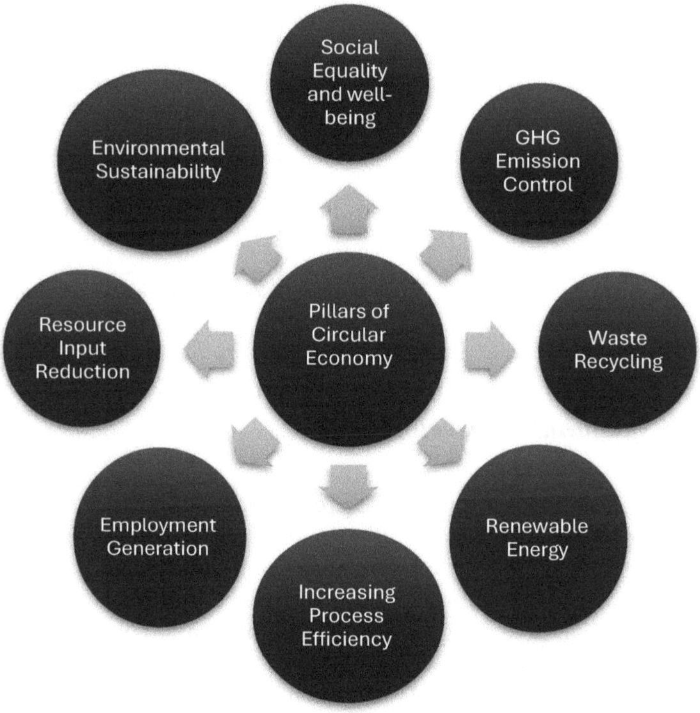

Fig. 2 The Principles of the circular economy for waste management

Sustainable development, while subject to varying interpretations, fundamentally balances societal environmental, social, and economic needs. Development often addresses immediate needs without considering long-term impacts. The consequences of this approach are now evident, ranging from financial crises stemming from irresponsible banking to global climate change driven by fossil fuel dependence [18]. The more we pursue unsustainable development, the more frequent and severe its consequences become. Therefore, we must strive for sustainable development from this day forward.

To date, sustainable development has been a key focus in numerous global conventions addressing crucial aspects of sustainability. Recycling and sustainable development are directly linked, as the limitation of natural resources and environmental challenges necessitate recycling as a prominent strategy for environmental preservation and sustainable growth.

Developing the recycling supply chain—One of the most critical strategies is for advancing recycling the development of a robust and efficient supply chain. This chain includes the collection, separation, recycling, and restoration of recyclable materials. To achieve this, collection and recycling systems must be improved, and innovations in recycling technologies should be promoted [19].

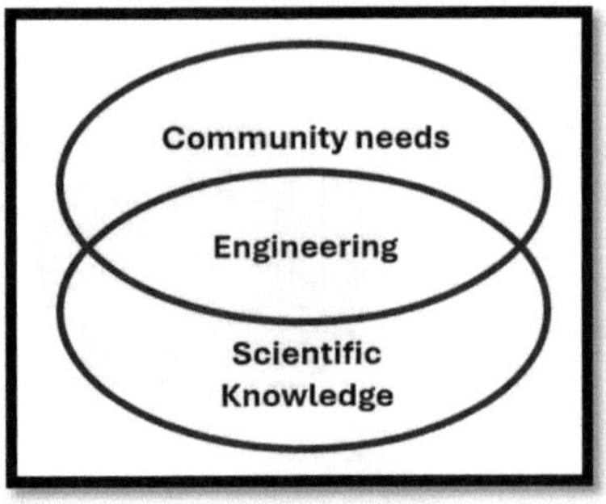

Fig. 3 The relationship between society's needs, engineering and scientific knowledge

2.1 *The Connection of Engineering Knowledge with Sustainable Development*

As discussed, all aspects of human life are changing, and we are witnessing the interconnection of various elements of human existence, including the economy, environment, culture, and politics. Today, sustainable development is a global movement in which many countries are striving to achieve this important goal and protect it. Conserving resources as long as possible has proven to be a societal challenge that affects everything from individual lifestyles to transportation. They have been the intermediary between the needs of society and scientific knowledge (Fig. 3). Engineers must possess characteristics enabling them to bridge the gap between society and the environment. While technical skills are essential, engineers, as the link between science and technology, require a broader skill set. They are key to sustainable development and navigating the relationship between science, technology, and ethics.

2.2 *The Significance of Circular Economy Approach in Sustainability*

The circular economy is crucial for sustainability, primarily due to its potential to mitigate environmental degradation. By reducing raw material extraction and waste generation, it lessens pressure on natural resources and ecosystems. This is vital, considering the growing populations and consumption levels that strain the planet's finite resources. Furthermore, the circular economy helps mitigate climate change.

By improving resource efficiency and reducing greenhouse gas emissions from production and disposal, it supports efforts to mitigate climate change. This is essential, given the urgent need to curb emissions and limit global warming and its impacts [20].

In addition to the environmental benefits, the circular economy approach also has a significant impact on economic development. This approach can create new opportunities for job creation and economic growth by strengthening innovation in product design, production processes, and business models. In addition, by promoting more efficient use of resources, businesses' costs decrease, and their competitiveness increase in a rapidly changing global market [21].

From a social perspective, the circular economy approach has important implications for sustainability. By promoting a more equitable distribution of resources and opportunities, this approach can help solve the problems of poverty and inequality. Furthermore, by encouraging collaboration between different stakeholders—including businesses, governments, non-profit organizations, and consumers—it can foster a more inclusive approach to sustainability that takes into account diverse perspectives and priorities [22].

Therefore, the circular economy approach represents a fundamental change in the current approaches to economic development, which is necessary to achieve the goals of sustainable development. By improving resource efficiency, reducing waste generation, mitigating environmental degradation, fostering economic growth, and addressing social inequalities, this approach has the potential to create a more flexible and equitable future for all. Similarly, policymakers, businesses, civil society organizations, and individuals should collaborate to adopt this approach as the foundation of sustainable development efforts worldwide [23]. There are different approaches to implementing the circular economy in sustainable development. These approaches can be categorized into three main areas: product design and innovation, business models and supply chain management, and policy and regulation.

Product design and innovation are key to implementing a circular economy. Rethinking product design to minimize waste and maximize resource efficiency is essential. Designing for durability, repairability, recyclability, or biodegradability extends product lifespans and reduces environmental impact. Incorporating renewable and recycled materials further closes material loops within supply chains [24].

Innovative business models, prioritizing resource efficiency over traditional consumption, are another key approach to implementing the circular economy. These models include renting/sharing products (e.g., car-sharing), product-as-a-service offerings (e.g., pay-per-use appliances), and take-back programs (e.g., electronics recycling). Furthermore, supply chain practices such as reverse logistics facilitate closed-loop systems, enabling the recovery of materials from end-of-life products [25]. Political interventions are a crucial factor; governments can implement regulations that encourage sustainable manufacturing practices (such as environmental design standards), promote producer responsibility (for example, requiring manufacturers to take back end-of-life products), or support investment in recycling infrastructure or rebuilding facilities [26].

3 Intelligent Systems

3.1 Smart Building

The built environment is central to human life, with people in developed nations spending 80–90% of their time indoors [27]. Our lives are shaped by this constructed world, whether in urban or rural settings. While "Intelligent Buildings" is an established concept, "smart" is increasingly prevalent, reflecting the adoption of smart technologies like sensors, materials, and meters for high-performance buildings across the sector [28]. It is important to understand how the building is being "designed" as this will affect the relationship one has to the design team, the owner and the construction company (Fig. 4).

According to the design and construction process using smart buildings reduce the costs of building in comparison with traditional buildings (Table 2; Fig. 5).

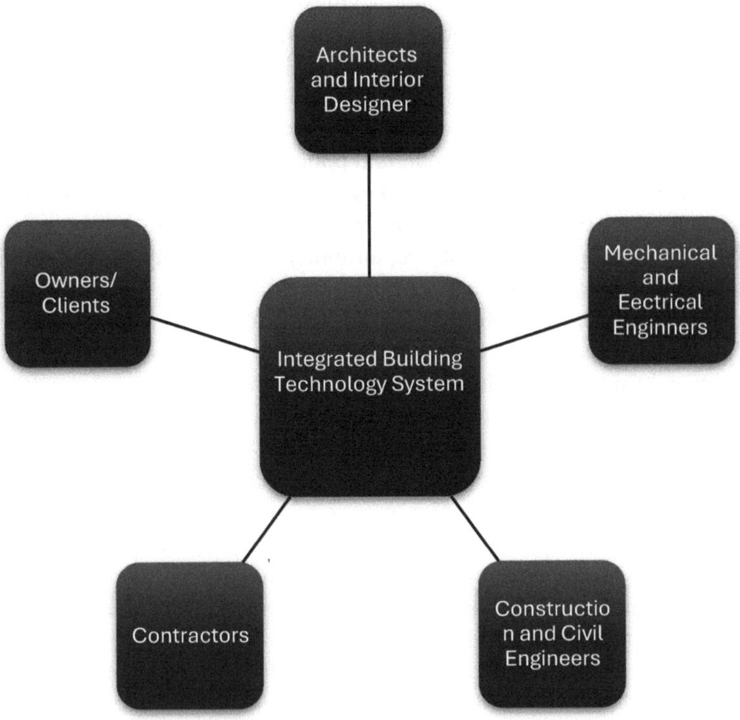

Fig. 4 Collaborative expertise in building design

Table 2 Smart Buildings vs. traditional Buildings

Amount (%)	Costs
56	Less in first cost
32	Less in annual costs
82	Less in annual maintenance
10	Less in utility costs
16	Less in cabling costs
44	Less in labour hours

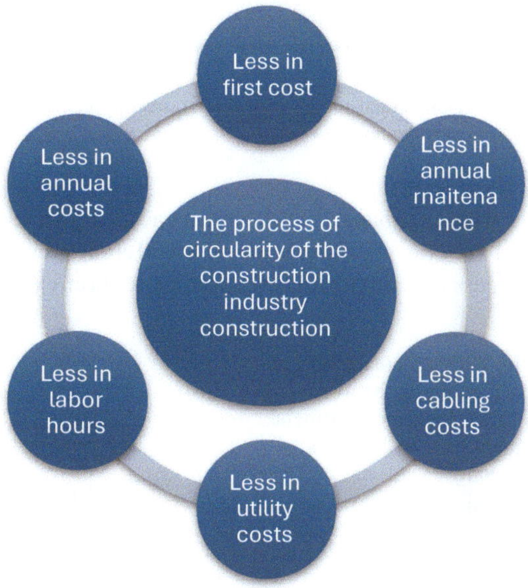

Fig. 5 The advantage of modern construction over traditional construction

3.2 Artificial Intelligence (AI)

Artificial intelligence (AI) and the circular economy (CE) can contribute to improving waste management systems in cities by providing innovative solutions.

The circular economy emphasizes minimizing waste, reusing, and recycling resources, while AI offers tools to optimize and automate waste management processes [29]. The integration of these two concepts enhances resource efficiency and reduces environmental footprints [30].

The application of AI in waste management enables the precise utilization of data and the analysis of big data, facilitating accurate predictions of waste generation and identifying consumption and production patterns [31]. These data-driven insights contribute to improved waste management and landfill reduction (Fig. 6).

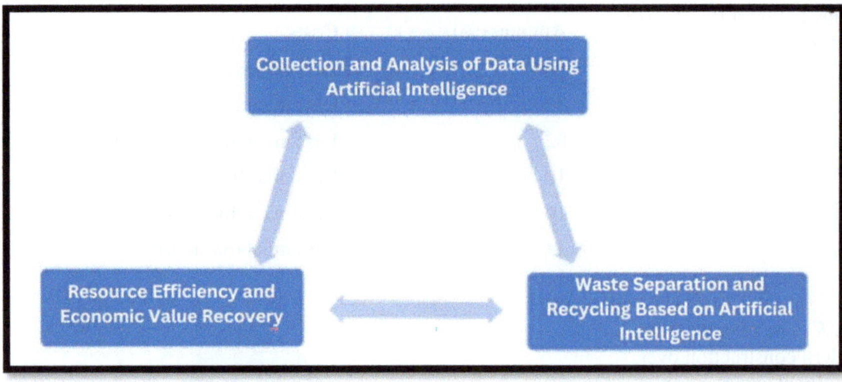

Fig. 6 Conceptual model of the relationship between waste management, artificial intelligence, and circular economy

AI also enhances waste sorting accuracy in recycling facilities, as machine vision and robotic algorithms identify different waste types, streamlining the recycling process and supporting the circular economy (Fig. 7).

Fig. 7 Conceptual model of the relationship between waste management, artificial intelligence, and circular economy

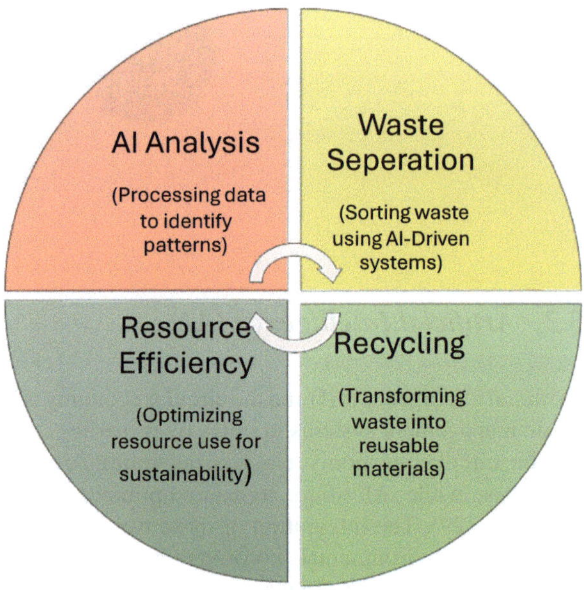

4 Strategies for Implementing a Circular Economy in Waste Management

A fundamental component of the circular economy is considering strategies for Implementing a Circular Economy in Waste Management [32].

4.1 Sustainable Product Design and Durability

A fundamental component of the circular economy is designing products for durability and recyclability. In this model, products are initially designed to have a longer lifespan and be fully recyclable or reusable. For example, manufacturers incorporate high-quality recyclable materials into their designs to ensure that products remain functional even after multiple uses. This approach not only reduces waste generation but also creates new opportunities for product reuse and repair.

4.2 Recycling and Material Recovery

The circular economy places significant emphasis on recycling and material recovery. Instead of discarding waste, this model aims to reintegrate materials into the production cycle as raw materials. This process helps to reduce waste volume while also decreasing the need for raw material extraction. Recycling plastic, metal, glass, and paper is a key application of this concept, enabling industries to source essential raw materials while reducing dependence on virgin resources [33].

4.3 Reducing Landfill Dependence and Environmental Protection

Traditional landfill waste disposal methods pose significant environmental challenges, often resulting in groundwater, soil, and air pollution. A primary goal of the circular economy is to minimize dependence on landfills. By promoting recycling and reuse, the volume of landfill waste is significantly reduced, thereby mitigating environmental pollution. This is particularly crucial for countries facing land scarcity for waste disposal [34, 35].

4.4 Economic Value Creation and Job Opportunities

Beyond environmental benefits, the circular economy also holds significant economic potential. Through recycling and material recovery, industries can utilize previously discarded resources, reducing production costs. Furthermore, this model fosters job creation in recycling, repair, remanufacturing, and sustainable design sectors. Many European countries have successfully developed recycling industries and incentivized material recovery, leading to the creation of stable employment opportunities and economic growth.

4.5 Enhancing Resource and Energy Efficiency

The circular economy contributes to improved resource and energy efficiency. Since materials are reused and recycled instead of being discarded, the need for raw material extraction and energy-intensive production processes is minimized. For example, producing aluminum from recycled materials requires approximately 90% less energy than producing aluminum from raw materials. This energy and resource conservation significantly lowers production costs and reduces environmental impact.

4.6 Promoting a Culture of Sustainable Consumption

A key role of the circular economy in waste management is fostering a culture of sustainable consumption and curbing excessive consumption habits. This model educates individuals and society to place greater value on products and materials, encouraging reuse and recycling over indiscriminate consumption [36]. Raising public awareness and implementing financial and social incentives for sustainable consumption can help shift societal attitudes, laying the foundation for sustainable waste management and long-term environmental preservation (Fig. 8).

One of the biggest obstacles to implementing a circular economy is the issue of policies and regulations. Achieving the benefits of a circular economy aligned with sustainability requires changes in legislation; therefore, policy-driven programs have also evolved in accordance with changes in the circular economy:

- The Paris Climate Agreement
- The United Nations Sustainable Development Goals (SDG 12, 11, 9, 13)
- The European Green Circular Economy Action Plan
- The 2019 Durban City Declaration in Africa
- China's Five-Year Circular Economy Plan
- Circular economy strategies in Latin American countries

Fig. 8 Consumer Behavior

Given the high costs associated with linear economic models, governments have begun to recognize the advantages of a circular economy.

5 Policy Making, Key Regulatory and Policy Barriers for Circular Economy

Policymaking is crucial for the circular economy, as evidenced by the agreement among governments, industries, and academics on the role of government in policy development and implementation. Incentive policies influencing consumer behavior are key to achieving circular economic goals. Resource efficiency policies, central to the circular economy, originated in Japan and Germany (1980s–1990s), focusing on closed-loop systems. In the last 15 years, numerous countries have implemented circular economy principles, often resulting in sustainable resource consumption, optimized raw material use, increased recycling, and reduced waste. The circular economy mitigates the conflict between economic development and emissions by addressing pollution and resource scarcity, while also enhancing Chinese business competitiveness and overcoming environmental trade barriers [37]. Growing Sino-German research collaborations demonstrate China's ability to leverage successful initiatives. While Germany leads in waste management, it has yet to translate these principles into a broader industrial economic model. China may succeed where Germany has not. Although East Asian nations (excluding Japan) initially dismissed eco-industrial reform as unnecessary for industrial development, they now view concepts like low-carbon technologies, renewable energy, and industrial symbiosis as potential sources of competitive advantage.

Evidence of the implementation of eco-industrial design principles is also found in Singapore, particularly in the Jurong petrochemical complex. It is only a matter of time before industrial symbiosis becomes a core industrial policy in South Korea and other East Asian nations, driven by tax exemptions, subsidies, and carbon credit incentives [38]. Policy-level challenges remain. China's corporate VAT (Value-Added Tax) system sometimes discourages the use of recycled materials. While recycled materials may be cheaper, their higher value-added ratio results in increased VAT, incentivizing the use of virgin materials.

Furthermore, environmental law enforcement is hampered by insufficient personnel and funding. Local authorities prioritizing short-term economic gains often overlook environmental concerns. Consequently, companies frequently discharge waste directly, neglecting by-product markets or pre-treatment equipment, thus disincentivizing environmentally friendly technologies and products [38]. Given these examples, it is evident that effective policies promoting green production, sustainable technologies, and responsible consumption are still lacking in China.

5.1 Key Challenges Hindering Circular Economy Implementation in China (Fig. 9)

- Lack of reliable information
- Absence of advanced technologies
- Weak enforcement of regulations
- Insufficient economic incentives
- Poor management and leadership
- Low levels of public awareness

Governments typically act as the intermediary linking energy companies and local communities through policymaking. Depending on the project's scale, a governing body may range from a local government to an international organization such as the European Union. A national government usually establishes a dedicated department with the necessary authority to systematically address obstacles [39].

Policymakers must recognize the need for intervention when market failures occur or when their actions enhance social welfare. Governments have the power to mandate cost-benefit assessments, control assets and resources, enforce long-term commitments, and coordinate the development of waste-to-energy (WTE) plants with efficient energy distribution infrastructure. Municipal authorities can also reduce costs by partnering with energy service companies (ESCOs) to link public investments. When institutions fail to regulate or govern effectively, the anticipated benefits of a policy become unattainable. Key stakeholders in policy implementation must adhere to action plans and performance criteria outlined within policies [40].

Comprehensiveness, coherence, and flexibility are critical attributes that must be maintained for successful policymaking. A policy achieves comprehensiveness by incorporating insights from multiple disciplines and accounting for external effects, ensuring synergy among its mechanisms. Coherence is maintained by setting clear objectives and following up with assessments of implemented actions. Policies with ambiguous goals can lead to the failure of projects, rendering them ineffective and ultimately unsuccessful. Conversely, well-defined objectives and measures facilitate

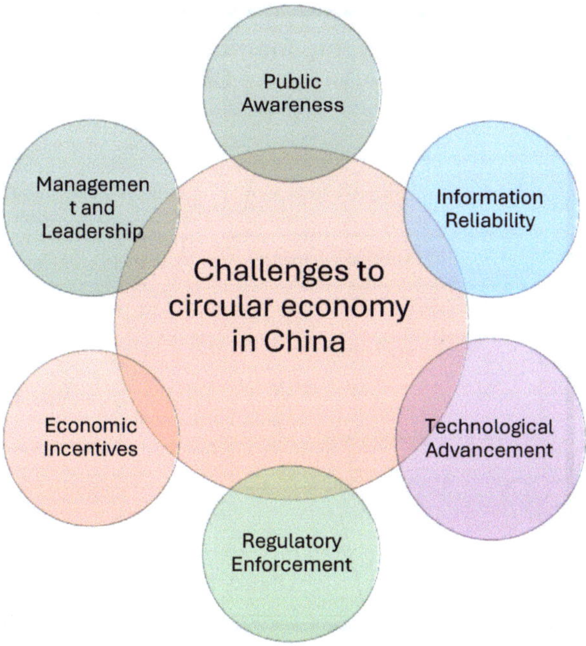

Fig. 9 Challenges to circular economy in China

project implementation, ease access to funding, and simplify progress evaluation. Another aspect of coherence is that policymakers should minimize the frequency and extent of changes. Constant policy shifts can deter investors and limit progress during implementation [41].

A supportive policy environment is crucial for effective cleaner production. Interagency collaboration is crucial for promoting cleaner production, resource efficiency, and waste minimization. A provincial-level roundtable involving stakeholders and senior management can ensure equitable consideration of all aspects, facilitate information exchange and feedback, secure resources, and enable negotiations [42].

Finally, institutional channels should be established to encourage public participation in voicing concerns and actively promoting cleaner production. For instance, direct communication with the Deputy Mayor responsible for cleaner production through a dedicated hotline, as well as engagement with local people's congress representatives and members of the local political consultative conference, could expedite responses to concerns about industrial pollution [43].

Policymaking is crucial for a circular economy, as evidenced by the consensus of governments, industry, and academia. Incentive policies influencing consumer behavior are key to achieving circular economy goals. Resource optimization policies, the foundation of a circular economy, first emerged in Japan and Germany in the 1980s and 1990s, focusing on closed-loop systems. Over the past 15 years, various nations have implemented circular economy principles, often witnessing direct results, including stabilized primary resource consumption, optimized raw material use, increased recycling, and reduced waste [44]. In the following sections,

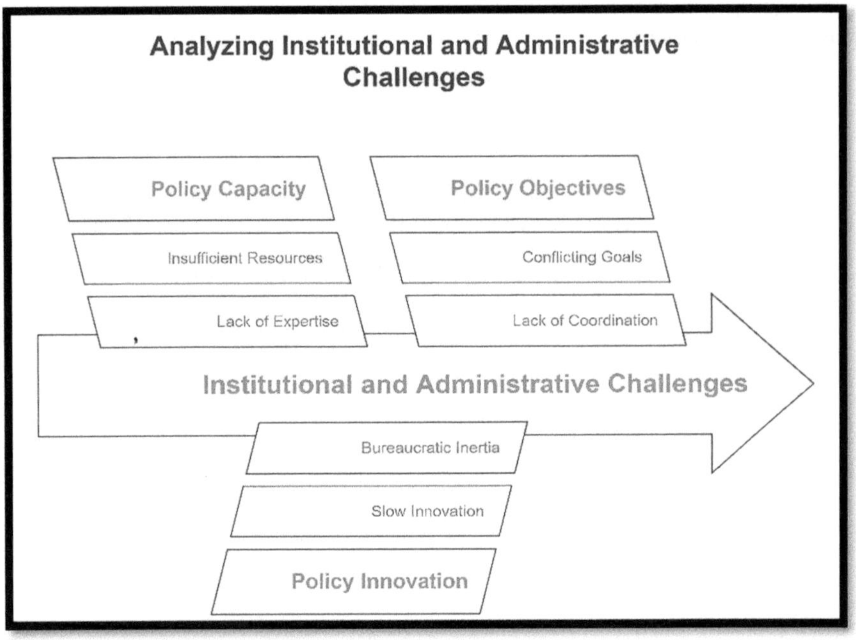

Fig. 10 Analyzing institutional and administrative challenges

we discuss the experiences of various countries with the concept of a circular economy. These policies encompass a range of simple waste management programs to support the transition to a circular economy (Fig. 10).

Less developed countries, such as Kyrgyzstan, Gambia, Kenya, Uganda, Rwanda, and Costa Rica, due to difficult geographical conditions and severe economic constraints, have enacted laws related to waste management in the long term and have attempted to develop comprehensive waste management systems at the national level. Some of these comprehensive waste management systems are electronic and smart government systems [45]. However, some construction or toxic waste is not recycled due to its specific complications and is sent to waste landfills with little change [46].

More developed countries, like France, pay attention to the volume of material consumption and to preventing product obsolescence. Policies are designed to extend the lifespan of various infrastructures and products, thereby minimizing depreciation as much as possible. Likewise, Spain has set a goal of reusing or recycling 50% of municipal waste in its comprehensive 6-year waste management plan (2016–2022). More comprehensively, the European Union Commission and its environmental agencies are planning Circular economy models in various industries, especially in the construction industry [47]. The purpose of these models, which will eventually become law, is to encourage industries to optimize material consumption and maximize the use of recycled materials, thereby promoting reuse. Considering that a quarter of the production wastage in the European Union is due to the construction industry, various models offer various suggestions, including the following [48]:

- Using asphalt chips and concrete chips or even steel, glass, aluminium, and wood in road construction.
- Use of usable materials obtained in the process of demolishing buildings.
- Improving the quality of consumables and increasing their lifespan.
- Using versatile materials for different uses that are compatible with the surrounding environment in the construction process.

Due to the rapid pace of industrialization and increasing urbanization, China has implemented various policies to promote circularity across its various industries, aiming to address the problems resulting from industrialization. The Chinese government explains its circular economy policies, which aim to reduce the consumption of raw materials. It encourages various industries, especially the construction industry, to minimize raw material consumption by designing optimally. China follows its policies by imposing restrictions on the consumption of raw materials [49].

South Korea has also explained the 10-year plan (2018–2027) for circularity of material consumption, and by setting national goals and various policies, it is moving in the direction of reducing wastage production and recycling material consumption and reusing them, identifying valuable wastage, separating recyclable and reusable materials from the rest of the wastage and encouraging industries to deliver recyclable materials and reducing the amount of wastage [50].

Policies can also target higher education, incorporating sustainability principles and skills development to cultivate a workforce capable of advancing sustainability. Circular economy training, within the broader sustainability framework, creates specialized personnel who can significantly reduce construction and operational costs by facilitating the transition from a linear to a circular economy [51].

One approach to relevant policies is the concept of a resistance economy, which was first proposed in September 2009 [51]. Among the strategic goals of the resistance economy in line with the road and construction industry, the following can be mentioned [52].

- Attention to entrepreneurs and producers in the field of the economy
- More use of domestic capacities in foreign markets
- Increased attention and focus on the real sector of the economy
- Establishing stability in the economy and ensuring its continuous growth in the long term

Therefore, the principles and concepts of the circular economy are related to the economy of resistance, and their application can help achieve the goals of the economy of resistance.

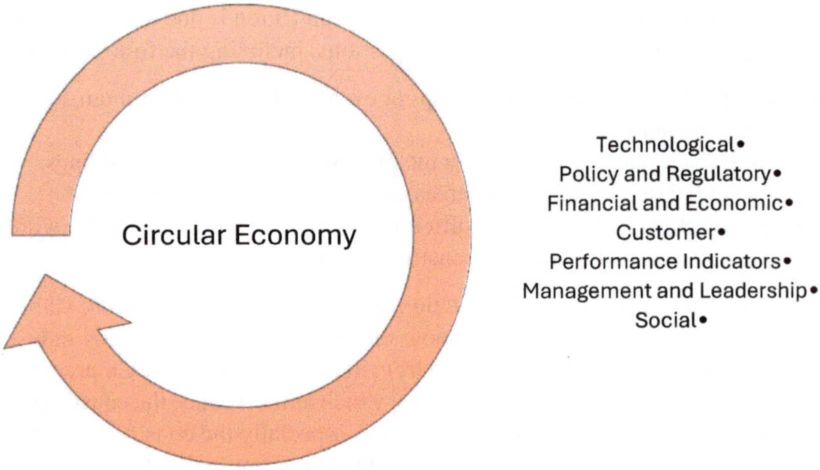

Fig. 11 Circular economy main barriers (overcoming barriers to progress)

6 Future Recommendation

By adhering to the principles of reduction, reuse, and recycling, the circular economy contributes to project sustainability and supports sustainable development goals through its strong economic, environmental, and social interconnections. Researchers must focus on addressing policy barriers to solving circular economy challenges, especially in developing countries. Additionally, collaboration between industry, government, and civil society is essential to facilitate the flow of recyclable materials and establish a strong and sustainable market for recycled products. The main barriers identified in the article were: (1) technological, (2) policy and regulatory, (3) financial and economic, (4) managerial, (5) performance indicators, (6) customer, and (7) social (Fig. 11). Policy barriers are one of the primary obstacles to the circular economy in developing countries, such as those in the Middle East, and researchers must address them.

7 Conclusion

Among the proposed approaches, AI has the potential to be more operational and practical throughout the project lifecycle, effectively reducing both economic and environmental costs. To adopt the circular economy as the economic model of the future, significant efforts are required to reform existing practices and formulate more comprehensive policies. As demonstrated by international experiences, economic instruments remain among the most effective means of environmental protection and resource conservation. The Chinese government must leverage policy-making to enhance economic incentives that guide corporate and citizen behaviours in line with circular economy principles.

Furthermore, the government must continue supporting key technologies essential for the circular economy. This necessitates accurately identifying priority technology areas and projects that align with both immediate and long-term economic needs. Additionally, research and development (R&D) investments at the corporate and university levels can enhance domestic technological innovation capacity.

To increase public awareness and engagement, periodic activities such as television campaigns, newsletters, achievement exhibitions, and workshops should be organized. These initiatives provide platforms for objective assessments of experiences from different countries and institutions. Moreover, corporate information exchange fosters mutual understanding and greater collaboration, laying a solid foundation for further promoting the circular economy.

Improving law enforcement and governmental management is also critical. This requires reforms in judicial management mechanisms, transparent oversight, and auditing practices. Additionally, the government should develop standardized methods for data collection, calculation, and reporting to enable accurate assessments of circular economy development.

The essential and urgent transition from goods production to services, and from a traditional to a digital, circular economy, cannot be delayed. Inaction is not an option, given the escalating accumulation of household and industrial waste. However, Fourth Industrial Revolution technologies and advancements in the digital economy offer potential solutions. Government, environmental organizations, industry associations, and leading companies must prioritize a focused, rapid, and responsible transition to a digital circular economy. This offers hope for reducing pollution, conserving resources, and promoting the reuse of non-renewable resources in developing countries.

The development and adoption of innovative technologies, along with advancements in research and development in areas such as e-commerce, reverse logistics, AI, big data analytics, and the Internet of Things, are crucial for achieving circular economy goals. Sufficient support and extensive investment in these areas can facilitate the implementation of such innovations within national industrial value chains, enabling the transformation from a linear to a circular value chain. In essence, supporting digital innovations that underpin new business models aligned with social responsibility—aimed at reducing environmental impacts and improving long-term societal well-being—enhances economic prosperity. Furthermore, leveraging domestic capacities, particularly digital capabilities, can significantly contribute to the successful development of a circular economy.

Despite the numerous benefits of AI in urban waste management, its application still faces challenges, including inadequate data availability, technical issues, integration complexities within urban systems, high costs, economic constraints, ethical concerns, and theoretical criticisms of the circular economy concept.

Nevertheless, AI applications in urban waste management and the circular economy can lead to cost reductions, increased efficiency, environmental improvements, and enhanced quality of life for citizens. With continued advancements in

AI, it is expected that this trend will persist, further optimizing waste management processes and promoting the development of a circular economy.

References

1. Choudhury M, Rajpal A, Goswami S, Chakravorty A, Raghavan V (eds) (2024) Analytical case studies on municipal and biomedical waste management: perspectives on sustainable development goals. CRC Press. https://doi.org/10.1201/9781003499695
2. Rajpal A, Choudhury M, Goswami S, Chakravorty A, Raghavan V (eds) (2024) Waste management and treatment: advances and innovations. CRC Press. https://doi.org/10.1201/9781003258377
3. Akramila N, Mappasere FA, Mahsyar A (2025) Towards a circular economy: government policy in waste management based on the 3R concept in Makassar City, Indonesia. J Govern Public Pol 12(1):1–17
4. Singh S, Singh P, Sharma A, Choudhury M (eds) (2022) Agriculture waste management and bioresource: the circular economy perspective. Wiley. https://doi.org/10.1002/9781119808428
5. Sun S, Li W, Sun X, Lin X (2024) The impact of public procurement on the adoption of circular economy practices. J Purch Supply Manag 30(4):100907
6. Sahamies K, Welinder O (2025) Orchestrating sustainability: government platforms for material circulation. Admin Soc 57(1):100–128
7. Bello AO, Isa RB, Oke AE, Arogundade S, Lewis JMO (2024) Circular economy implementation in the construction industry: an examination of the barriers in a developing country. Int J Build Pathol Adapt
8. Feldman J, Seligmann H, King S, Flynn M, Shelley T, Helwig A, Burey PP (2024) Circular economy barriers in Australia: how to translate theory into practice? Sustain Prod Consump 45:582–597
9. Gallego-Schmid A, López-Eccher C, Muñoz E, Salvador R, Londono NAC, Barros MV et al (2024) Circular economy in Latin America and the Caribbean: drivers, opportunities, barriers and strategies. Sustain Prod Consump
10. Mubarik MS, Kontoleon A, Shahbaz M (2024) Beyond the hurdles: exploring policy obstacles in the path to circular economy adoption. J Environ Manag 370:122667
11. Dey S, Santra M, Choudhury M, Ghosh AR, Samanta P (2024) Food waste generation and its industrial utilization: an overview. Environ Sci Pollut Res. https://doi.org/10.1007/s11356-024-34252-3
12. Gajanayake A, Iyer-Raniga U (2025) If there is waste, there is a system: understanding Victoria's circular economy transition from a systems thinking perspective. Ecol Econ 227:108395
13. Çimen Ö (2021) Construction and built environment in circular economy: a comprehensive literature review. J Clean Prod 305:127180
14. Choudhury M, Adhikari MD, Agarwal S, Samanta P, Sharma A, Kundu D, Kumar S (2025) Pharmaceuticals and personal care products in soil: sources, impacts and myco-remediation strategies. Emerg Contam 2025:100488. https://doi.org/10.1016/j.emcon.2025.100488
15. Roy and Choudhury (2022) Bioeconomy and green plastic production. In: Baskar C, Ramakrishna S, Daniela La Rosa A (eds) Encyclopedia of green materials. Springer, Singapore. https://doi.org/10.1007/978-981-16-4921-9_189-
16. Hossain MU, Ng ST, Antwi-Afari P, Amor B (2020) Circular economy and the construction industry: existing trends, challenges and prospective framework for sustainable construction. Renew Sustain Energy Rev 130:109948
17. Ogunmakinde OE, Egbelakin T, Sher W (2022) Contributions of the circular economy to the UN sustainable development goals through sustainable construction. Resour Conserv Recycl 178:106023

18. Kumar A, Thakur AK, Gaurav GK, Klemeš JJ, Sandhwar VK, Pant KK, Kumar R (2023) A critical review on sustainable hazardous waste management strategies: a step towards a circular economy. Environ Sci Pollut Res 30(48):105030–105055
19. Baghchesaraei OR, Baghchesaraei A, Cheshnasar SP (2020) Assessment of forecasting energy consumption in residential buildings based on local climate of major cities in Australia. In: 2020 International conference on smart grids and energy systems (SGES). IEEE, pp 729–733
20. Alnafrah I, Alqararah K, AlJaradin M (2025) From linear to circular: evaluating the effects of circular economy on sustainable development progress in the EU economies. J Environ Manag 382:125383
21. Schroeder P, Anggraeni K, Weber U (2019) The relevance of circular economy practices to the sustainable development goals. J Ind Ecol 23(1):77–95
22. Khajuria A, Atienza VA, Chavanich S, Henning W, Islam I, Kral U et al (2022) Accelerating circular economy solutions to achieve the 2030 agenda for sustainable development goals. Circ Econ 1(1):100001
23. Knäble D, de Quevedo Puente E, Pérez-Cornejo C, Baumgärtler T (2022) The impact of the circular economy on sustainable development: a European panel data approach. Sustain Prod Consump 34:233–243
24. Campoli JS, Kodama TK, Nagano MS, Burnquist HL (2025) Advancing circular economy: G20 nations' path towards 12th sustainable development goal. Circ Econ Sustain 5(1):1–24
25. Vann Yaroson E, Chowdhury S, Mangla SK, Dey P, Chan FT, Roux M (2024) A systematic literature review exploring and linking circular economy and sustainable development goals in the past three decades (1991–2022). Int J Prod Res 62(4):1399–1433
26. D'Adamo I, Favari D, Gastaldi M, Kirchherr J (2024) Towards circular economy indicators: evidence from the European Union. Waste Manag Res 42(8):670–680
27. Mandpe A, Paliya S, Gedam VV, Patel S, Tyagi L, Kumar S (2023) Circular economy approach for sustainable solid waste management: a developing economy perspective. Waste Manag Res 41(3):499–511
28. Baghchesaraei OR, Baghchesaraei A (2016) Smart buildings: design and construction process. J Eng Appl Sci 11:783–787
29. Choudhary R, Kumar A, Narayan S, Mishchenko I, Choudhury M, Bharadwaj LK et al (2025) Integrating fuzzy logic into life cycle assessment for sustainable municipal solid waste management in metro cities. Ecol Eng Environ Technol 26(7):316–331. https://doi.org/10.12912/27197050/205500
30. Yu Y, Junjan V, Yazan DM, Iacob ME (2022) A systematic literature review on Circular Economy implementation in the construction industry: a policy-making perspective. Resour Conserv Recycl 183:106359
31. Baghchesaraei A, Baghchesaraei O (2014) The importance of infrastructures in the development of modern methods of construction. Int J Appl Eng Res 9(21):11689–11692
32. Tutore I, Parmentola A, di Fiore MC, Calza F (2024) A conceptual model of artificial intelligence effects on circular economy actions. Corporate Soc Respons Environ Manag 31(5):4772–4782
33. Choudhury M, Roy P (2025) Challenges with microplastic pollution in the regime of UN sustainable development goals. World Dev Sustain 6:100216. https://doi.org/10.1016/j.wds.2025.100216
34. Mondal T, Choudhury M, Kundu D, Dutta D, Samanta P (2023) Landfill: an eclectic review on structure, reactions and remediation approach. Waste Manag 164:127–142. https://doi.org/10.1016/j.wasman.2023.03.034
35. Paul S, Choudhury M, Deb U, Pegu R, Das S, Bhattacharya SS (2019) Assessing the ecological impacts of ageing on hazard potential of solid waste landfills: a green approach through vermitechnology. J Clean Prod 236:117643. https://doi.org/10.1016/j.jclepro.2019.117643
36. Roberts H, Zhang J, Bariach B, Cowls J, Gilburt B, Juneja P et al (2024) Artificial intelligence in support of the circular economy: ethical considerations and a path forward. AI Soc 39(3):1451–1464

37. Fitch-Roy O, Benson D, Monciardini D (2021) All around the world: assessing optimality in comparative circular economy policy packages. J Clean Prod 286:125493
38. Camilleri MA (2020) European environment policy for the circular economy: implications for business and industry stakeholders. Sustain Dev 28(6):1804–1812
39. de Melo TA, de Oliveira MA, de Sousa SR, Vieira RK, Amaral TS (2022) Circular economy public policies: a systematic literature review. Procedia Comput Sci 204:652–662
40. Geng Y, Fu J, Sarkis J, Xue B (2012) Towards a national circular economy indicator system in China: an evaluation and critical analysis. J Clean Prod 23(1):216–224
41. Herrador M, Cho Y, Park PH (2020) Latest circular economy policy and direction in the Republic of Korea: room for enhancements. J Clean Prod 269:122336
42. Ayar P, Baghchesaraei OR, Khademi A, Baibordy A (2023) A descriptive-comparative study of incorporating the sustainability approach in civil engineering education. Majallah-iAmuzih-iMuhandisi-i Iran 24(96):93–95
43. Nunes AMM, Coelho Junior LM, Abrahão R, Santos Júnior EP, Simioni FJ, Rotella Junior P, Rocha LCS (2023) Public policies for renewable energy: a review of the perspectives for a circular economy. Energies 16(1):485
44. Bourdin S, Jacquet N (2025) Closing the loop at the local scale: investigating the drivers of and barriers to the implementation of the circular economy in cities and regions. Ecol Econ 231:108542
45. Goswami S, Choudhury M, Agarwal S (eds) (2025) Electronic waste: impact on health, animals, and the environment. CRC Press. https://doi.org/10.1201/9781003582311
46. Salmenperä H, Pitkänen K, Kautto P, Saikku L (2021) Critical factors for enhancing the circular economy in waste management. J Clean Prod 280:124339
47. Parajuly K, Fitzpatrick C, Muldoon O, Kuehr R (2020) Behavioral change for the circular economy: a review with focus on electronic waste management in the EU. Resour Conserv Recycl 6:100035
48. Grafström J, Aasma S (2021) Breaking circular economy barriers. J Clean Prod 292:126002
49. Kirchherr JW, Hekkert MP, Bour R, Huijbrechtse-Truijens A, Kostense-Smit E, Muller J (2017) Breaking the barriers to the circular economy
50. Gedam VV, Raut RD, de Sousa Jabbour ABL, Tanksale AN, Narkhede BE (2021) Circular economy practices in a developing economy: barriers to be defeated. J Clean Prod 311:127670
51. Geng Y, Xinbei W, Qinghua Z, Hengxin Z (2010) Regional initiatives on promoting cleaner production in China: a case of Liaoning. J Clean Prod 18(15):1502–1508
52. Durdyev S, Koc K, Tleuken A, Budayan C, Ekmekcioğlu Ö, Karaca F (2025) Barriers to circular economy implementation in the construction industry: causal assessment model. Environ Dev Sustain 27(2):4045–4081

Longevity and Sustainability in Design: Approaches to Reducing Waste Through Durable Products

Prasann Kumar, Amit Raj, Velugolu Venkata Suneel Kumar, Ugyen Tshering, and Daza

Abstract Growing concern about the world's waste problem has led to increasing efforts towards achieving environmental sustainability. One of the most promising approaches to reducing waste throughout a product's lifecycle is designing durable, resource-efficient, and environmentally friendly products. In this chapter, we discuss the synergies of advanced manufacturing technology, new material development, and design techniques that can cooperatively prolong product lifetimes. Utilizing case studies and literature, it assesses key strategies, including modular design, repairability, and circular economy integration, that enable prolonged product use and reduced material degradation. The study also examines the facilitating influence of industry partnerships, regulatory policies, and consumer behaviors on the market penetration of sustainable durables. This paper holds theoretical interest and policy relevance, as it integrates technical, economic, and social analyses to inform the transition to sustainable production and consumption. It is further evidence of the need to design for durability as a lever to attain global sustainability objectives and decrease ecological land demand.

Keywords Resource-efficient · Waste · Sustainability · Repairability · Durability

P. Kumar (✉) · A. Raj · V. V. Suneel Kumar
Department of Agronomy, School of Agriculture, Lovely Professional University, Phagwara, Punjab, India

U. Tshering
Agriculture Research and Development Centre, Mongar, Bhutan

Daza
Stanley College of Diploma, Perth, WA, USA

S. Majumdar et al. (eds.), *Sustainable Urban Future*, Urban Sustainability,
https://doi.org/10.1007/978-981-95-2976-6_8

1 Introduction

The emission of waste has become one of the most pressing environmental issues worldwide, primarily due to current factors such as industrialization, population growth, and increased consumption. The current consumption patterns of non-biodegradable products and those with very short life spans have significantly contributed to high rates of waste products, which have polluted the world's dump-sites, water sources, and air, thereby exacerbating climate change [1]. The production of solid waste is estimated at 2.24 billion metric tons per year, and if not correctly disposed of, it has adverse effects on the environment and human health [2]. In the first option, organic waste decomposes in landfills, producing methane (CH_4), a greenhouse gas. In contrast, the second option involves burning the trash, which releases carbon dioxide (CO_2) and toxic hydrocarbons, including dioxins [3].

Nonetheless, high tendencies to acquire things based on fleeting trends and a culture of disposability have led to short product life cycles and, consequently, high rates of product replacement and sector waste [4]. For instance, the use of graphene-enhanced coatings in electronic devices has enhanced their wear and corrosion resistance, thereby increasing their lifespan. Embracing the circular economy, which is defined as a new economic model that departs significantly from the traditional linear economy focused on a 'make-use-dispose' system, involves concepts such as product refurbishment and remanufacturing [5]. Governments and environmental organisations have initiated awareness campaigns by advocating for repair cafes and sharing economies to enhance product durability [6].

Considering the effects of the traditional and long-term oriented conception of durability as a means of preserving resources are decreasing the carbon footprint, and avoiding the exacerbation of the landfill problem [7]. This discussion encompasses various industries, including consumer electronics, apparel, packaging, and automotive sectors. It compares and contrasts the traditional and modern perspectives on the durability of waste reduction. Testimonials of use and real-life examples will be used to demonstrate the effectiveness of rightsizing durables in meeting the waste challenge [8]. This chapter utilises the progress in science, industry knowledge, and policies to present the need to shift society's focus from short-term achievements and heedlessly chosen goals to durability as the new economic model for sustainable development and reducing negative environmental impacts.

2 Concept of Durability in Product Design

Product durability can be enhanced through materials science, product and component design, and redesign, as well as multiple-criteria design and multi-attribute product development [9]. Long-life products also yield economic benefits in terms of cost savings from reduced replacements and repairs, as well as the conservation of the earth's natural resources and raw materials, all while minimising waste sent

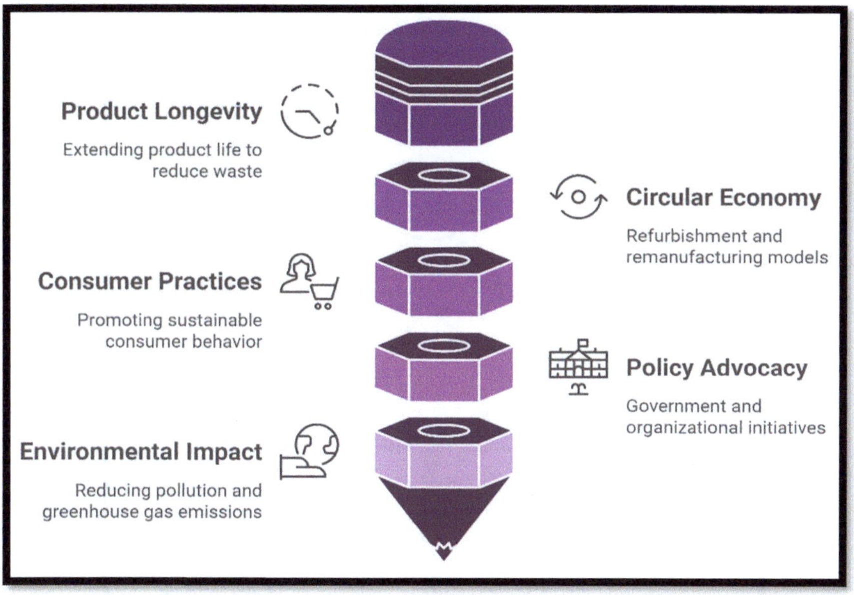

Fig. 1 Strategies for sustainable waste management

to landfills [10, 11]. The effective use of materials, design for sustainability, and a circular economy model contribute to the development of sustainable product design [12]. As new technologies emerge with the means to create long-lasting or even ever-lasting products affordable to the masses, it becomes necessary for manufacturers to adopt an ecological and economic approach toward sustainability. It also caters to consumer demands for durable, high-quality products, which are essential in a world where resources are rapidly depleting (Fig. 1).

2.1 Definition and Importance of Durable Products

2.1.1 Defining Durability in Design

Durability refers to a product's ability to withstand wear, pressure, or damage while maintaining its intended function throughout its intended lifespan [13].

2.1.2 Economic and Environmental Significance

The concept of sustainability is not limited to the consumer level but is also pivotal for an organisation's success. Thus, using durable products relieves pressure on raw

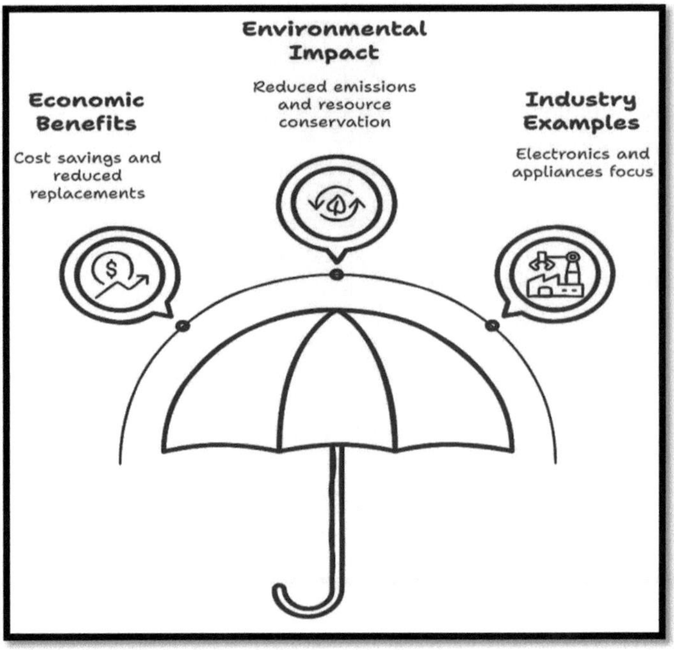

Fig. 2 Sustainability's multifaceted impact

materials and manufacturing, thereby decreasing energy consumption and ultimately reducing greenhouse gas emissions [14]. Additionally, such economies help save costs and preserve the environment by minimising the associated impact of product disposal. For example, in the electronics or home appliances industry, where products quickly wear out, the policy will open doors for cost-effectiveness and minimise electronic waste at some time in the future (Fig. 2).

2.2 Characteristics of Durable Materials and Designs

2.2.1 Material Properties and Selection

One attribute that defines the product's lifespan is the use of materials resistant to degradation. These comprise high-grade stainless steel, which offers good resistance to corrosion, wear, and thermal deterioration, as well as engineered polymers and advanced ceramics [15].

2.2.2 Structural and Functional Design Principles

While material selection plays a crucial role in developing durability, it is also highly dictated by the design work done on any structure. When designing a strong and reliable product, stress distributions, redundancy, and modularity are sound principles [16].

2.2.3 Aesthetic and Ergonomic Considerations

Durability in product design encompasses not only the durability of the product itself but also the durability of attractive post-design shapes and usability characteristics [17]. This is because a product whose look, feel, and texture have not deteriorated due to regular usage will likely be considered pricey and offer quality services to users and consumers.

2.3 Lifecycle Analysis and Sustainability Metrics

2.3.1 Understanding Lifecycle Analysis (LCA)

Life Cycle Analysis (LCA) is a method used to assess the environmental impacts of a product throughout its entire life cycle, from cradle to grave. At each stage, designers are in a position to understand what can be done to reduce power usage, minimize waste, and increase the recycling rate. LCA provides tangible measures in terms of carbon impact, water consumption, and resource drawdown, which are essential for evaluating the sustainability of a product [18].

2.3.2 Sustainability Metrics and Economic Impact

Such measurements as energy-to-weight, emissions-to-weight, or recycling rate can compare various sustainability solutions. Integrating durability into product design will help minimise the rate at which products need replacement, thereby reducing the overall environmental impact of manufacturing new products [19].

2.3.3 Integrating Durability with Circular Economy Principles

A sustainable approach to product design considers not only durability but also incorporates concepts of the circular economy. This means that one needs to design easily disassembled, recyclable, or reusable products, and the materials are kept in the loop [20, 21]. By aiming to make products more durable and recyclable simultaneously, it is possible to lengthen the valuable product's lifetime and, as a result, minimise the environmental impacts on the product life cycle [22]. This integration fosters innovation and sustainability in resource utilisation.

3 Strategies for Enhancing Product Durability

Improving the durability of a product is essential through the use of better materials, design strategies, and advanced manufacturing techniques, which collectively constitute sustainable means of creating superior products. Durable polymers and eco-friendly substitutes offer the most desirable characteristics, fostering the development of stable products [23].

3.1 Material Innovations

3.1.1 High-Performance Materials for Extended Lifespan

In sustainable product design, the choice of material with high mechanical strength, corrosion resistance, and compatibility with various other rigours that the product will likely undergo during its useful life cannot be overemphasised. For instance, polymer matrix composites (PMCs), such as carbon fibre-reinforced polymers (CFRPs), are acknowledged in the market for their high strength and lightweight properties, which make them suitable for use in aerospace parts, sporting equipment, and other applications [24].

3.1.2 Biodegradable and Renewable Options

However, over the past few years, there has been an increasing demand for biomorphic and bio-based materials to apply for high-performance materials with circular economy strategies. These include polylactic acid (PLA) derived from cornstarch or sugarcane, as well as bio-based composite blends with natural fibres like hemp or flax [25].

3.1.3 Advanced Sustainable Packaging

It probably cannot get any better than envisioning a packaging material that utilises a high-performance biodegradable polymer base and natural fibre reinforcements. It provides the required reinforced baggage and safeguards that are mandatory for transportation and storage, and returns to nature at the time of disposal, thus reducing the amount of packaging waste in landfills [26].

3.2 *Design Approaches*

3.2.1 Modular and Repairable Product Designs

Modularity in product design refers to the ability to create a product that allows for easy and quick replacement of a part later on. This approach helps reduce resource consumption for maintaining the systems and minimize output waste by replacing only the destructive components rather than the entire system [27].

3.2.2 Adaptive and Multipurpose Functionality

Additionally, designing modular and multi-purpose products is another way to enhance durability. Thus, such designs enable products to adapt to users' needs and anticipate future developments. For instance, one can customise an adaptive product to perform a given task and many others; thus, purchasing additional equipment decreases inventories and reduces energy demand [28].

3.2.3 Modular Consumer Electronics

A multifunctional smartphone produced using high-quality, strong-built subassemblies that can be repaired or updated singly. This design enables users to upgrade the device's camera, battery, or processor without replacing the entire unit [29].

3.3 *Manufacturing Processes*

3.3.1 Advanced Fabrication Technologies

High levels of technological development in the production process can reduce the rate of product breakdown. These processes minimise defects while lowering energy and material consumption and potentially create geometries that enhance a product's mechanical properties [30].

3.3.2 Quality Control and Precision Engineering

Ensuring quality and dimensional accuracy is crucial when constructing products that meet the desired level of durability and reliability. Production efficiency is secured with the help of automated inspection systems, non-destructive testing methods, and other statistical process control techniques that detect defects [31].

3.3.3 Durable Automotive Components

As technology in the fabrication processes has improved in the automotive industry, mechanical components have been produced with such high precision that they can withstand the harshest environments for extended periods. For instance, by periodically arranging elements made of lightweight materials, such as metals bonded with high-strength alloys, it is possible to design and produce engine parts that are both lightweight and durable [32]. The improved quality control mechanism ensures that every part meets the required quality standards, thereby enhancing the vehicle's performance, safety, and durability (Table 1).

4 Role of Technology in Durability

These innovations enhance the mechanical and design aspects of products, ensuring that systems retain their functions over time [33]. At the same time, AI is used in design to assist with the accurate modelling, simulation, and optimisation of product design for durability and strength [34].

4.1 Integration of Smart Technologies for Predictive Maintenance

4.1.1 Real-Time Monitoring and Condition-Based Maintenance

Complex systems are automatically monitored for various stressors, such as temperature, vibration, or other vital indicators, by integrating a wide range of sensors [35].

4.1.2 Predictive Analytics and Data-Driven Insights

The ambient data collected by smart sensors is analyzed with the help of advanced statistical tools, such as predictive modeling. For instance, infinitesimal variations in vibration characteristics or temperature fluctuations can classify the early stage of component deterioration [36].

Table 1 The application of these measures creates awareness of sustainable ways to design durable, repairable, and adaptable products, which, in a way, helps minimize wastage and conserve the environment

Strategy	Description	Benefits	Examples	References
Self-Healing Materials	Materials that can autonomously repair damages, extending their functional life	Reduces maintenance needs and prolongs product lifespan	Self-healing metals that repair fatigue cracks	[23]
High-Performance Alloys	Alloys are engineered for superior strength and corrosion resistance	Enhances durability in demanding environments	Stainless steel in the construction and automotive industries	[24]
Biodegradable Polymers	Polymers that decompose naturally, minimising environmental impact	Reduces long-term pollution from discarded products	Polylactic acid (PLA) is used in packaging materials	[25]
Renewable Resource-Based Materials	Materials derived from sustainable sources, such as bamboo or hemp	Lowers carbon footprint and promotes sustainability	Bamboo flooring and hemp-based textiles	[26]
Modular Design	Designing products in interchangeable modules for easy upgrades and repairs	Simplifies maintenance and extends product usability	Modular smartphones allow component replacements	[27]
Repairable Product Architecture	Creating products with accessible components to facilitate repairs	Encourages fixing over discarding, reducing waste	Appliances are designed with easily replaceable parts	[28]
Adaptive Functionality	Products capable of evolving functionalities to meet changing user needs	Increases product relevance and lifespan	Smart home devices receive software updates for new features and functionality	[29]
Multipurpose Design	Designing products to serve multiple functions, reducing the need for additional items	Decreases overall consumption and waste	Sofa beds serve as both seating and sleeping arrangements	[30]

(continued)

Table 1 (continued)

Strategy	Description	Benefits	Examples	References
Advanced Fabrication Technologies	Utilising cutting-edge manufacturing techniques, such as 3D printing, for precise and efficient production	Enhances product quality and durability	Aerospace components are manufactured using additive manufacturing	[31]
Precision Engineering	Apply meticulous design and manufacturing processes to ensure the highest quality products	Reduces defects and extends product life	Swiss-made watches are known for their precision and longevity	[32]
Quality Control Systems	Implementing rigorous testing protocols during manufacturing to detect and correct defects early	Ensures consistent product reliability and durability	Automotive industry standards for safety and performance testing	[23]
Redundancy in Design	Incorporating backup systems or components to maintain functionality in case of failure	Enhances reliability and extends operational life	Dual-engine designs in aircraft for safety and reliability	[24]
Predictive Maintenance Technologies	Utilising sensors and AI to anticipate and mitigate potential failures before they occur	It prevents unexpected breakdowns and extends product lifespan	Industrial machinery equipped with IoT sensors for real-time monitoring and control	[25]
Corrosion-Resistant Coatings	Applying protective layers to materials to prevent degradation from environmental exposure	Extends the life of products exposed to harsh conditions	Epoxy coatings are applied to pipelines to prevent rust	[26]

(continued)

Table 1 (continued)

Strategy	Description	Benefits	Examples	References
Heat Treatment Processes	Techniques such as annealing or quenching are used to enhance material properties, including strength and durability	Improves resistance to wear and fatigue	Hardened steel tools for industrial applications	[27]
Ergonomic Design	Creating comfortable and efficient human-use products reduces wear and tear from misuse	Enhances user satisfaction and product longevity	Ergonomically designed office chairs that promote proper posture	[28]
Standardisation of Components	Using uniform parts across different product models to simplify repairs and replacements	Facilitates maintenance and prolongs product life	Universal charging ports for electronic devices	[29]
Environmental Stress Screening	Tested products under extreme conditions to identify potential failures before releasing them to the market	Ensures product robustness and reliability	Consumer electronics are subjected to temperature and humidity tests	[30]
User Education and Maintenance Guides	Providing consumers with information on proper use and upkeep to prevent premature failure	Empower users to maintain products effectively, extending their lifespan	Instruction manuals detailing maintenance routines for home appliances	[31]
Sustainable Supply Chain Practices	Ensuring that materials and components are sourced responsibly, with a focus on quality and durability	Enhances overall product reliability and supports environmental sustainability	Electronics companies source conflict-free minerals and emphasise long-lasting components	[32]

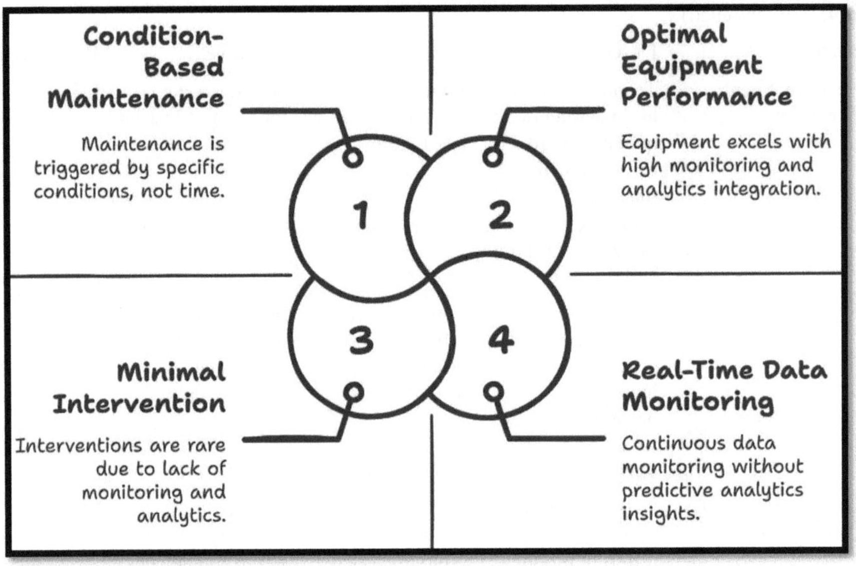

Fig. 3 Smart maintenance usually involves a combination of the following technologies

4.1.3 New Example

Suppose a sizable industrial piece of equipment is equipped with intelligent sensors to monitor process data continuously. Due to constant monitoring and timely maintenance services, the machine performs optimally for a longer period compared to its previous periods of use, making the physical implementation of intelligent systems essential for optimizing performance and potentially reducing operating expenses [37] (Fig. 3).

4.2 Internet of Things (IoT) and Connected Devices for Longevity

4.2.1 Seamless Connectivity and Data Integration

IoT has an enabling characteristic of extending the lifespan of products by connecting devices, sensors, and machinery. IoT systems enable the implementation of continuous and interoperable data exchange between system components and the necessary information regarding a product, its usage, and its performance [38].

4.2.2 Smart Infrastructure and Remote Management

This concept enhances maintenance by encompassing various connected devices, systems, and structures. In the production, transportation, and energy industries, IoT-based platforms enable remote monitoring and control [39].

4.2.3 New Example

A set of interrelated agriculture irrigation networks composed of IoT sensors for monitoring the soil's moisture, the pump's efficiency, and the pressures within the system. Information from these sensors is transmitted to a virtual system that provides a remote-control method and water supply feedback [40].

4.3 Role of Artificial Intelligence in Designing Durable Systems

4.3.1 Computational Modelling and Simulation

AI is transforming product design as a whole, as well as the industries involved in product design, by enabling the computational modelling and simulation of material behaviour under various conditions [41].

4.3.2 Optimization of Design and Adaptive Control

Computer-aided design optimisation utilises data from past performances and future performance simulations to achieve better designs. Techniques like generative design can generate numerous designs and identify configurations that satisfy strength, weight, and durability requirements [42].

4.3.3 New Example

The final product demonstrates that AI can improve durability by forecasting outcomes and designing changes to satisfy clients, thereby maintaining the product's long-term functionality [43].

5 Consumer Behaviour and Education: Shifting Consumption Patterns Toward Sustainable, Durable Products

The suggestion to enhance consumer knowledge about the long-term returns associated with claims made about durable goods will shift demand from low-quality, short-life-cycle products to those with longer life cycles [44]. The invention of tools like the Digital Sustainability Index demonstrates how technology can effectively empower consumers while also encouraging the market to develop ideas that promote sustainability worldwide [45].

5.1 Role of Consumer Awareness in Supporting Durable Products

5.1.1 Theoretical Foundations of Consumer Awareness

In the case of durable products, awareness encompasses not only the appreciation of a product's long durability but also the understanding of the economic benefits of purchasing products that will last for both the consumer and the economy as a whole [46].

5.1.2 Communication of Sustainability Metrics

By quantifying these aspects, companies could make some statistics private, which consumers can use to assess products, not only by price [47].

5.1.3 Consumer Perception and Trust

Providing information on third-party certifications, environmental labels, and transparent and accurate supply chain information will help strengthen consumer confidence [48]. Based on the perception that a product is durable and environmentally friendly, consumers adjust their behaviour towards making purchases in a market that is not geared towards the large-scale production of inferior-quality goods (Fig. 4).

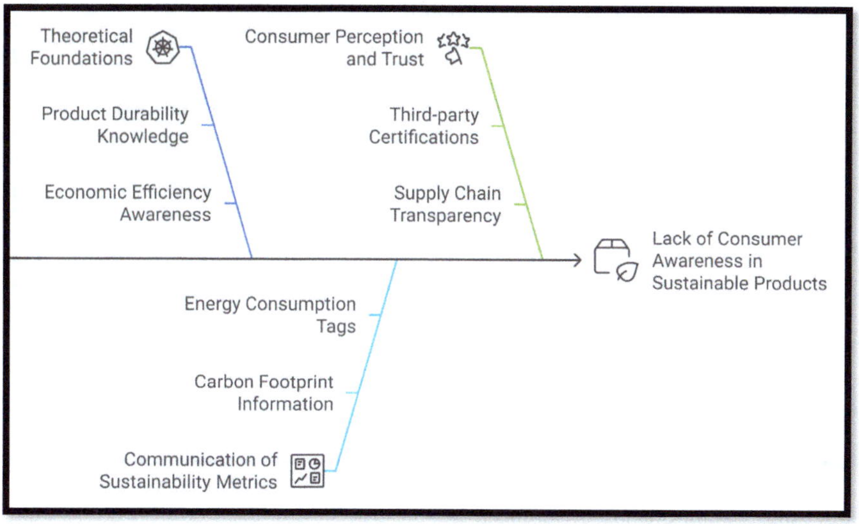

Fig. 4 Role of consumer awareness in supporting durable products

5.2 Strategies for Shifting Consumption Patterns toward Sustainability

5.2.1 Educational Campaigns and Outreach Programs

To achieve this, it is essential to counter some social myths and beliefs, for instance, the idea that the durability of a product is by default a costly affair, yet the focus should be put on the value of the product over the lifetime rather than the cost of depreciation [49].

5.2.2 Incentivizing Sustainable Choices

Other measures mainly encourage consumers to buy durable products through incentives based on their economic value [50]. They enable reduced capital investment at the onset and foster a market that encourages longevity and sustainability (Fig. 5).

5.2.3 Leveraging Digital Platforms and Social Influence

With technological advances, other online platforms, such as e-commerce sites and mobile device applications, are also having a growing impact on consumer behaviour. Through such sustainability characteristics, these platforms incorporate essential and

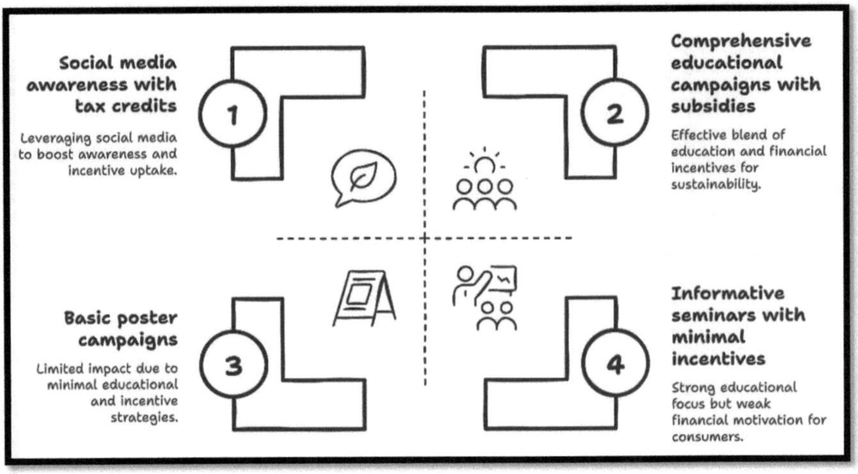

Fig. 5 Strategies for sustainable consumption

valuable information that encourages consumers to shop for products with sustainable bills of materials [51].

5.3 *Hypothetical Digital Sustainability Index*

A product with high DSI can, therefore, be considered strong because it lasts longer and uses resources more efficiently, with minimal impact on the surrounding environment [52].

6 Policy and Regulatory Frameworks

Governments and industries have developed policies and standards to sustain products, enhancing their longevity in the market. Like others, such frameworks also encourage first-of-its-kind and high-quality manufacturing solutions. However, in addition to that, they extend to the element of environmental compliance throughout the whole product's lifecycle [53]. At the same time, the advent of vendor-specific industry standards and certification programs presents other key factors that help build consumer confidence while also creating competitive market and product distinctions [54].

6.1 Government Initiatives and Policies to Encourage Durability

6.1.1 Regulatory Incentives and Economic Levers

Additionally, policies that focus on extended producer responsibility refer to obligations that require producers to take full responsibility for each product, from its production to its recycling or disposal at the end of its useful life [55].

6.1.2 Legislative Frameworks and Environmental Standards

Besides monetary considerations, legal stimuli set the comprehensive quality and Sustainability Standards of any product up for use in the market. The federal standards for product life, durability, and endurance under various environmental conditions, as well as their capability for repair and refurbishment, ensure that only high-quality products reach the market [56].

6.1.3 New Example

This policy approach is beneficial in improving the quality and sustainability of products in the market and incentivises the innovation of other durable technologies [57].

6.2 Industry Standards and Certifications

6.2.1 Development and Adoption of Durability Standards

Organizations such as the International Organization for Standardization (ISO) and regional bodies develop standards that specify material properties, testing methodologies, and performance benchmarks [58].

6.2.2 Certification Programs and Market Transparency

Certification programs play a crucial role in verifying compliance with established durability standards. Independent certification bodies conduct rigorous assessments of products, awarding certifications that signal to consumers that the product has been designed with longevity and sustainability in mind [59].

6.2.3 New Example

A recently introduced certification program for consumer electronics has introduced a "Durability Seal." This seal, awarded after comprehensive testing of product resilience and repairability, has rapidly gained recognition among environmentally conscious consumers [60]. Its adoption by leading manufacturers has raised industry standards and spurred investments in materials research and design innovations to extend product lifespans (Fig. 6).

7 Economic and Environmental Benefits

The focus on using durable goods offers numerous benefits from both economic and environmental perspectives, which are significant for sustainable development in global markets [61]. It promotes and facilitates the circular economy by reducing the extraction of resources and preventing and repurposing solid waste [62].

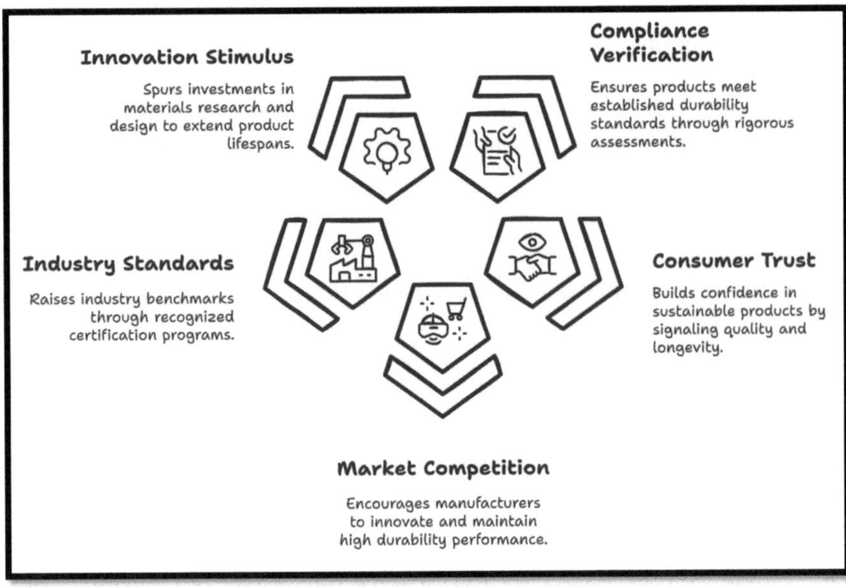

Fig. 6 Driving sustainability through certification and innovation

7.1 *Cost-Effectiveness of Durable Products*

7.1.1 Reduced Lifecycle Costs

Buyers are led to spend less on replacement parts and services, while producers of goods and service providers quickly secure future orders by offering quality products [63].

7.1.2 Enhanced Performance and Reliability

High-quality products result in minimal production loss in industrial markets, as they are less prone to breakdowns commonly associated with machines [64].

7.1.3 Advanced Consumer Electronics

These solid-state devices feature a high degree of artistry and utilize the finest quality materials, offering exceptional durability. Thus, such designs not only bring down the cost for users but also help develop the utilisation and stewardship of products [65].

7.2 *Contribution to Circular Economy and Resource Conservation*

7.2.1 Reducing Resource Extraction and Waste Generation

When product wear and durability periods are extended, the demand for new materials is significantly reduced, as are the detrimental effects of mining, manufacturing, and transportation [66].

7.2.2 Fostering Recycling and Upcycling Initiatives

Additionally, upcycling enterprises recycle products, create other valuable products, and promote economic benefits while minimising environmental pollution [67].

7.2.3 Sustainable Automotive Components

This can be easily dismantled and recycled, and in the process, all the metal and polymers used in the casing manufacturing are retrieved [68]. It reduces the need for new resources to be sourced from primary sources and lowers the environmental impact of car manufacturing companies, thereby promoting a circular economy (Table 2).

8 Challenges and Barriers

The following challenges must be addressed to increase the rate of change: market resistance, primarily due to high initial costs of durability, technical barriers, innovation gaps, and a lack of cooperation [69].

8.1 Market Resistance to Higher Initial Costs

8.1.1 Economic Trade-Offs and Consumer Perception

The administrative challenges include economic trade-offs, as stakeholders must compare the immediate costs of implementing Green IT with the potential long-term cost savings and benefits of this technique [70].

8.1.2 Investment in Innovation Versus Short-Term Gains

Enormous initial costs often render the practice beyond the reach of SMEs, as these companies typically lack sufficient capital for its implementation.

8.1.3 New Example

Concentrate on a case of a manufacturer creating a new, improved version of a household appliance made from cutting-edge and long-lasting materials.

8.2 Technical Limitations and Innovation Gaps

8.2.1 Material and Process Constraints

There are high-performance materials that might show some good characteristics when tested in the laboratory but are prone to some form of weakness or deterioration when exposed to conditions within the environment and other forms of usage that were not envisaged when the materials were produced [71].

Table 2 Implementing durable products offers significant economic and environmental benefits, contributing to cost savings and promoting a circular economy

Aspect	Description	Economic benefits	Environmental benefits	References
Reduced Replacement Frequency	Durable products require less frequent replacement, leading to lower long-term costs	Decreases consumer expenditure over time	Reduces waste generation and resource extraction	[61]
Lower Maintenance Costs	High-quality, durable products often need less maintenance	Saves on repair and upkeep expenses	Minimises environmental impact from maintenance activities	[62]
Enhanced Resale Value	Durable goods maintain value longer, offering higher resale potential	Provides financial returns through second-hand markets	Extends product lifecycle, reducing the need for new resources	[63]
Energy Efficiency	Durable products are often designed with energy efficiency in mind	Reduces energy bills for consumers	Lowers greenhouse gas emissions associated with energy production	[64]
Resource Conservation	Long-lasting products decrease the demand for raw materials	Stabilises material costs and supply chains	Preserves natural resources and biodiversity	[65]
Support for the Circular Economy	Durable products facilitate reuse, refurbishment, and recycling, thereby reducing waste and promoting environmental sustainability	Opens new business models and revenue streams	Reduces landfill usage and environmental pollution	[66]
Reduced Manufacturing Demand	A less frequent need for replacement decreases manufacturing pressure	Lowers production costs and capital expenditure	Diminishes industrial emissions and waste	[67]
Improved Brand Reputation	Companies that offer durable products tend to gain the trust of consumers	Leads to increased sales and customer loyalty	Encourages industry-wide sustainable practices	[68]
Compliance with Regulations	Durable products often meet stringent environmental standards	It avoids fines and reduces regulatory compliance costs	Aligns with global sustainability goals	[61]

(continued)

Table 2 (continued)

Aspect	Description	Economic benefits	Environmental benefits	References
Innovation and Competitive Advantage	Investing in durability drives innovation	Differentiates products in the market, potentially allowing premium pricing	Promotes sustainable product development	[62]
Reduced Disposal Costs	Longer product lifespans mean fewer disposal requirements	Saves on waste management expenses	Lessens landfill burden and associated environmental hazards	[63]
Consumer Satisfaction	Durable products enhance user experience and satisfaction	Increases customer retention and positive word-of-mouth	Encourages sustainable consumption patterns	[64]
Economic Stability	Durable goods contribute to stable consumption patterns	Mitigates economic volatility from overproduction	Balances resource use and environmental impact	[65]
Job Creation in Maintenance and Repair	Emphasis on durability boosts the repair and maintenance sectors	Generates employment opportunities	Promotes skills development and sustainable services	[66]
Extended Product Warranties	Durable products often come with more extended warranties	Provides financial security to consumers	Encourages manufacturers to maintain high-quality standards	[67]
Lower Transportation Emissions	Fewer replacements reduce transportation needs	It cuts logistics and shipping costs	Decreases carbon footprint associated with product distribution	[68]
Material Efficiency	Durable products utilise materials more effectively	Optimises production costs	Reduces mining and extraction activities	[61]
Support for Sustainable Policies	Aligns with governmental sustainability initiatives	Access to subsidies or tax incentives	Advances national and global environmental objectives	[62]

(continued)

Table 2 (continued)

Aspect	Description	Economic benefits	Environmental benefits	References
Educational Value	Promote consumer awareness about sustainability	Encourages informed purchasing decisions	Fosters a culture of environmental responsibility	[63]
Long-Term Investment	Durable products serve as long-term investments for consumers	Provides better return on investment over time	Reduce the overall environmental footprint of consumer goods	[64]

8.2.2 Research and Development Bottlenecks

Product development gaps occur when the existing state of technology cannot effectively support the product, rendering it unsustainable and short-lived. Due to funding restrictions, a lack of coordination in research, and insufficient interdisciplinary research in the technological field, it takes time to develop new materials and production processes.

8.2.3 New Example

Let us assume that engineers want to design a biodegradable material for use in consumer electronics applications. The initial trials of the new polymer-coated products under limited conditions have demonstrated good readability and biodegradability.

8.3 Need for Industry-Wide Collaboration

8.3.1 Standardization and Shared Best Practices

The lack of standardization and compliance standards across different industries may hinder the ability to adopt everlasting technologies. Commitment at all levels of the value chain is necessary to establish standards for the entire industry, ensuring that the products produced are safe and environmentally friendly [72].

8.3.2 Cross-Sector Partnerships and Policy Support

This also involves strengthening partnerships between private corporations, academic institutions, and government agencies.

8.3.3 New Example

A group of leading companies cooperating to design a common platform for reasonable manufacturing. By having a collective set of durability standards and information, these companies reduce the time it takes to research and select better materials, ultimately lowering costs.

9 Future Directions and Recommendations

As the call for sustainability and the long-term expansion of its application to industries and markets increases, long-term product innovation will inevitably redirect markets and mitigate environmental impacts. So far as material science, digital tooling, and manufacturing technologies are concerned, new opportunities are continuously emerging. Still, research and interdisciplinary collaborations will remain key to advancing the field and ensuring that tangible improvements in structural integrity are attained. From an outlook on material, design and research, durability product development is, therefore, set to be driven by the new and emerging interfaces created by advanced materials, digital designs, interdisciplinary research, government and regulatory policies and active market, all of which form the key components [73].

9.1 Emerging Trends in Durable Product Innovation

9.1.1 Advanced Materials and Nanotechnology

Recent advancements in materials science have enabled the development of strong and lightweight materials with extended lifespans. New materials, such as carbon nanotubes, advanced composites, and self-healing polymers, have been developed at the molecular level to exhibit high wear resistance, fatigue resistance, and environmental damage tolerance.

9.1.2 Digital Design and Simulation

Virtual testing, real-time performance simulations, computer-aided simulations, and 3D visualization and modeling are some of the numerous sub-processes that are rapidly transforming product design through the best practice use of digital technologies. These technologies aim to create models that test how products would fare under relevant usage conditions over several years. These simulation tools accelerate innovation cycles, enabling the creation of high-quality products while conserving resources [74].

9.2 Call for Interdisciplinary Research and Development

9.2.1 Integration of Diverse Scientific Disciplines

From the proper perspective, breakthrough ideas toward a more sustainable world would need input from specialised fields like materials science, engineering, computer science, and environmental sciences. Joint research in these fields enables the development of approaches to enhance durability that span a range of time frames and length scales, from the atomic to the integrated system level.

9.2.2 Innovative Methodologies and Data-Driven Insights

Multidisciplinary work yields the creation of new scientific research, particularly through the combined use of experimental and computational methods. Researchers can gain knowledge through testing prototypes and numerical models, which enhances the durability of the models' foresight.

9.2.3 New Example

Suppose a multidisciplinary team of experts is tasked with designing and launching a new range of consumer electronics products with extended durability.

9.3 Framework for Global Adoption of Durable Product Strategies

9.3.1 Harmonization of Standards and Regulatory Support

The regulation of durable products must align with the products to gain market acceptance.

9.3.2 Market Incentives and Consumer Education

Incentives encompass both monetary and non-monetary elements, including subsidies, tax reductions, and preferential procurement, which have a positive impact on the adoption of durable products. The integration of digital technologies and social media into educational practices can enhance people's awareness of sustainability and durability, which, in turn, will influence international purchasing habits [75].

9.3.3 New Example

Suppose an international collaboration can establish a standard for the durability of consumer electronics.

10 Conclusion

Driven by rapid industrialization, unsustainable consumer behaviour, and the proliferation of disposable products, the global waste crisis has evolved into a critical and alarming situation. The excessive generation of waste has far-reaching effects, including severe land degradation from landfilling, ecosystem pollution, and increased greenhouse gas (GHG) emissions that hasten climate change. The rising levels of electronic waste (e-waste) and plastic pollution, which now predominantly contaminate soil and water, are particularly concerning. These persistent toxins not only disrupt the natural balance but also pose serious risks to human health, affecting food safety, water quality, and overall well-being. Addressing this complex challenge requires a paradigm shift toward responsible consumption and sustainable production, safeguarding both public health and the environment.

References

1. Rajpal A, Choudhury M, Goswami S, Chakravorty A, Raghavan V (eds) (2024) Waste management and treatment: advances and innovations. CRC Press. https://doi.org/10.1201/9781003258377
2. Cárdenas I, Koeva M, Davey C, Nourian P (2023) Solid waste in the virtual World: a digital twinning approach for waste collection planning. In: International 3D GeoInfo conference. Springer Nature Switzerland, Cham, pp 61–74. https://doi.org/10.1007/978-3-031-43699-4_4
3. Bano A (2025) Direct environmental pollution from solid waste. In: Waste-to-energy. Elsevier, pp 249–269. https://doi.org/10.1016/B978-0-443-22356-3.00011-7
4. Stephan A, Athanassiadis A (2018) Towards a more circular construction sector: estimating and spatialising current and future non-structural material replacement flows to maintain urban building stocks. Resour Conserv Recycl 129:248–262. https://doi.org/10.1016/j.resconrec.2017.09.022

5. Pesce D, Franzè C, Paolucci E (2024) Meeting the closed-loop challenge: the "Who-What-How" strategic choices for a successful remanufacturing approach. Bus Strateg Environ 33(6):5661–5680. https://doi.org/10.1002/bse.3770
6. Hernandez RJ, Miranda C, Goñi J (2020) Empowering sustainable consumption by giving back to consumers the 'right to repair'. Sustainability 12(3):850. https://doi.org/10.3390/su12030850
7. Bualuang T, Jitsangiam P, Chusai N, Suwan T, Rattanasak U, Chindaprasirt P (2024) Utilization of dumped coal ash from power-plant landfills for carbon footprint reduction in sustainable pavement base construction. Constr Build Mater 441:137462. https://doi.org/10.1016/j.conbuildmat.2024.137462
8. Zorpas AA (2024) The hidden concept and the beauty of multiple "R" in the framework of waste strategies development reflecting to circular economy principles. Sci Total Environ 952:175508. https://doi.org/10.1016/j.scitotenv.2024.175508
9. Zhao R, Su H, Chen X, Yu Y (2016) Commercially available materials selection in sustainable design: an integrated multi-attribute decision making approach. Sustainability 8(1):79. https://doi.org/10.3390/su8010079
10. Mondal T, Choudhury M, Kundu D, Dutta D, Samanta P (2023) Landfill: an eclectic review on structure, reactions and remediation approach. Waste Manag 164:127–142. https://doi.org/10.1016/j.wasman.2023.03.034
11. Paul S, Choudhury M, Deb U, Pegu R, Das S, Bhattacharya SS (2019) Assessing the ecological impacts of ageing on hazard potential of solid waste landfills: a green approach through vermitechnology. J Clean Prod 236:117643. https://doi.org/10.1016/j.jclepro.2019.117643
12. Mestre A, Cooper T (2017) Circular product design. A multiple loops life cycle design approach for the circular economy. Des J 20(Suppl 1):S1620–S1635. https://doi.org/10.1080/14606925.2017.1352686
13. Bhong M, Khan TK, Devade K, Krishna BV, Sura S, Eftikhaar HK et al (2023) Review of composite materials and applications. Mater Today Proc. https://doi.org/10.1016/j.matpr.2023.10.026
14. Correa JP, Montalvo-Navarrete JM, Hidalgo-Salazar MA (2019) Carbon footprint considerations for biocomposite materials in sustainable product development: a review. J Clean Prod 208:785–794. https://doi.org/10.1016/j.jclepro.2018.10.099
15. Michla JJ, Nagarajan R, Krishnasamy S, Siengchin S, Ismail SO, Prabhu TR (2021) Conventional and additively manufactured stainless steels: a review. Trans Indian Inst Metals 74(6):1261–1278. https://doi.org/10.1007/s12666-021-02305-7
16. Kharrazi A, Yu Y, Jacob A, Vora N, Fath BD (2020) Redundancy, diversity, and modularity in network resilience: applications for international trade and implications for public policy. Curr Res Environ Sustain 2:100006. https://doi.org/10.1016/j.crsust.2020.06.001
17. Spreafico C, Landi D (2022) Using product design strategies to implement circular economy: differences between students and professional designers. Sustainability 14(3):1122. https://doi.org/10.3390/su14031122
18. Pfister S, Boulay AM, Berger M, Hadjikakou M, Motoshita M, Hess T et al (2017) Understanding the LCA and ISO water footprint: a response to Hoekstra (2016) "A critique on the water-scarcity weighted water footprint in LCA". Ecol Indic 72:352–359. https://doi.org/10.1016/j.ecolind.2016.07.051
19. Rivera JL, Lallmahomed A (2016) Environmental implications of planned obsolescence and product lifetime: a literature review. Int J Sustain Eng 9(2):119–129. https://doi.org/10.1080/19397038.2015.1099757
20. Goswami S, Choudhury M, Samanta P (eds) (2025) Micro-nano plastics exposure, environmental degradation and public health crisis. Springer Singapore. https://doi.org/10.1007/978-981-96-5047-7
21. Roy and Choudhury (2022) Bioeconomy and green plastic production. In: Baskar C, Ramakrishna S, Daniela La Rosa A (eds) Encyclopedia of green materials. Springer, Singapore. https://doi.org/10.1007/978-981-16-4921-9_189-

22. Iraldo F, Facheris C, Nucci B (2017) Is product durability better for environment and for economic efficiency? A comparative assessment applying LCA and LCC to two energy-intensive products. J Clean Prod 140:1353–1364. https://doi.org/10.1016/j.jclepro.2016.10.017
23. Mukherjee C, Varghese D, Krishna JS, Boominathan T, Rakeshkumar R, Dineshkumar S et al (2023) Recent advances in biodegradable polymers—properties, applications and future prospects. Eur Polym J 192:112068. https://doi.org/10.1016/j.eurpolymj.2023.112068
24. Das TK, Ghosh P, Das NC (2019) Preparation, development, outcomes, and application versatility of carbon fiber-based polymer composites: a review. Adv Compos Hybrid Mater 2:214–233. https://doi.org/10.1007/s42114-018-0072-z
25. Pokharel A, Falua KJ, Babaei-Ghazvini A, Acharya B (2022) Biobased polymer composites: a review. J Compos Sci 6(9):255. https://doi.org/10.3390/jcs6090255
26. Walsh JSP, Gorman AC, Castaño P (2022) Postorbital discard and chain of custody: the processing of artifacts returning to Earth from the International Space Station. Acta Astronaut 195:513–531. https://doi.org/10.1016/j.actaastro.2022.03.035
27. Oldfrey B, Barbareschi G, Morjaria P, Giltsoff T, Massie J, Miodownik M, Holloway C (2021) Could assistive technology provision models help pave the way for more environmentally sustainable models of product design, manufacture and service in a post-covid world? Sustainability 13(19):10867. https://doi.org/10.3390/su131910867
28. Bonvoisin J, Halstenberg F, Buchert T, Stark R (2016) A systematic literature review on modular product design. J Eng Des 27(7):488–514. https://doi.org/10.1080/09544828.2016.1166482
29. Vermeşan H, Tiuc AE, Purcar M (2019) Advanced recovery techniques for waste materials from IT and telecommunication equipment printed circuit boards. Sustainability 12(1):74. https://doi.org/10.3390/su12010074
30. Benedetti M, Du Plessis A, Ritchie RO, Dallago M, Razavi N, Berto F (2021) Architected cellular materials: a review of their mechanical properties for fatigue-tolerant design and fabrication. Mater Sci Eng R Rep 144:100606. https://doi.org/10.1016/j.mser.2021.100606
31. Jaber A, SattarpanahKarganroudi S, Meiabadi MS, Aminzadeh A, Ibrahim H, Adda M, Taheri H (2022) On smart geometric non-destructive evaluation: inspection methods, overview, and challenges. Materials 15(20):7187. https://doi.org/10.3390/ma15207187
32. Zhao Q, Sun Q, Xin S, Chen Y, Wu C, Wang H et al (2022) High-strength titanium alloys for aerospace engineering applications: a review on melting-forging process. Mater Sci Eng A 845:143260. https://doi.org/10.1016/j.msea.2022.143260
33. Goworek H, Oxborrow L, Claxton S, McLaren A, Cooper T, Hill H (2020) Managing sustainability in the fashion business: challenges in product development for clothing longevity in the UK. J Bus Res 117:629–641. https://doi.org/10.1016/j.jbusres.2018.07.021
34. Elahi M, Afolaranmi SO, Martinez Lastra JL, Perez Garcia JA (2023) A comprehensive literature review of the applications of AI techniques through the lifecycle of industrial equipment. Discov Artif Intell 3(1):43. https://doi.org/10.1007/s44163-023-00089-x
35. Suslu B, Ali F, Jennions IK (2023) Understanding the role of sensor optimisation in complex systems. Sensors 23(18):7819. https://doi.org/10.3390/s23187819
36. Moshawrab M, Adda M, Bouzouane A, Ibrahim H, Raad A (2022) Smart wearables for the detection of occupational physical fatigue: a literature review. Sensors 22(19):7472. https://doi.org/10.3390/s22197472
37. Wang J, Gao RX (2022) Innovative smart scheduling and predictive maintenance techniques. In: Design and operation of production networks for mass personalization in the era of cloud technology. Elsevier, pp 181–207. https://doi.org/10.1016/B978-0-12-823657-4.00007-5
38. Noura M, Atiquzzaman M, Gaedke M (2019) Interoperability in internet of things: taxonomies and open challenges. Mobile Netw Appl 24:796–809. https://doi.org/10.1007/s11036-018-1089-9
39. Mishra P, Singh G (2023) Energy management systems in sustainable smart cities based on the internet of energy: a technical review. Energies 16(19):6903. https://doi.org/10.3390/en16196903
40. García L, Parra L, Jimenez JM, Lloret J, Lorenz P (2020) IoT-based smart irrigation systems: an overview on the recent trends on sensors and IoT systems for irrigation in precision agriculture. Sensors 20(4):1042. https://doi.org/10.3390/s20041042

41. Mozaffar M, Liao S, Xie X, Saha S, Park C, Cao J et al (2022) Mechanistic artificial intelligence (mechanistic-AI) for modeling, design, and control of advanced manufacturing processes: current state and perspectives. J Mater Process Technol 302:117485. https://doi.org/10.1016/j.jmatprotec.2021.117485
42. Barbieri L, Muzzupappa M (2022) Performance-driven engineering design approaches based on generative design and topology optimisation tools: a comparative study. Appl Sci 12(4):2106. https://doi.org/10.3390/app12042106
43. Wang L, Liu Z, Liu A, Tao F (2021) Artificial intelligence in product lifecycle management. Int J Adv Manuf Technol 114:771–796. https://doi.org/10.1007/s00170-021-06882-1
44. Prata JC, Silva ALP, Da Costa JP, Mouneyrac C, Walker TR, Duarte AC, Rocha-Santos T (2019) Solutions and integrated strategies for the control and mitigation of plastic and microplastic pollution. Int J Environ Res Public Health 16(13):2411. https://doi.org/10.3390/ijerph16132411
45. Tay LY, Tai HT, Tan GS (2022) Digital financial inclusion: a gateway to sustainable development. Heliyon 8(6). https://doi.org/10.1016/j.heliyon.2022.e09766
46. Lin ST, Niu HJ (2018) Green consumption: environmental knowledge, environmental consciousness, social norms, and purchasing behavior. Bus Strateg Environ 27(8):1679–1688. https://doi.org/10.1002/bse.2233
47. Shooshtarian S, Caldera S, Maqsood T, Ryley T (2020) Using recycled construction and demolition waste products: a review of stakeholders' perceptions, decisions, and motivations. Recycling 5(4):31. https://doi.org/10.3390/recycling5040031
48. Maranesi C, De Giovanni P (2020) Modern circular economy: corporate strategy, supply chain, and industrial symbiosis. Sustainability 12(22):9383. https://doi.org/10.3390/su12229383
49. Fallahi S, Mellquist AC, Mogren O, Listo Zec E, Algurén P, Hallquist L (2023) Financing solutions for circular business models: exploring the role of business ecosystems and artificial intelligence. Bus Strateg Environ 32(6):3233–3248. https://doi.org/10.1002/bse.3297
50. Zhang X, Bai X, Shang J (2018) Is subsidized electric vehicles adoption sustainable: consumers' perceptions and motivation toward incentive policies, environmental benefits, and risks. J Clean Prod 192:71–79. https://doi.org/10.1016/j.jclepro.2018.04.252
51. Jacobson J, Harrison B (2022) Sustainable fashion social media influencers and content creation calibration. Int J Advert 41(1):150–177. https://doi.org/10.1080/02650487.2021.2000125
52. Verma SK, Prasad A, Katiyar V (2024) State of art review on sustainable biodegradable polymers with a market overview for sustainability packaging. Mater Today Sustain 2024:100776. https://doi.org/10.1016/j.mtsust.2024.100776
53. Konstantinidis FK, Myrillas N, Tsintotas KA, Mouroutsos SG, Gasteratos A (2023) A technology maturity assessment framework for industry 5.0 machine vision systems based on systematic literature review in automotive manufacturing. Int J Prod Res:1–37. https://doi.org/10.1080/00207543.2023.2270588
54. Mittal S, Khan MA, Romero D, Wuest T (2018) A critical review of smart manufacturing & Industry 4.0 maturity models: implications for small and medium-sized enterprises (SMEs). J Manuf Syst 49:194–214. https://doi.org/10.1016/j.jmsy.2018.10.005
55. Ameli M, Mansour S, Ahmadi-Javid A (2019) A simulation-optimization model for sustainable product design and efficient end-of-life management based on individual producer responsibility. Resour Conserv Recycl 140:246–258. https://doi.org/10.1016/j.resconrec.2018.02.031
56. Charter M, Gray C, Clark T, Woolman T (2017) The role of business in realising sustainable consumption and production. In: System innovation for sustainability 1. Routledge, pp 56–79
57. Khan N, Ray RL, Kassem HS, Hussain S, Zhang S, Khayyam M et al (2021) Potential role of technology innovation in transformation of sustainable food systems: a review. Agriculture 11(10):984. https://doi.org/10.3390/agriculture11100984
58. Chen Z, Han C, Gao M, Kandukuri SY, Zhou K (2022) A review on qualification and certification for metal additive manufacturing. Virtual Phys Prototyp 17(2):382–405. https://doi.org/10.1080/17452759.2021.2018938

59. Ghisellini P, Quinto I, Passaro R, Ulgiati S (2024) Exploring environmental and social performances of circular start-ups: an orientation and certification assessment. Bus Strateg Environ 33(4):3222–3241. https://doi.org/10.1002/bse.3620
60. Sonego M, Echeveste MES, Debarba HG (2022) Repair of electronic products: consumer practices and institutional initiatives. Sustain Prod Consump 30:556–565. https://doi.org/10.1016/j.spc.2021.12.031
61. Korhonen J, Honkasalo A, Seppälä J (2018) Circular economy: the concept and its limitations. Ecol Econ 143:37–46. https://doi.org/10.1016/j.ecolecon.2017.06.041
62. Yang M, Chen L, Wang J, Msigwa G, Osman AI, Fawzy S et al (2023) Circular economy strategies for combating climate change and other environmental issues. Environ Chem Lett 21(1):55–80. https://doi.org/10.1007/s10311-022-01499-6
63. Khan MA, Mittal S, West S, Wuest T (2018) Review on upgradability–A product lifetime extension strategy in the context of product service systems. J Clean Prod 204:1154–1168. https://doi.org/10.1016/j.jclepro.2018.08.329
64. Mohan TR, Roselyn JP, Uthra RA, Devaraj D, Umachandran K (2021) Intelligent machine learning based total productive maintenance approach for achieving zero downtime in industrial machinery. Comput Ind Eng 157:107267. https://doi.org/10.1016/j.cie.2021.107267
65. Bennett NJ, Whitty TS, Finkbeiner E, Pittman J, Bassett H, Gelcich S, Allison EH (2018) Environmental stewardship: a conceptual review and analytical framework. Environ Manag 61:597–614. https://doi.org/10.1007/s00267-017-0993-2
66. Mohajerani A, Burnett L, Smith JV, Kurmus H, Milas J, Arulrajah A et al (2019) Nanoparticles in construction materials and other applications, and implications of nanoparticle use. Materials 12(19):3052. https://doi.org/10.3390/ma12193052
67. Pandey S, Pandit P, Pandey R, Pandey S (2020) Marketing strategies for upcycling and recycling of textile and fashion. In: Recycling from waste in fashion and textiles: a sustainable and circular economic approach, pp 253–275. https://doi.org/10.1002/9781119620532.ch12
68. Gu F, Guo J, Zhang W, Summers PA, Hall P (2017) From waste plastics to industrial raw materials: a life cycle assessment of mechanical plastic recycling practice based on a real-world case study. Sci Total Environ 601:1192–1207. https://doi.org/10.1016/j.scitotenv.2017.05.278
69. Gupta H, Kusi-Sarpong S, Rezaei J (2020) Barriers and overcoming strategies to supply chain sustainability innovation. Resour Conserv Recycl 161:104819. https://doi.org/10.1016/j.resconrec.2020.104819
70. Stornelli A, Ozcan S, Simms C (2021) Advanced manufacturing technology adoption and innovation: a systematic literature review on barriers, enablers, and innovation types. Res Policy 50(6):104229. https://doi.org/10.1016/j.respol.2021.104229
71. Wang JX, Burke H, Zhang A (2022) Overcoming barriers to circular product design. Int J Prod Econ 243:108346. https://doi.org/10.1016/j.ijpe.2021.108346
72. Rueda X, Garrett RD, Lambin EF (2017) Corporate investments in supply chain sustainability: selecting instruments in the agri-food industry. J Clean Prod 142:2480–2492. https://doi.org/10.1016/j.jclepro.2016.11.026
73. Stier SP, Kreisbeck C, Ihssen H, Popp MA, Hauch J, Malek K et al (2024) Materials acceleration platforms (MAPs): accelerating materials research and development to meet urgent societal challenges. Adv Mater 36(45):2407791. https://doi.org/10.1002/adma.202407791
74. Dangelico RM, Pujari D, Pontrandolfo P (2017) Green product innovation in manufacturing firms: A sustainability-oriented dynamic capability perspective. Bus Strateg Environ 26(4):490–506. https://doi.org/10.1002/bse.1932
75. Rosati F, Faria LGD (2019) Business contribution to the Sustainable Development Agenda: organizational factors related to early adoption of SDG reporting. Corporate Soc Respons Environ Manag 26(3):588–597. https://doi.org/10.1002/csr.1705

Cutting-Edge Recycling and Reuse: Exploring Advanced Technologies for Urban Environments

Reza Mirzaei, Elnaz Lakizadeh, and Swarnava Dey

Abstract Advanced technologies are recognized as the primary drivers of transformation in urban environments, significantly enhancing the quality of life for citizens. This chapter explores the impact of emerging technologies on cities, with a particular focus on the Internet of Things (IoT) and its applications in urban management. The Internet of Things, through the use of sensors and monitoring systems, facilitates the optimization of resource consumption and the intelligent management of infrastructure. Additionally, big data is presented as a vital tool for effective decision-making in urban management, enabling precise analysis and efficient forecasting. Additionally, renewable energy and green technologies are proposed as solutions to mitigate pollution and reduce energy consumption in urban areas. This chapter also examines smart transportation as a crucial aspect, encompassing intelligent public transportation systems and autonomous vehicles. Cybersecurity issues are receiving significant attention due to the growing reliance of cities on technology, and strategies for addressing cyber threats are proposed.

Additionally, urban design is explored as a crucial factor in developing innovative and efficient urban spaces, with successful global examples in this field highlighted. This chapter emphasizes the importance and necessity of utilizing advanced technologies to create smart and sustainable cities. It also addresses the challenges and opportunities that lie ahead in this field.

Keywords Smart City · Internet of Things (IoT) · Big data · Renewable energy · Smart transportation

R. Mirzaei (✉) · E. Lakizadeh
Faculty of Architecture and Urban Planning, Iran University of Art, Tehran, Iran
e-mail: Eng.RezaMirzaeii@gmail.com

S. Dey
Department of Architecture, Jadavpur University, Kolkata, West Bengal, India

S. Majumdar et al. (eds.), *Sustainable Urban Future*, Urban Sustainability,
https://doi.org/10.1007/978-981-95-2976-6_9

1 Introduction

In the present era, technological advancements are recognized as the primary drivers of transformation in urban environments. These developments, particularly with the emergence of the Internet of Things (IoT) and its applications in urban management, have a profound impact on the quality of life for citizens [1]. The Internet of Things, through the utilization of sensors and monitoring systems, aids in the optimization of resource consumption and the intelligent management of infrastructures, and, most importantly, during an era of climate change and the crises associated with it. Emerging technologies can be considered a driving force in enhancing climate resilience [2–4]. Intelligent transportation is also regarded as a key aspect of these transformations, encompassing innovative public transportation systems and autonomous vehicles. In this context, big data emerges as a vital tool for effective decision-making in urban management, enabling precise analysis and efficient forecasting [5, 6]. However, the increasing dependence of cities on technology has turned cybersecurity issues into serious challenges, necessitating the development of strategies to address cyber threats [7].

Alongside these factors, urban design is examined as a critical element in the development of innovative and efficient urban spaces, with successful global examples introduced as models for implementation in other regions [8]. Given these transformations, further research is necessary in the field of integrating advanced technologies into urban management to identify and analyze existing challenges and opportunities effectively. Therefore, this research investigates the impact of advanced technologies on urban management and the quality of life for citizens, seeking to answer this key question: How can modern technologies are leveraged to create innovative and sustainable cities while managing the associated challenges.

2 Bibliometric Analysis

The bibliometric analysis, based on data retrieved from the Web of Science (2015–2024), highlights the rapid growth and interdisciplinary nature of research on advanced technologies in recycling and reuse, especially in urban environments. Publications increased significantly from 536 in 2015 to 2384 in 2024 (Fig. 1), reflecting an intensified global interest in innovative, resilient, and sustainable approaches to urban waste management. Prominent contributions from authors such as Liu Y, Li Y, and Wang Y underscore the pivotal role of East Asian researchers in advancing this domain (Fig. 2). Subject category analysis highlights a diverse disciplinary foundation, with Environmental Sciences (27.81%) and Environmental Studies (26.15%) forming the most significant knowledge clusters, followed by Urban Studies (16.68%) and Green Sustainable Science & Technology (13.42%) (Fig. 3), which emphasizes are the environmental and technological imperatives of urban transformation. Overall, the bibliometric analysis reflects a dynamic and evolving research landscape that underpins the development of innovative, efficient, and sustainable cities through the strategic deployment of advanced technologies.

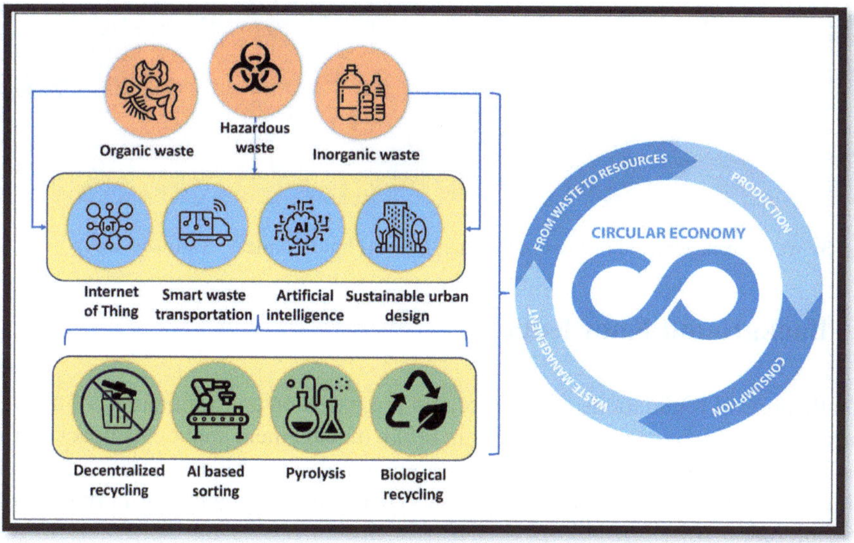

Fig. 1 Circular economy concept with various wastes and technology application

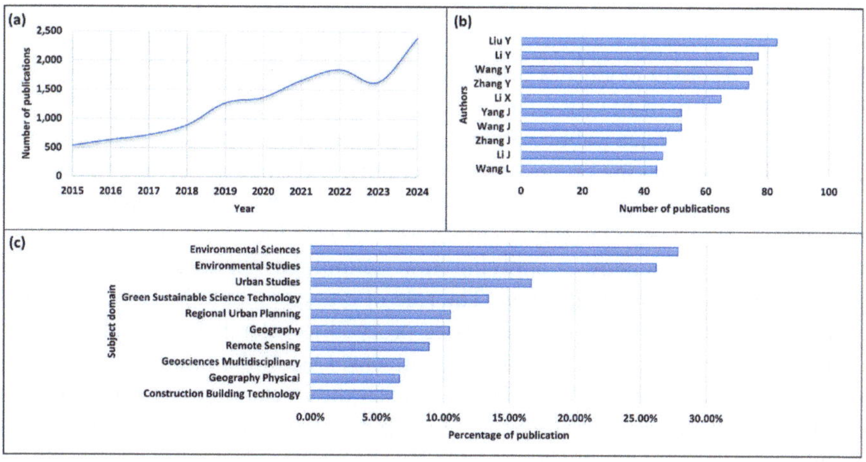

Fig. 2 Bibliometric trends in research on advanced urban technologies in recycling and reuse (2015–2024) based on Web of Science data: (**a**) Annual distribution of publications; (**b**) Top 10 most productive authors; (**c**) Distribution of publications across major subject domains

Fig. 3 Waste management practices in Kolkata, India. (Photo Courtesy: Swarnava Dey)

3 Background Knowledge

The theoretical foundations of this research are based on an examination of the key components of advanced technologies in urban management and their impact on the quality of life for citizens. In this section, operational experiences in various domains related to smart cities, the Internet of Things (IoT), and intelligent transportation will be explored.

3.1 Smart Cities and Quality of Life

Smart cities, as novel models of urban management, aim to optimize resources and enhance the quality of life for citizens. According to theorists such as Hollands [9], smart cities can enhance public services, reduce costs, and increase citizen participation in decision-making processes by leveraging information and communication technologies (ICT). This, in turn, leads to an improvement in the quality of life and public satisfaction. One important aspect that becomes evident in the managerial and executive structure of applying new technologies in urban management is the necessity of citizen engagement. The role of citizens should be defined through legal frameworks to ensure that processes proceed as intended and that managers can leverage the participatory capacity of citizens. One of the foundational international documents in the realm of citizens' rights is the Declaration of Cultural Rights, which highlights the participatory role of the public that managers must consider in the cultural policymaking of smart cities [10].

In essence, smartening up is an idea aimed at enhancing and improving the quality of life, which has become a key concept in urban planning and has also been accompanied by a significant increase in urban population in contemporary times. Urban growth has influenced the development of urban facilities and services. Therefore, to enhance the quality of life in urban spaces, a comprehensive reevaluation across various domains is necessary. In this context, one of the most important factors for

Table 1 Explaining the areas and dimensions of smart cities

Domain	Dimensions
Smart Governance	Integrating services and interactions between the government, people, municipality and related organizations, increasing the quality of urban life, increasing efficiency and effectiveness in the performance and provision of public and social urban services based on technology and with environmental orientations
Smart Mobility	Establishing an integrated information and communication technology infrastructure, empowering the smart transportation system, reducing air pollution
Smart Environment	Smart and renewable energy network, monitoring and controlling pollution, renovating buildings and facilities, green buildings and urbanization, optimizing energy consumption and resources
Smart Living	Promoting a healthy, safe and refreshing lifestyle, diverse cultural facilities in the city, quality housing and social cohesion and capital
Smart Economy	Increasing productivity, creating a platform for entrepreneurship, creating innovative products, and finally, new business models and services based on information and communication technology and also with environmental orientations in line with the use of clean energy
Smart Citizen	Continuous learning of people, using smart platforms to improve the level of education and creativity, and not having negative resistance to the use of technology in business and life decisions
Smart Technology	Sensors, radars, data, data analytics, broadband, fiber optics, user interface

Source: Based on the authors' studies

achieving the desired objectives is the utilization of the potential of urban environments and spaces in employing smart facilities and creating interactions with their users as a crucial element in the intelligence of cities, enabling goals such as efficiency, responsiveness of urban spaces, and efforts to increase both the qualitative and quantitative presence of people to be realized [11–13] (Table 1).

3.2 *Internet of Things (IoT)*

The Internet of Things serves as a key infrastructure in smart cities, facilitating the collection and analysis of data through sensors and Internet-connected devices. Zaman et al. [14] emphasize that IoT can assist in optimizing energy consumption, managing traffic, and monitoring environmental quality. This technology empowers urban managers to make data-driven decisions and enhance the efficiency of public services.

3.3 Intelligent Transportation

Intelligent transportation, as a vital component of smart cities, encompasses smart public transportation systems and autonomous vehicles. Litman [15] states that these systems are can contribute to reducing traffic congestion, improving safety, and increasing access to public services. Furthermore, the establishment of appropriate infrastructure for intelligent transportation can lead to decreased pollution and improved quality of life for citizens.

Advancements in transportation systems, driven by technological innovations, have evolved into intelligent transportation systems, which establish intelligent communication links among vehicles, between vehicles and traffic infrastructure, and between vehicles and humans [16]. Within this transportation framework, to ensure a safe travel experience, optimize routes, provide parking solutions, implement bright lighting, and identify accidents through extensive coordination among traffic control systems, a collection of data and information related to travel is gathered, generated, and processed. This interconnected network facilitates the dissemination of helpful information and effective utilization in management, control, monitoring, and travel guidance, thereby innovatively reshaping transportation [17–19]. By tracking intelligent transportation systems, the efficiency of urban mobility networks is ensured, as they encompass a network of interconnected stakeholders and actors, fostering a technical and social understanding within the urban transportation network. This factor paves the way for participatory management, particularly the collaborative management of intelligent devices and networks [20–22]. Moreover, the overarching goal of integrated intelligent transportation management is to unify the constituent elements of the system, including computing technology, communication technology, image processing, and the understanding and control of online public opinion, thereby facilitating the combination of information and communication technology with one another and enabling the distribution of traffic flows among various modes of transportation. The relationship between transit-oriented development and intelligent transportation can be explored within the shared framework of innovative transit-oriented development (SMART-TOD), which will be discussed further. Transit-oriented development is grounded in the core domains of intelligent transportation and can play a significant role in enhancing the management of intelligent transportation systems. The definition of SMART-TOD can be anchored in two pillars: the public transportation axes of transit-oriented development, including metro, bus, and taxi services equipped with information and communication technology, and an integrated definition derived from their amalgamation. This definition positions S-TOD as a fundamental component of urban transportation, traffic planning, and management. Here, a contradiction in the definition is highlighted, which can be addressed in two ways. The question arises whether S-TOD refers to a TOD within a smart city or whether it pertains to the intelligence of the TOD within a city. In response to this inquiry, it can be stated that the characteristics of the city in question form the core of the answer. If the city in question is a smart city equipped with intelligent technologies, it demonstrates S-TOD in the form of transit-oriented development that exists within a smart city. Conversely, if the city is not smart, S-TOD is illustrated as the intelligence of the TOD within a city. The emergence of S-TOD is driven by guiding urban development through the integration of technology and transportation to address the needs of urban residents, relying on planning and urban design based on TOD principles and research related to traffic forecasting, objects, and cloud computing. Spatial awareness is a crucial aspect considered for tracking activities through various tools, such as mobile phones and cameras. Another important point is the spatial areas and nodes that are the targets of TOD. The structure of S-TOD encompasses support areas (planning and design tools, processing capabilities, development technologies, internet terminals, support facilities, virtual space, and traffic space), the core area (internal station space, external station space, drop-off areas inside and outside, online space), and the coverage area (areas with significant urban functions, traffic space, virtual space, and areas lacking significant urban functions) [23].

The increase in traffic volume and the consequent rise in fuel consumption have prompted urban traffic managers to implement innovative solutions, leading to the activation of new streams such as GIS- and GPS-based applications within the transportation system, culminating in complete system integration and the provision of comprehensive capabilities. The GPRS infrastructure provides cellular data and related messages for optimal communications to the TOD system. GIS database producers supply map and traffic information to the TOD system. In this context, public transportation (such as metro, bus, and taxi) is connected to information and communication technology support systems. In general, it can be stated that within this structure, the TOD system connects to intelligent distribution centers in the form of operation-support centers (routing, monitoring, coordination, and management) both outside and inside the vehicle (security, authentication, positioning, traffic information). Within this framework, the TOD interfaces with operators and platforms. The operational efficiency of the TOD fleet is enhanced through intelligent transportation management, and the fleet capacity is increased with automated dispatch processes [24] (Table 2).

4 Methodology

The present study is conducted with a reliance on qualitative methodology. Here, the experiences of Iran, the United States, Italy, India, and China in the realm of emerging technologies are examined through the analysis of documents, laws, regulations, and plans from these countries. Initially, official documents and reports related to government policies and their executive programs in the field of emerging technologies were scrutinized. These documents include government reports, guidelines, long-term strategies, and development plans in areas such as the Internet of Things (IoT) in coastal cities, innovative Transit-Oriented Development (TOD), and quality of life in smart cities. Subsequently, their experiences in each domain were analyzed. The results of these analyses can help policymakers and executive managers in various

Table 2 Description of smart transportation areas, regions and indicators

Area	Region	Index
Software infrastructure (software infrastructure applies to all infrastructure)	Backup Area	Possibility of online ticket charging, timer indicating the time remaining to the destination, application indicating the location of public transport vehicles, possibility of online ticket purchase, intelligent traffic signal control system, mobile internet, data transmission equipment (energy consumption, vehicle behavior, speed limit, number of passengers, etc.) to the main network, data analysis platform, multi-modal electronic service access platform
	Main Area	Internet terminals including computers, mobile phones and other smart devices, electronic payment systems, authentication mechanisms, Bluetooth, cellular network, Wi-Fi, Internet of Things, big data, Internet facilities in public transport, user guidance (information about queues and congestion, etc.), service information (arrival time, departure time, etc.)
	Coverage Area	Virtual space for online human activities relying on the Internet
Physical infrastructure	Backup Area	Parking lot, traffic-receiving space
	Main Area	• *Furnishings*: Lighting, furniture, signs and panels, drinking fountains, toilets • *Services*: Shops, markets, ancillary and welfare services • *Parking*: Car and bicycle parking, open areas • *Accessibility*: Interchange with other modes of transport, average time to reach the metro station
	Coverage Area	Physical-spatial value of the neighborhood: Public pedestrian traffic in the vicinity of the metro station, existing spatial structure around the metro station, streetscape, concentration and diversity within the station area, ease of movement in different parts of the city, retail activities around the station, average distance from the station to medical, residential, commercial, residential, educational and shopping centers and stores, access opportunities from work to the station
Land use pattern	Backup Area	Commercial, administrative, medical facilities
	Main Area	–

(continued)

Table 2 (continued)

Area	Region	Index
	Coverage Area	• *Mixed use*: Establishment of diverse activities and services within walking distance of residence and employment, 24-h activity and vitality of public space, mixed commercial-residential use, integration of land use with public transportation, area of use (commercial, residential, educational and medical), number of hotels • *Distribution and dispersion*: The desired distribution and dispersion of area use • *Building density*: Relative density of residence and supporting activities and in interaction with public transportation, local services and active public space, adaptation of access to public transportation services to the density of activity, size of building blocks, number of units per hectare, horizontal density (ratio of number of plots to land area per hectare)
Public transportation infrastructure	Backup Area	• Parking support facilities, street connections in the form of short-distance pedestrian and bicycle paths compared to the vehicular route • Pedestrian capability (safe pedestrian network, integrated pedestrian network, vitality of the pedestrian area, climatic comfort, short (compared to the vehicular route), direct and diverse, wide sidewalks in the vicinity of the metro station, favorable pedestrian access to the station area, route length), density and number of pedestrian intersections • Cycling capability, safe cycling network, integrated cycling network, short (compared to the vehicular route), direct and diverse, route length, density and number of bicycle intersections
	Main Area	Access to metro station, access to bus station, access to taxi station, 24-h traffic of vehicles, user-friendliness of transportation facilities (passenger safety at the station, basic facilities at the station, information display system)
	Coverage Area	Virtual space for online human activities relying on the internet
Traffic infrastructure (facilities and equipment)	Backup Area	Various surveillance cameras (static surveillance cameras at traffic intersections, highway surveillance cameras, roadside cameras, drone cameras, and dashboard cameras), real-time road management system, road passenger transportation system on various routes

(continued)

Table 2 (continued)

Area	Region	Index
	Main Area	Monitoring the stability of tunnels, corridors, interfaces, traffic lights, warning systems, intelligent information systems, intelligent incident management systems, intelligent parking management systems, intelligent pavement management systems, location tracking systems, smart cards, signs, signs
	Coverage Area	Virtual space for online human activities relying on the internet
Socio-economic infrastructure	Backup Area	Density of commercial establishments (number of commercial establishments/square kilometer), private sector investment
	Main Area	Social interactions inside and outside the station, number of passengers in the metro station
	Coverage Area	Economic value of land, population density (average distance from the station to residential area (number of people/square kilometer), population of statistical blocks)
Legal infrastructure	Backup Area	Government evidence-based policymaking, laws and regulations, implementation of advanced technologies in the smart transport ecosystem, national documents, education, local and regional traffic management programs, public participation
	Main Area	Station scale and design details
	Coverage Area	Land use policies, urban design and street layouts suitable for active travel and public transport

Source: Based on the authors' studies

countries worldwide leverage the successful experiences of other nations to design and implement effective policies and projects.

5 Urban Infrastructure for Recycling and Reuse

Urban infrastructure for recycling and waste reuse plays a pivotal role in achieving sustainable development and a circular economy. With the increasing waste production in large cities and the necessity to reduce pressure on the environment, innovative technologies and waste management systems have gained paramount importance. In this context, the use of decentralized recycling facilities, digital innovation centers, and the recovery of construction waste, as complementary approaches, has opened new horizons in waste recycling and reuse. Decentralized recycling facilities, such as small-scale anaerobic digestion systems, convert organic waste into energy and fertilizers while minimizing the need for waste transportation. Digital platforms, by connecting individuals and organizations, enhance participation in recycling and foster a culture of waste separation. The recycling of construction waste also constitutes a significant portion of waste and requires innovative technologies, as well as collaborative efforts in terms of both legal and operational aspects. The integration of these technologies and infrastructures is key to achieving sustainable and smart cities that embed recycling and reuse into daily life.

5.1 Decentralized Recycling Facilities

Decentralized recycling facilities (or small-scale anaerobic digestion systems) represent a novel and flexible technology for managing food waste, designed with the aim of energy production and carbon emission reduction. Unlike large centralized factories located at a single site, these systems are installed locally near waste generation sources (such as food distribution centers or neighborhoods) and typically have processing capacities of less than 1000 tons per year or electrical outputs below 15 kW. These systems convert organic waste into biogas (methane) and nutrient-rich fertilizers. In many countries, large-scale versions of these systems have been utilized in a centralized manner at urban or regional scales. However, recent attention has shifted towards small-scale and decentralized facilities due to their lower installation and maintenance costs, which reduce expenses and pollution associated with waste transportation, while also increasing social participation since their outputs (electricity, heat, and fertilizers) are directly returned to the local community. Decentralized anaerobic digestion facilities, particularly small-scale models, offer a viable solution for managing food waste, which can significantly contribute to carbon emission reduction while generating energy. Consequently, the use of these systems is more effective in reducing carbon dioxide emissions compared to landfilling or incinerating waste. In a real-world example, a small decentralized facility was established at a food supply center in Singapore, processing 300 kg of food waste daily. Within 75 days, approximately 1895 kg of waste were decomposed, producing 173 cubic meters of biogas with a methane content of 53%. The efficiency of organic material decomposition was about 88% [25, 26].

5.2 Innovation Centers and Reuse Sharing Platforms

- Digital platforms, by serving as intermediaries among citizens, municipalities, and waste management companies, contribute to the regeneration of the recycling value chain. These platforms leverage technologies such as barcodes, GPS, databases, and artificial intelligence to facilitate multi-directional communications, knowledge exchange, and active stakeholder participation. Their key functions can be summarized in four areas: connection, information dissemination, mobilization, and integration. Through these four functions, citizens transition from passive roles to active participants in the design and implementation of waste policies.

- Incentive mechanisms, such as gamification, scoring, and real-time feedback, create an engaging experience that fosters a culture of source separation, reduces operational costs, and increases recycling rates.
- However, a critical challenge lies in the digital literacy gap; without equitable access to technology, universal benefits cannot be achieved. Therefore, adopting a human-centered approach and implementing digital empowerment programs are prerequisites for success.

The combination of platforms with policies such as pay-as-you-throw (PAYT) can reform the waste tax structure and strengthen data-driven urban governance. Ultimately, as the user base expands, direct and indirect network effects proliferate, gradually institutionalizing sustainable behaviors and steering towards the achievement of Sustainable Development Goals (SDGs). Digital platforms enhance recycling practices within a framework of collaborative production by linking citizens, municipalities, and waste companies. This model, grounded in the literature on value co-creation, defines the platform as a circular intermediary among stakeholders with complementary objectives, capable of influencing the processes and outcomes of participation [27, 28].

5.3 Recovery of Construction and Demolition Waste

The recovery of construction and demolition waste (C&DW), accounting for nearly 40% of total global waste, plays a vital role in achieving sustainable development and a circular economy. The quantitative forecasting of recyclable, reusable, and landfillable waste has been facilitated through a hybrid approach combining Building Information Modeling (BIM) and Machine Learning (ML). The input data includes characteristics such as floor area, volume, structural type, and building usage, with the type of structure and its usage being the most influential factors. The primary challenges include the lack of initial waste segregation at project sites, weak regulatory frameworks, insufficient inter-agency collaboration, and execution issues. Addressing these challenges necessitates macro-level policy-making, the formulation of new regulations, enhanced project management awareness, and the allocation of adequate resources.

Furthermore, strengthening recycling industries and developing cost-effective business models for waste management are essential. The utilization of digital technologies such as BIM, ML, the Internet of Things, blockchain, and robotics enables the optimization of waste management, reduction of design waste, and enhancement of resource efficiency. Big data and geographic analysis facilitate the discovery of hidden patterns and precise tracking of waste, aiding in the prevention of illegal dumping. Geographically, Asian countries face more significant challenges in management and regulation, while European countries focus on developing laws and standardization. A shift in attitude towards built assets and sustainable collaborations among companies is crucial for realizing a circular economy in the construction industry. This hybrid framework offers a precise tool for informed decision-making prior to demolition, paving the way for optimizing resource consumption, reducing waste, and advancing towards sustainable construction [29–31].

6 Examination of Practical Experiences in the Use of New Technologies in Urban Environments

One of the critical and central factors in managing contemporary cities is the utilization of new technologies across various domains. Through these technologies, the efficiency of urban infrastructures is also enhanced, thereby improving the quality of life for citizens. To achieve an efficient smart city, it is essential to establish coordination and synergy among these technologies. The development and expansion of the necessary infrastructure for this coordination, particularly in the areas of communications, cybersecurity, and support for research and development, are vital for realizing the vision of smart cities. The following sections will examine the experiences of Iran, the United States, India, and China to inform the strategies of other countries.

6.1 Bushehr/Iran

For the first time in the country, coinciding with National Persian Gulf Day, the project for designing the smart city of Bushehr was initiated through collaboration among governmental agencies, the executive body of Bushehr Province, and the Center for the Development of Information Technology (MAGFA). In this context, the Bushehr Port Municipality has been working since 2016 on drafting the Smart City Development Document in three phases: foundational studies, designing the conceptual model of the smart city of Bushehr, and formulating the strategic document for the smart city of Bushehr. In the final phase, strategic policies, significant projects, and key actions for the city to become smarter have been outlined. Among the actions taken by the Bushehr Port Municipality in this regard, one can mention the design and implementation of over 39 km of fiber optic network, the establishment of a Wi-Fi umbrella along the city's coastal strip, efforts towards designing and implementing a centralized citizen card, and the creation of dedicated bike lanes and stations, among others. The overarching goals of this project include enhancing the quality of life, improving and intelligently managing urban services, optimizing resource consumption, and fostering the growth and flourishing of scientific and innovative capacities in Bushehr. By leveraging the expertise of leading international consultants in this field, the project aims to serve as the first smart city initiative in Iran, facilitating the transfer of technology and localization of innovative city concepts, thereby paving the way for the development of smart cities nationwide. This project was launched with the support of the Bushehr Governorate and in collaboration with the Center for the Development of Information Technology (MAGFA), along with all scientific, executive, and public agencies in the province, including the municipality, the General Directorate of Communications and Information Technology of the province, the Persian Gulf University, the Persian Gulf Science and Technology Park, and other stakeholders of the smart city, marking it as the first smart city in Iran.

6.2 New York/United States

"BIM" is the industry-accepted acronym for "Building Information Modeling." This guide provides guidance and requirements for projects by engineers in the New York area that utilize Building Information Modeling (BIM). The primary documents include the Army Design Process Manual and submission requirements prepared by the U.S. Army Corps of Engineers (CELRL) and the Center of Standardization (CoS) for Army Reserve Centers [32]. This document is structured as follows:

Document Objectives [26]

1. Utilize technology to meet expectations.
2. Demonstrate the constructed model.
3. Provide a model with future usability capabilities.
4. Manage facilities/infrastructure.
5. Facilitate future development.

The benefits of using BIM are outlined as follows [32]

- Increased site safety (safety hazards will be identified in the early stages before issues arise).
- Faster approvals (3D images and real-time models enhance communication between field inspectors and office supervisors, aiding them in resolving issues immediately).
- Better service and fewer office visits (site safety plans submitted through this process are entirely reviewed, revised, and approved via the website and online document management tools).
- Improved decision-making processes.
- Reduced requests for information.
- Generation of superior designs.
- Cost reductions.
- Enhanced quality and overall construction scheduling.

In 2014, this system was implemented in New York zoning, providing a comprehensive 3D modeling system. Policies such as Building Information Management and Cadastre have been introduced. In the realm of Building Information Management, as referenced in the New York Housing Document, this initiative was operationalized in 2014, introducing a three-dimensional zoning framework. Additionally, all necessary information is available on its website in accessible file formats, and all building information has been thoroughly documented with the assistance of various engineering branches. Furthermore, policies for the preservation and maintenance of housing are visible in this document, with part of the document's objectives focused on protective approaches for building management. The cadastre has also been prepared at the state level and incorporated into a comprehensive system (nyc.gov).

The New York State GIS Coordination Body (CB) and the Office of Cybersecurity and Critical Infrastructure Coordination (now the Office of Cybersecurity or OCS) have identified the development of a statewide layer as one of the highest priorities stemming from the 2008 strategic plan. This business plan outlines the steps necessary to achieve this goal. In addition, this plan supports the vision of the National Spatial Data Infrastructure (NSDI) by establishing a process that collects local and authoritative cadastres into a statewide dataset, which can subsequently be integrated at the national level. The national-level benefits reflect those gained at local and regional levels, as NSDI reduces redundant efforts and ensures that the best available cadastral information is utilized in decision-making (33).

In this context, BIM and GIS are two key technologies for enhancing the smartness of cities. These technologies enhance the quality of life for citizens and improve the safety and efficiency of infrastructure. The integration of information at the national level, utilizing the NSDI (National Spatial Data Infrastructure), enhances the efficiency of urban planning. Smart cities, due to their high dependence on data, require robust cybersecurity strategies, which are addressed in this document.

6.3 Turin, Italy

In the management structure of Transit-Oriented Development (TOD) in Turin, effective relationships and collaborations among various stakeholders—such as municipal authorities, private developers, public transportation agencies, and local communities—are crucial. These partnerships ensure that urban planning and development efforts are well-coordinated, sustainable, and aligned with broader economic, social, and environmental goals. The smart TOD framework of Turin promotes a comprehensive approach to urban mobility and infrastructure growth by strengthening open communication and shared objectives, thereby enhancing the overall quality of life for its residents (Table 3).

The management structure of Transit-Oriented Development (TOD) in the city of Turin relies on collaboration among several stakeholders, each playing a role in enhancing urban mobility and sustainability. The Municipality of Turin plays a pivotal role, accounting for approximately 35% of the influence through the formulation of policies and regulations [34]. The municipality coordinates with the Turin Transport Company (GTT), which manages public transportation operations, including buses, trams, and the metro, exerting 30% of its influence through the

Table 3 SMART-TOD management structure in Turin

Type of relationship	Stakeholders involved	Natures of relationship	Description
Vertical—Public Sector	City of Torino and Public Sector Agencies	Policy-Making and Oversight	The City of Torino sets policies, regulations, and strategic frameworks for urban development and transportation. This includes budget planning, resource allocation, and overseeing implementation
Vertical—Public Sector and Civil Society	City of Torino, GTT, and Civil Society	Service Provision and Accountability	The City of Torino and GTT (Torino Transport Company) provide transport services and development projects to meet community needs, addressing feedback and ensuring accountability
Citizen to Government and Public Sector	Citizens and Public Sector Entities	Participation and Oversight	Citizens engage through public consultations, providing input on transportation services and urban planning projects, influencing decision-making processes
Horizontal—Public Sector and Private Sector	City of Torino, GTT, and Private Companies	Collaboration in Projects	Public sector entities collaborate with private companies and technology providers for implementing infrastructure projects, smart technologies, and service innovations. This includes PPPs and investment partnerships
Government and Citizens	City of Torino, Citizens, and Civil Society Organizations	Consultation and Interaction	The City of Torino interacts with citizens and civil society through meetings, surveys, and consultations, ensuring public input is considered in TOD strategies and projects
Civil Society and Private Sector	Civil Society Organizations and Private Companies	Social Responsibility and Impact	Civil society organizations partner with private entities to promote social responsibility and sustainability through joint initiatives, campaigns, and educational programs

Source: Based on the authors' studies

integration of transportation systems with urban development's [35]. The regional government of Piedmont also contributes by providing regional planning support and funding, aligning local TOD projects with broader strategies, thus accounting for 20% of the influence [36]. The private sector and technology providers contribute approximately 10% of the impact through public-private partnerships (PPPs) and the implementation of innovative technologies designed to enhance transportation services. Public engagement, which accounts for 5% of the influence, strengthens public support and aligns with community needs by involving residents in the planning process. Overall, the relationships among these stakeholders have created a cohesive and efficient structure for managing TOD in Turin, enabling the city to implement successful urban development strategies, improve public transportation, reduce reliance on private vehicles, and support sustainable urban growth [37]. The intelligent transportation infrastructure in Turin, Italy, exemplifies the integration of Intelligent Transportation Systems (ITS) in urban mobility. Turin has adopted a multi-layered approach to smart transportation, focusing on real-time data collection, traffic management, and integrating public transport with digital services [38].

1. *Real-Time Data and Traffic Management*: Turin employs advanced ITS technologies for more effective traffic flow management. The city has deployed sensors and connected devices throughout its transportation network, enabling the real-time collection of data on vehicle movement, road congestion, and public transport availability. This data is analyzed to optimize traffic signals, reduce congestion, and enhance public transportation planning [39].
2. *Integrated Public Transport Systems*: Turin has developed a platform that integrates various modes of public transport, including buses, trams, and metro systems, into a single digital system. Passengers can access real-time information about vehicle arrival times, route changes, and other transport updates through mobile applications and digital kiosks installed throughout the city. This approach ensures that users have easy access to the most up-to-date information, thereby enhancing overall efficiency and user experience [40].
3. *Smart Ticketing and Payment Systems*: The city's public transport system also benefits from innovative ticketing solutions. Turin has implemented contactless payment options, allowing passengers to pay for their journeys using mobile apps or smart cards. This reduces the need for physical tickets and streamlines the payment process for travelers [41].
4. *Environmental Sustainability*: Turin's smart transit infrastructure is designed with environmental sustainability in mind. The city promotes the use of electric buses and low-emission vehicles to reduce the carbon footprint of its public transportation system. Additionally, data collected from the ITS network is utilized to promote more efficient driving routes and reduce unnecessary fuel consumption [42].

6.4 China

The strategic position of China's coastal regions is of paramount importance, as evidenced by the fact that urban development centers at the national level are gravitating towards coastal cities, with urban areas near the shoreline undergoing significant growth. The final report of the China project on spatial planning in coastal areas—Disaster Prevention and Sustainable Development: Phase 3—constitutes one of the principal documents in this domain. China boasts over 18,000 km of coastline, encompassing 11 provinces (autonomous regions and municipalities directly under the central government) within the coastal zone. By 2010, approximately 300 million people resided in the coastal region, accounting for 22% of China's total population. Consequently, the management of crises related to coastal incidents holds considerable significance. The Gross Domestic Product (GDP) of coastal areas represents 36.8% of the national total. The coastlines of Guangdong, Guangxi, and Hainan belong to the South China Sea region and constitute more than half of China's total coastline [43]. The coastal area is the most developed and rapidly growing region in China, which has consequently led to numerous environmental challenges for these coastal zones. The foremost issue is water quality pollution. In 2010, marine water quality at nearly a quarter of all monitoring stations was identified as being around or below Class IV of the national standard. The second problem pertains to the degradation of coastal wetlands and habitats such as coral reefs, mangroves, and seagrass beds. Moreover, China's coastal region is vulnerable to climate change and marine disasters, with frequent storms and increasing seawater intrusion into rivers exacerbating the challenges faced by economic activities and daily life [44].

Based on improvements in administrative structure, another critical issue is the establishment of relevant laws or regulations to facilitate the integrated management of coastal areas [45]. Given the conditions, there is an urgent need to enhance capacity building for coastal spatial planning and integrated coastal management (ICM). Therefore, China is actively participating in the Sida project "Spatial Planning in Coastal Areas—Disaster Prevention and Sustainable Development." This project is implemented in three phases: Phase 1—preparation of regional resource documentation and a guiding document for integrating new concepts such as climate change adaptation, disaster risk reduction, ecosystem-based management, and outcomes-based policy and procedures for coastal spatial planning (complete); Phase 2—national consultation and planning for the national adaptation of the regional resource document; and Phase 3—capacity building, national adaptation, and demonstrations [46].

The overarching goal of implementing this project is to promote sustainable development in China's coastal areas through the use of spatial planning for integrated coastal zone management (ICZM) and ecosystem-based management (EBM), while also mitigating and preventing the impacts of natural disasters, climate change, and rising sea levels. China will undertake further actions to follow up on the results of the Sida-COBSEA project [45]:

- Accelerating the official release of the national resource document following review and improved refinement.
- Enhancing training and capacity building at the national level across local levels and various governmental branches.
- Continuing to utilize the ideas and methods transferred by the "Spatial Planning in Coastal Areas—Disaster Prevention and Sustainable Development" project in China and the region through joint projects.

- Promoting improvements in the legal and institutional environment in China to facilitate the implementation of integrated coastal management.

New concepts and technologies, such as ecosystem-based management and treating land and sea as a unified entity in coastal planning, have been applied in the coastal environmental program of the (Ministry of Ecology and Environment (MEP) for the Twelfth Five-Year Plan period and are involved in several other related national research projects [47]).

The outline of the Fourteenth Five-Year Plan (2021–2025) for national economic and social development and the vision for 2035 of the People's Republic of China addresses coastal economic development in a separate chapter. Harmonious development is key. To safeguard the health and integrity of marine ecosystems and ensure coordination between people and the sea while simultaneously fostering marine economic development, coordination between land and maritime activities is essential. The plan outlines the following discussions [48]:

- Strengthening and expanding marine engineering equipment and the marine biomedical industry.
- Promoting seawater desalination and large-scale utilization of marine energy.
- Promoting and developing cultural marine tourism.
- Optimizing the distribution of green aquaculture in coastal areas.
- Identifying several demonstration zones for high-quality marine economy development and marine industrial clusters.
- Strengthening the development of the three main marine economic rings in the north, east, and south comprehensively.
- Deepening maritime cooperation with neighboring countries based on coastal economic belts.

It should also be noted that in recent years, the development of intelligent monitoring systems and early warning platforms for tsunamis has gained significant attention in China. These platforms have been independently developed by the National Marine Environmental Forecasting Center (NMEFC). Global seabed earthquake monitoring, global water level monitoring, a GPU-based parallel tsunami prediction model for the Pacific Ocean, and a tsunami scenario are integrated by these systems, which are based on the Internet of Things (IoT). In this IoT framework, the system serves as part of the early warning system. This approach aims to create a modern network of infrastructure that is efficient, functional, intelligent, environmentally friendly, and capable of managing environmental crises [49].

6.5 India

The land of India features an extensive coastline, providing a significant foundation for coastal city initiatives such as Sagarmala. Sagarmala has introduced new dimensions to the coastal economy. The country's coastal economy is being bolstered through the Sagarmala program, which involves not only the modernization of old ports but also the construction of new ones. India asserts that ports serve as gateways to development. Consequently, the government has launched the Sagarmala program to modernize connections and infrastructure along its shores. The Sagarmala program is a flagship initiative of the Ministry of Ports, Shipping, and Waterways, aimed at promoting port development in the country. It leverages India's 7500-km coastline, 14,500 km of potentially navigable waterways, and its strategic position along major maritime trade routes [50].

India's blue economy encompasses the entire system of resources in the ocean, as well as the economic infrastructure in marine, coastal, and terrestrial areas within the country's legal jurisdiction. India's extensive maritime interests position it as a significant potential contributor to the nation's economic growth. The Indian government's vision of a New India highlights the blue economy as one of the ten key dimensions of growth. The blue economy is articulated as the sixth dimension of this vision, emphasizing the necessity of a coherent policy that integrates numerous sectors to enhance the livelihoods of coastal communities while accelerating development and employment. According to the draft policy framework, India's approach to maximizing the socio-economic potential of the blue economy should include a framework for: appropriately measuring blue economy activities and their contribution to national income, ensuring the welfare, safety, and livelihoods of fishers in coastal areas, fostering innovation to ensure zero waste, adopting low-carbon technologies, and guaranteeing balanced international engagement. India has demonstrated interest in strengthening all three pillars of the blue economy: security, sustainability, and economic prosperity [51].

1. Coastal Economy Development
   - Integration of the coastal economy with ports through coastal development
   - Establishment of economic zones and projects synergistic with coastal industrial corridors [50].
   - Development of smart port-based cities and other urban infrastructure to enhance living standards [50].
   - Implementation of skill development/livelihood projects for coastal community development, such as coastal tourism development projects (lighthouses and land islands) and fisheries sector development [50].
2. Marine Development and Infrastructure
   - Modernization and expansion of existing port capacities and establishment of green ports to alleviate future growth bottlenecks [52].
   - Development of port discharge (road/rail and inland waterway) and logistical infrastructures to reduce overall logistics costs and enhance cargo movement in and out of the country [52].
   - Development of the marine sector, such as shipbuilding and repair clusters [52].

In summary, the following projects are being pursued by India

- Port-based industrialization projects: Infrastructure development, maintenance in industrial/urban areas near ports, establishment of industrial parks [50]
- Port-based urbanization projects: Greenfield and brownfield urbanization projects to enhance quality of life, development of smart port cities [50].
- Coastal community development projects: Infrastructure and non-infrastructure projects related to coastal community development, such as skill development and production, health projects [51].
- Coastal shipping projects: Infrastructure development and maintenance projects to promote coastal shipping, construction of jetties for coastal shipping, dredging [50].
- Modernization of existing ports and development of new ports: Projects aimed at improving existing port infrastructures and configurations, increasing green ports [52].
- Renewable marine energy projects: Renewable energy projects for the sustainable development of ports and coastal areas, solar and wind energy projects [52].
- Storage platform projects [52].

The Integrated Coastal Surveillance System (ICSS) is a system developed to protect India's coastal territory, enabling the rapid identification, location, and monitoring of suspicious vessels and motorized and traditional boats within a specified controlled area, operating around the clock without physical presence. In India, IoT has been extensively utilized to facilitate operations without requiring physical attendance. This system comprises coastal radar stations, optical and electro-optical sensors, and Automatic Identification Systems (AIS) [53].

7 Discussion and Implications

The world is in a constant state of transformation, and humanity has always strived to overcome existing challenges and create an environment conducive to living, where the highest level of satisfaction with the quality of life can be achieved. One of the new concepts for addressing the current challenges of cities is innovative city development, which has garnered significant attention in recent years. Given the increasing trend of population growth in metropolitan areas, the establishment of smart cities in today's advanced world is an inevitable necessity. The planning and utilization of electronic tools and communication systems aimed at enhancing the quality of life for city citizens determine the level of intelligence of that city. To effectively implement modern technologies across various domains, it is essential to be familiar with the challenges associated with each one (Table 4).

Table 4 Areas, challenges and solutions for applying new technologies towards the realization of smart cities

Area	Challenge	Solution
SMART-TOD	Lack of coordination and consistency of documents and programs with international laws	The need to develop integrated policies in line with international documents
	Limited public participation and awareness	Implement comprehensive public awareness campaigns and educational programs to inform citizens about the benefits of smart transportation, hold public meetings
	Financial constraints and resource limitations hinder the implementation and maintenance of intelligent transportation systems	Attract private sector investments through public-private partnerships (PPPs), green bonds and urban development funds
	Lack of environmental impact assessment of intelligent transportation projects	Carry out rigorous environmental impact assessments (EIA), promote the use of electric vehicles, increase green public transportation options
	Slow adoption and implementation of new technologies in intelligent transportation projects	Support research and development in smart transportation technologies, integrate the Internet of Things (IoT), artificial intelligence (AI) and big data analytics in the transportation network
	Inefficiency in coordination and cooperation between various management and executive agencies	Establish centralized and specialized institutions to oversee the management and implementation of SMART-TOD projects. Strengthen inter-sectoral coordination through clear communication channels
Coastal cities	Old and inefficient infrastructure	Upgrading infrastructure to prepare it for the application of new technologies
	Emerging disaster forecasting and early warning systems in early development	Increasing forecast accuracy with the help of physical and digital presence
	Data-related problems and limited access to up-to-date data	Accessing accurate and up-to-date data
	High cost and financial constraints	Private financing, investing in strengthening communication infrastructure and data networks
Quality of life and Smartness	Limited public participation	Recognize cultural rights in public participation documents and encourage people to play an effective role

(continued)

Table 4 (continued)

Area	Challenge	Solution
	Concerns about privacy and data security	Restrict access to Internet of Things (IoT)-based surveillance systems and smart sensors
	Lack of clear rules and regulations	Establish clear laws and regulations for the use of new technologies and emphasize data security and privacy
	Unequal access to technology in some places and the creation of a digital divide	Develop Internet and communications infrastructure in underserved areas

Source: Based on the authors' studies

8 The Urban Waste Challenges

The management of urban waste falls under the jurisdiction of local governments, the formal and informal private sectors, and civil society. The issues of social and environmental justice are intricately linked to the increasing volume of urban waste and its management, closely correlating with the Sustainable Development Goals (SDGs). This management faces challenges such as high volume, diverse compositions, geographical dispersion, and significant social and environmental consequences. The importance of this issue in achieving sustainable development goals is profoundly evident, encompassing a wide range from poverty alleviation and hunger reduction to health promotion, gender equality, climate change mitigation, ecosystem protection, and the advancement of social justice and participation. Improper waste disposal leads to the emission of greenhouse gases and pollution of air, soil, and water, jeopardizing human health and biodiversity. Ultimately, the best strategy is to prevent waste generation and improve management practices to safeguard community health and mitigate environmental pressures.

8.1 Rising Waste Volumes

The unprecedented growth of population and the rapid development of cities, coupled with limited land resources, have resulted in an exponential increase in waste production, marking one of the most serious management challenges facing urban and environmental decision-makers in the twenty-first century. The accelerating growth of waste necessitates meticulous planning, continuous monitoring, and targeted policymaking to address this issue effectively. Global estimates indicate that the volume of waste produced in 2017 was approximately 20 billion tons, and if current trends persist, this figure could exceed 46 billion tons by 2050. This implies that within less than four decades, the waste volumes will more than double. Among this, urban waste, which includes household, commercial, and service waste, although constitutes a smaller share of total global waste, holds a significant position in terms of its direct impact on urban environments and public health. In 2019 alone, approximately 2.7 billion tons of urban wastes were generated, representing a 30–50% increase compared to previous years. Projections suggest that this amount could reach between 2.89 and 4.54 billion tons by 2050, indicating an intensification of the crisis.

Alongside the relentless increase in waste production, the quality of its management faces numerous challenges. Approximately one-third of urban waste is not collected at all, and a significant portion of collected waste is disposed of through unsustainable and improper methods, such as illegal dumping or uncontrolled burning; actions that carry severe environmental and health repercussions. Furthermore, the absence of global standards in defining, categorizing, and measuring waste has resulted in statistics that are non-comparable, incomplete, and inconsistent. In many countries, particularly developing nations, formal waste registration systems operate poorly, with a substantial portion of waste either managed by the informal sector or not recorded at all. Reports indicate that globally, only about 58% of urban waste is appropriately managed (such as through sanitary landfilling or recycling), while the remaining 42% is either incinerated or sent to unauthorized disposal sites. These unauthorized locations affect approximately 3.5–4 billion people, with around 64 million individuals—equivalent to the population of France—directly exposed to the associated hazards. It is anticipated that without intervention, hundreds of millions more will join this group. The health and environmental consequences of this dire situation are catastrophic. For instance, in 2010, approximately 8.7 million people in three countries—India, Indonesia, and the Philippines—were exposed to direct contact with toxic pollutants, such as lead and hexavalent chromium, at 373 unauthorized disposal sites. Moreover, it is estimated that between 400,000 and one million deaths annually are linked to poor waste management. Unauthorized burning of waste at disposal sites or homes has also been reported as a cause of premature death for around 270,000 individuals each year. According to ISWA reports, fewer than 750 direct deaths due to improper waste management at disposal sites were recorded in less than a year. In the face of this crisis, prioritizing waste reduction at the source and enhancing recycling processes presents a key opportunity to reduce final waste volume and move towards sustainable resource management. Adopting a circular economy approach, focusing on reduction, reuse, and recycling, can play a pivotal role in alleviating environmental pressures and achieving sustainable development goals. However, a fundamental prerequisite is the development of comprehensive and coordinated measurement and reporting systems at local, national, and global levels [54, 55].

8.2 Complex Waste Streams

In recent decades, the excessive consumption of plastics and the inadequacy of efficient recycling systems have transformed the plastic waste crisis into one of the most complex and serious environmental issues. The COVID-19 pandemic exacerbated this problem by increasing the use of disposable protective equipment, such as surgical masks, significantly raising the volume of unrecyclable waste [56]. In this context, chemical recycling technologies, particularly pyrolysis, have garnered widespread attention as an effective and sustainable alternative to traditional methods such as incineration. Pyrolysis is a thermal process that converts plastics into gas, oil, and biochar in the absence of oxygen, and life cycle assessments indicate that this method, especially at around 500 °C for surgical masks, is environmentally much less impactful than direct incineration; for instance, its greenhouse gas emissions are over ten times lower than those from burning, and it performs better in terms of human toxicity, ecosystem damage, and air pollution indicators. However, pyrolysis faces challenges such as high energy consumption and the production of products with a broad molecular distribution that require technological optimizations. In this regard, catalytic pyrolysis has been developed using catalysts such as ZSM-5 and β-Zeolite, which enhance reaction selectivity, reduce the necessary temperature, and improve the quality of final products. Additionally, integrating plastic pyrolysis with biomass can produce oils of higher quality and lower oxygen content; however, the differing optimal temperatures of these two materials present a significant operational challenge. In contrast, the incineration of plastic waste, especially that containing chlorine or nitrogen, generates highly toxic gases such as dioxins, furans, and carcinogenic aromatic compounds, which have detrimental effects on human health, biodiversity, and soil and air quality [57, 58]. Moreover, incomplete combustion generates soot and toxic ash, the disposal of which necessitates costly secondary processes. Economically, pyrolysis is considered one of the most cost-effective recycling methods, with an approximate net profit of $427 per ton of plastic. In contrast, mechanical recycling often proves inefficient in many areas due to high collection and sorting costs. To address the plastic waste crisis, a multi-tiered approach is required, encompassing the development of bioplastics, the enhancement of chemical recycling technologies, the formulation of incentivizing and restrictive policies, and the strengthening of collaboration among academia, industry, and government [59]. The combination of these technologies with renewable energy and the utilization of comprehensive life cycle assessments can pave the way for intelligent and sustainable management of plastic waste. On the other hand, the issue of electronic waste recycling has also become a significant challenge due to the rapid growth of production and low recycling rates [60]. Over 53 million tons of electronic waste is generated annually, with projections suggesting this could rise to approximately 75 million tons by 2030. These wastes represent rich and complex sources of metals, and their recycling is complicated by the diversity of elements and the need for costly selective processes. However, the production of high-entropy alloys (HEAs) from these wastes presents a novel opportunity in the fields of recycling and industrial development. HEAs, composed of multiple different metals, exhibit excellent mechanical properties, high corrosion resistance, and desirable thermal and electrical characteristics; however, their industrial development is often limited by reliance on critical and expensive metals. Utilizing metals found in electronic waste for the fabrication of these alloys eliminates the need for selective recovery of individual elements and enables the direct production of multi-component alloys. Recent studies have demonstrated that by precisely combining recycled metals from devices such as smartphones, laptops, CPUs, and lithium-ion batteries, HEAs with diverse phase structures (single-phase or dual-phase) and desirable mechanical properties can be produced. These findings, corroborated by microscopic analyses and thermodynamic calculations, highlight the significant potential of electronic waste as a strategic source for the production of engineering alloys. Thus, the innovative recycling of electronic waste not only helps reduce dependence on critical resources but also provides an effective and sustainable solution for managing this type of complex waste, contributing to industrial development in the field of advanced materials [57, 58]. Therefore, the intelligent and sustainable management of complex plastic and electronic waste streams necessitates the integration of novel recycling technologies, intelligent policymaking, intersectoral collaboration, and a comprehensive approach that ensures both environmental and human health while generating economic added value [61, 62].

8.3 Environmental and Economic Impacts

Urban waste has widespread and detrimental effects on the environment, including the pollution of water and soil resources, increased greenhouse gas emissions, and challenges associated with improper waste disposal. The implementation of sound and efficient waste management practices, such as source segregation and effective recycling, plays a crucial role in mitigating these harms and can significantly contribute to environmental health. Among these, marine debris, particularly plastic waste, represents one of the most significant and complex environmental challenges of today's world, posing a serious threat to the sustainability of marine ecosystems, as well as to the health of marine organisms and humans. This waste, a substantial portion of which consists of single-use plastics and microplastics, enters the oceans from two primary sources: terrestrial and marine [63] Due to their durability and buoyancy, they can accumulate across a wide range of marine environments, from surface waters and coastlines to the ocean depths and even polar regions. Plastic pollution not only disrupts environmental health but also inflicts considerable economic damage on sectors such as fisheries, tourism, and aquaculture. This issue has gradually intensified since the 1980s, with reports indicating that between 1990 and 2015, approximately 26–91 million tons of waste entered the seas annually, a significant portion of which is plastic. It is projected that this figure could reach 48 million tons per year by 2030. Weak waste management, excessive plastic consumption, the absence of effective policies, and a lack of global cooperation are key factors exacerbating this crisis [64].

Furthermore, the direct impacts of this waste on ecosystems include toxic pollution, ingestion by marine organisms, blockage of respiratory passages, and damage to the food chain, ultimately threatening human health as well. In response to this crisis, international organizations such as the United Nations Environment Programme (UNEP), the World Health Organization (WHO), and international conventions such as the United Nations Convention on the Law of the Sea (UNCLOS) have provided legal frameworks and guidelines to reduce plastic pollution and support marine environmental protection. Regional and national collaborations are also of paramount importance. The European Union has implemented comprehensive waste management strategies and community awareness initiatives, while Asian countries have adopted programs to promote recycling and reduce plastic consumption. African nations have enacted bans on plastic bags, serving as examples of these efforts. Tools such as plastic taxes and circular economy policies in some countries have contributed to reducing plastic consumption; however, challenges remain, including the lack of global coordination, binding standards, and comprehensive and reliable data. Effectively addressing marine waste requires a holistic approach in which all stakeholders—from governments and international organizations to the private sector and the general public—actively participate. The use of modern technologies for monitoring and managing waste, promoting responsible consumption, and reforming production and recycling systems will be key to achieving a clean ocean and sustainable ecosystems. These efforts align with the United Nations Sustainable Development Goals, particularly regarding ocean conservation and human health, charting a clear path toward a healthier future and a more sustainable environment [65, 66]. Economically, waste management in urban areas is of great significance, as population growth leads to increased waste volumes and, consequently, higher costs for collection, transportation, and disposal. In this study, four different waste management scenarios were examined, and the total costs of each scenario were calculated. The results indicate that landfill methods combined with methane recovery are the most economically viable option. Additionally, a combination of recycling, waste-to-energy (WTE) incineration, and controlled landfilling can not only reduce greenhouse gas emissions but also help decrease waste management costs. Therefore, the selection of an appropriate waste management method should consider technical, environmental, and economic aspects to achieve optimal efficiency and minimal costs [67].

9 Advanced Recycling Technologies

Advanced recycling technologies represent a pivotal advancement in waste management and environmental protection, leveraging cutting-edge scientific and technological achievements to enable the recovery of usable materials from complex waste streams. Among these, artificial intelligence and robotics, through the integration of deep learning and machine vision systems, automate the sorting and classification of waste with remarkable speed and precision. On the other hand, chemical recycling employs specialized methods such as pyrolysis and gasification to transform hard-to-recycle plastic waste into valuable feedstock, thereby reducing environmental burdens and contributing to a circular economy. Additionally, biological and enzymatic recycling utilizes microorganisms and specialized enzymes, offering a sustainable and eco-friendly approach to plastic degradation [68]. This approach, in conjunction with biotechnological advancements, is continuously improving in speed and efficiency. Thermal conversion technologies, including controlled incineration, pyrolysis, and gasification, not only reduce waste volume but also facilitate the production of sustainable energy. Collectively, these technologies provide a novel horizon for effective recycling and sustainable development, significantly aiding the achievement of environmental and economic goals.

9.1 Artificial Intelligence and Robotics in Waste Sorting

The research presents an intelligent system for waste sorting that significantly enhances the level of automation and accuracy in the separation process by integrating deep learning and robotic arm technology. By employing data augmentation techniques, the generalization capability of the deep learning model DenseNet121 has been improved, thereby enhancing its ability to identify various types of complex waste. The robotic arm, equipped with a machine vision system, is capable of automatically and accurately identifying, locating, and sorting waste, ensuring the system's effectiveness under real operational conditions. Experiments indicate that this system not only increases the speed and accuracy of sorting but also provides a robust foundation for the development of advanced and practical environmental technologies at scale. The utilization of this system represents an efficient and resilient alternative to traditional manual waste sorting methods [69].

9.2 Chemical Recycling

The chemical recycling process serves as a complementary approach to mechanical recycling, enabling the conversion of hard-to-recycle and end-of-life plastic waste into fundamental materials such as monomers, oligomers, and other valuable chemical compounds. This method employs various technologies depending on the type of polymer: solvolysis for PET, pyrolysis for polyolefins (such as PE, PP, and PS), and combined processes of dechlorination and gasification for PVC. Each of these methods offers specific advantages in restoring the molecular structures of plastics and producing secondary products with added value. Simultaneously, the recycling of unsorted or mixed plastics can also be achieved using staged or "one-pot" methods. Existing evidence suggests that chemical recycling, which can result in a significant reduction in environmental burden of up to 60% compared to mechanical methods, can contribute to sustainability goals, particularly when combined with renewable energy sources and optimized technologies. The success of these technologies necessitates the establishment of appropriate infrastructure, incentivizing policies, and the development of value-added business models, thereby redesigning the entire production and recycling chain of plastics within an integrated and sustainable framework [70].

9.3 Biological and Enzymatic Recycling

Biological and enzymatic recycling of plastic waste represents a novel, sustainable, and environmentally friendly approach that has fundamentally transformed the management of polymeric waste. This method leverages the biological capabilities of microorganisms, enzymes, and even insect microbiota to effectively degrade plastics such as PET, PE, PS, PU, and PLA. At the heart of this process are enzymes like PETase and MHETase, extracted from the bacterium *Ideonella sakaiensis*, which break down the ester bonds of PET, converting it into recyclable monomers such as terephthalic acid and ethylene glycol. The incorporation of complementary enzymes such as lipases, esterases, and even proteinase K, along with enzyme cocktails, enhances the rate and scope of degradation. Furthermore, microbial consortia composed of degrading fungi and bacteria, such as *Bacillus subtilis*, *Aspergillus flavus*, and *Pseudomonas* sp., have demonstrated significant reductions in the weight of target polymers in various environments, including soil, water, and marine settings, within short timeframes. In this context, research utilizing advanced tools such as protein engineering and data mining databases like ProthermaDB has led to the development of enzymes with high thermal stability and efficiency, including glycosylated enzymes in eukaryotic hosts, which exhibit significantly enhanced performance and durability under operational conditions [71].

Additionally, the application of analytical methods, such as FTIR, SEM, GC-MS, and HPLC, enables the precise monitoring of the degradation process, the identification of intermediate products, and the optimization of environmental conditions, including pH, temperature, and humidity. Within the framework of a circular economy, enzymatic recycling offers the opportunity to reintegrate high-purity monomers directly into the production cycle, as evidenced by the company CARBIOS. Degraded PET is reused in the production of high-quality products, including food containers. Similarly, the degradation of PLA into lactide is being developed through enzymes such as lipase and cutinase. Despite challenges such as the inherent resistance of synthetic polymers like PE and PVC to biological degradation, slow degradation rates at an industrial scale, and the costs associated with biotechnology, emerging trends such as utilizing unconventional biological resources (e.g., insects), designing specific enzymes, creating synergistic multi-enzyme systems, and advancing bioremediation technologies outline a pathway toward sustainability for the effective and industrial recycling of plastic waste.

Achieving this goal requires the integration of microbiology, biotechnology, environmental engineering, and intelligent policymaking to position biological and enzymatic recycling as a cornerstone of sustainable waste management [72, 73].

9.4 Thermal Conversion Technologies

Thermal conversion technologies for municipal solid waste (MSW) to energy are emerging as effective and sustainable solutions in waste management and energy supply, encompassing four main branches: incineration, pyrolysis, gasification, and hydrothermal processing. In the realm of incineration, one significant research trend is the utilization of circulating fluidized bed (CFB) reactors, which, despite their advantages—such as high efficiency and lower pollutant emissions compared to traditional methods—face challenges including lower operating temperatures and the need for precise fuel preparation. Recent advancements have led to the development of internal circulating fluidized bed (ICFB) boilers, which facilitate faster combustion reactions, more uniform temperatures, and reduced emissions; a notable example of this technology is the successful operation at the Jeonju Paper plant in South Korea, with a capacity of 33 mW. Additionally, at more minor scales, the development of household and industrial incinerators with capacities of around 2 tons per hour, enhanced by optimally injected air, recovery of hot gases, and the incorporation of the Organic Rankine Cycle (ORC), has increased the thermal-to-electric conversion efficiency to up to 60%. The ORC cycle utilizes low-boiling-point organic fluids, such as R245fa, toluene, and R134a, to enhance the energy recovery process. Recent research has also focused on miniaturizing household incinerators using fluidized reactors and HHO gas (a mixture of hydrogen and oxygen generated from water electrolysis), which can achieve waste volume reductions of up to 90% at temperatures exceeding 1000 °C. In the area of pyrolysis, energy recovery and waste volume reduction through thermal decomposition in the absence of oxygen have been enhanced; designs that separate the drying chamber from the main reactor and recover waste heat have enabled the processing of high-moisture waste (up to 60%). Moreover, the use of microwave radiation to increase heating speed and uniformity, alongside the addition of activated carbon as an adsorbent, represents emerging trends in this domain. In pyrolysis, the application of catalysts such as zeolites (ZSM-5 and Y-zeolite) to accelerate the breakdown of polymer chains and increase aromatic compounds in oil products, metal oxides (MgO, CaO) to reduce oxygen and enhance gas production, iron-based catalysts to lower activation energy and produce combustible gases, and Ni/ZSM-5 combinations to increase hydrogen content and aromatize liquid products are key trends.

Furthermore, to reduce costs, natural and recycled catalysts such as natural zeolite, bentonite, and shell waste have gained attention. In the field of gasification, the use of catalysts has improved syngas composition, increased process yield, and reduced tar production. Effective catalysts include dolomite (which reduces tar and converts carbon to syngas), active metals such as nickel and iron (which increase hydrogen and reduce carbon deposits), combined catalysts like Rh/CeO_2 (which perform well at low temperatures), and alkaline catalysts such as sodium carbonate (which enhance gasification yield and reduce activation energy). The advanced plasma gasification technology, which utilizes electric arcs to achieve extremely high temperatures (up to 2500 °C), is particularly applicable in military vessels and marine lines for managing specific waste types, producing syngas with a high calorific value of over 9 mJ per cubic meter. Finally, hydrothermal processing of waste (HTMSW), conducted in closed reactors with high-pressure steam at temperatures ranging from 180 to 235 °C, has successfully increased the energy density of waste, reduced chlorine content, and improved combustion characteristics, resulting in solid fuels suitable for power plants or cement factories; an industrial example of this technology in Indonesia has a daily processing capacity of 50 tons of waste. Collectively, these innovative trends demonstrate that thermal conversion of waste to energy can serve as an efficient and environmentally friendly tool in waste management and sustainable energy supply, provided that the physical and chemical characteristics of the waste are precisely controlled, effective catalysts are employed, and advanced technologies are utilized. However, it is worth noting that only the biological and organic fractions of waste are recognized as renewable energy sources. At the same time, other materials, such as plastics and rubber, are deemed non-renewable and sometimes considered polluting. These technologies can have extensive applications in the formulation of industrial reports, research, and policy-making in the fields of waste management and renewable energy [74].

10 Challenges and Considerations

In the waste recycling process, we encounter a set of multifaceted challenges that can hinder the achievement of environmental and economic objectives in this field. On one hand, the need for substantial investments in advanced equipment and ongoing maintenance costs imposes significant financial pressures on projects. On the other hand, the technical complexities and high energy consumption of certain technologies create limitations for efficient operation. Furthermore, deficiencies and ambiguities in legal and regulatory frameworks, along with behavioral and cultural barriers within society, pose serious challenges to the acceptance and success of recycling processes. Table 5 outlines these challenges and their characteristics, offering a clearer perspective on the obstacles ahead.

11 Recommendations

Given the technical and economic complexities of advanced recycling technologies, it is recommended that municipalities and responsible entities adopt collaborative models and collective investment to mitigate the high costs associated with

Table 5 Challenges, explanations and considerations

Challenges	Explanations	Considerations
(A) High Capital and Maintenance Costs	• Advanced recycling devices and technologies, especially new technologies such as artificial intelligence, robotics, enzyme therapy and thermal conversion processes, require significant initial investment • Costs related to specialized maintenance, equipment upgrades and periodic services constitute a major part of the ongoing costs • Lack of sustainable financial resources and lack of appropriate economic support may cause the activity to stop or the quality of the recycling process to decline	• Before selecting advanced technologies, a careful cost-benefit assessment should be conducted to ensure the economic feasibility of the project. It is essential to design sustainable financing models such as joint ventures, low-interest bank facilities, or the use of government and international support funds • Developing regular and preventive plans for equipment repairs and upgrades, along with specialized training of human resources, can prevent unwanted cost increases and extend the useful life of technologies • Creating financial and legal incentives to attract private sector investors and facilitate the entry of new technologies are key priorities

(continued)

Table 5 (continued)

Challenges	Explanations	Considerations
(B) Technical Limitations and Energy Use	• New technologies in recycling are usually technically complex and require high expertise and appropriate infrastructure • Energy consumption in processes such as pyrolysis, gasification and thermal conversion is high and if clean energy sources are not used, negative environmental impacts may increase • The quality of input raw materials and the level of contamination in waste can reduce the performance of the recycling system and produce low-quality products	• Training of specialized personnel and improving the technical capabilities of operational teams, along with the development of technological and communication infrastructure, are prerequisites for the successful implementation of advanced technologies • Integrating recycling technologies with renewable energy sources such as solar and wind and utilizing smart energy management systems can reduce environmental impacts and increase productivity • Establishing waste control and pre-processing systems to remove pollutants and increase the quality of input materials will ensure the optimal performance of recycling technologies and the production of quality products
(C) Regulatory and Behavioral Barriers	• Inadequate, complex or unclear regulations can limit the development of new technologies in recycling • The lack of supportive policies and financial incentives, including legal facilities and exemptions, reduces the motivation to adopt new technologies • Public culture, how people separate waste and their level of participation in the recycling process are of great importance, and a lack of awareness or social resistance can reduce the effectiveness of recycling programs	• Reviewing and simplifying laws and regulations related to recycling and supporting new technologies, along with clarifying implementation processes, will facilitate the development of this area • Incentive policies such as tax exemptions, subsidies, credit facilities, and accreditation of advanced recycling projects will increase the motivation necessary to adopt new technologies. • Extensive public education programs and active community participation in waste separation and environmental protection, along with media campaigns and cooperation between various institutions, can reduce social resistance and strengthen effective participation

Source: Based on the authors' studies

establishing and maintaining innovative systems. Such collaborations can take the form of joint funds or innovative waste management service contracts, involving partnerships with the private sector, non-governmental organizations, and research institutions. Furthermore, a particular emphasis on specialized training for technical personnel and the development of clean energy infrastructures adjacent to recycling facilities will not only reduce energy consumption and environmental impacts but also enhance the productivity and quality of recycled products. Additionally, the utilization of smart data and continuous monitoring systems in managing recycling processes will enable accurate forecasting of potential issues and optimization of system performance. To institutionalize an advanced recycling culture and promote sustainable reuse in large cities, comprehensive public education programs should be implemented in conjunction with the use of digital technologies and social networks, aiming to increase public participation and mitigate behavioral resistance. Leveraging digital platforms to facilitate source separation of waste, providing transparent information about the positive impacts of recycling, and offering financial or non-financial incentives (such as tax discounts or certificates of appreciation) are also effective strategies. Moreover, the application of advanced technologies in the recycling of construction and electronic waste, coupled with legal coordination and the establishment of a sustainable supply chain for recycled materials, can play a crucial role in achieving a circular economy and mitigating environmental pressures.

12 Conclusion

Today, urbanization and urban development represent a distinctive phenomenon that has transformed the ability of urban managers to provide services into a significant challenge. The complex economic and technical changes that followed the Industrial Revolution, laying the foundation for urbanization and the development of contemporary cities, date back to the eighteenth century. These changes have resulted in profound alterations in the dimensions of cities and the population ratios of their residents. Based on conducted studies and references to relevant articles, it can be stated that, initially, one of the most critical issues concerning the intelligence of cities is the application of indicators and components aimed at creating and developing healthier, safer, more vibrant, and more comfortable environments. Smartening is an idea aimed at enhancing and improving the quality of life, regarded as a key concept in urban planning and coinciding with the significant growth of urban populations in contemporary times. Urban growth has influenced the expansion of urban facilities and services.

Thus, in the pursuit of improving the quality of life in urban spaces, a comprehensive review across various domains is required. In this regard, one of the most important factors for achieving the desired goal is the utilization of the potential of urban environments and spaces in the implementation of smart facilities and the establishment of interaction with their users, as a crucial element in the intelligence of cities, enabling objectives such as efficiency, responsiveness of urban spaces, and efforts to enhance both the qualitative and quantitative presence of people. Based on the conducted investigations, advanced technologies, particularly the components discussed in this research, play a crucial role in the development of smart cities and in enhancing the quality of life. These components not only contribute to the optimization of resources and public services but can also mitigate the existing challenges in urban management.

Acknowledgements In this section, we would like to express our gratitude to Professor Majumdar and all the editors for providing us with this opportunity.

References

1. Kitchin R (2014) The data revolution: Big data, open data, data infrastructures and their consequences. SAGE Publications. https://books.google.com/books/about/The_Data_Revolution.html?id=GfOICwAAQBAJ/
2. Bertot J, Gorham U, Jaeger P, Sarin L, Choi H (2014) Big data, open government and e-government: issues, policies and recommendations. Inf Polity 19:5–16. https://doi.org/10.3233/IP-140328/
3. Mirzaei R, Lakizadeh E (2024a) Designing the institutional framework of hybrid special economic zones (HEZ). In: Siew G, Allam Z, Cheshmehzangi A (eds) Diversity as catalyst: economic growth and urban resilience in global cityscapes. Urban sustainability. Springer, Singapore. https://doi.org/10.1007/978-981-97-8776-0_2/
4. Mirzaei R, Lakizadeh E (2024b) Developing a climate resilience framework for cities with an emphasis on innovative technology applications. In: Ghosh S, Majumdar S, Cheshmehzangi A (eds) Cities of tomorrow: urban resilience and climate change preparedness. Urban sustainability. Springer, Singapore. https://doi.org/10.1007/978-981-97-9658-8_14/
5. Choudhary R, Kumar A, Narayan S, Mishchenko I, Choudhury M, Bharadwaj LK et al (2025) Integrating fuzzy logic into life cycle assessment for sustainable municipal solid waste management in metro cities. Ecol Eng Environ Technol 26(7):316–331. https://doi.org/10.12912/27197050/205500
6. Litman T (2025) Smart congestion relief. Comprehensive evaluation of traffic congestion costs and congestion reduction strategies. Victoria Transport Policy Institute. https://www.vtpi.org/cong_relief.pdf/
7. Moore T (2010) The economics of cybersecurity: principles and policy options. Int J Crit Infrastruct Protect 3(3):103–117. https://doi.org/10.1016/j.ijcip.2010.10.002/
8. Gehl J (2010) Cities for people. Island Press. https://islandpress.org/books/cities-people#desc/
9. Hollands R (2008) Will the real smart city please stand up? City 12:303–320. https://doi.org/10.1080/13604810802479126/
10. Mirzaei R, Purohit S, Lakizadeh E (2024) Development of the framework for the application of cultural rights in smart urban governance. J Glob Resour 11:10.46587/JGR.2025.v11i01.001/
11. Batty M, Axhausen KW, Giannotti F, Pozdnoukhov A, Bazzani A, Wachowicz M et al (2012) Smart cities of the future. Eur Phys J Special Top 214:481–518. https://doi.org/10.1140/epjst/e2012-01703-3/
12. Caragliu A, Del Bo C, Nijkamp P (2011) Smart cities in Europe. J Urban Technol 18(2):65–82. https://doi.org/10.1080/10630732.2011.601117/
13. Harrison C, Donnelly I (2011) A theory of smart cities. In: 55th annual meeting of the International Society for the Systems Sciences 2011. https://www.researchgate.net/publication/228428752_A_Theory_of_Smart_Cities/

14. Zaman M, Abdelwahed S, Zohrabi N (2024) A review of IoT-based smart city development and management. J Smart Cities 7:1462–1500. https://doi.org/10.3390/smartcities7030061/
15. Litman T (2024) Smart transportation emission reduction strategies. Identifying truly optimal ways to conserve energy and reduce emissions. Victoria Transport Policy Institute. https://www.vtpi.org/ster.pdf/
16. Balasubramaniam A, Paul A, Hong W-H, Seo H, Kim JH (2017) Comparative analysis of intelligent transportation systems for sustainable environment in smart cities. Sustainability 9(7):1120. https://doi.org/10.3390/su9071120/
17. Anbukkarasi S, Dhivyaa CR (2024) AI techniques for future smart transportation. In: Artificial intelligence for future intelligent transportation. Apple Academic Press, pp 243–268. https://www.taylorfrancis.com/chapters/edit/10.1201/9781003408468-10/ai-techniques-future-smart-transportation-anbukkarasi-dhivyaa/
18. Deveci M, Yaşar I, Pamucar D, Zaidan A, Wen X, Gupta BB (2023) Evaluation of Cooperative Intelligent Transportation System scenarios for resilience in transportation using type-2 neutrosophic fuzzy VIKOR. Transp Res A Policy Pract 172:103666. https://doi.org/10.1016/j.tra.2023.103666/
19. Oladimeji D, Gupta K, Kose NA, Gundogan K, Ge L, Liang F (2023) Smart transportation: an overview of technologies and applications. Sensors 23(8):3880. https://doi.org/10.3390/s23083880/
20. Bagga P, Das AK, Rodrigues J (2023) Bilinear pairing-based access control and key agreement scheme for smart transportation. Cyber Secur Appl 1:100001. https://doi.org/10.1016/j.csa.2022.100001/
21. Shi H, Wang H, Ma R, Hua Y, Song T, Gao H, Guan H (2022) Robust searching-based gradient collaborative management in intelligent transportation system. ACM Trans Multimed Comput Commun Appl 20. https://doi.org/10.1145/3549939/
22. Zhang J, Li S, Wang Y (2023) Shaping a smart transportation system for sustainable value co-creation. InfSyst Front 25:365–380. https://doi.org/10.1007/s10796-021-10139-3/
23. Bai L, Xie L, Li C, Yuan S, Niu D, Wang T, Yang Z, Zhang Y (2023) The conceptual framework of smart TOD: an integration of Smart City and TOD. Land 12(3):664. https://doi.org/10.3390/land12030664/
24. Philippopoulos PI, Koutkias V, Fafoutis X (2007) ToD: an intelligent location-based transport management system. In: 2007 IEEE intelligent vehicles symposium, Istanbul, Turkey, pp 793–798. https://doi.org/10.1109/IVS.2007.4290213/
25. Moheb-Alizadeh H, Sadeghi AH, Jaunich MK, Kemahlioglu-Ziya E, Handfield RB (2023) Reverse logistics network design to estimate the economic and environmental impacts of take-back legislation: a case study for E-waste management system in Washington State. https://arxiv.org/abs/2301.09792/
26. Tian H, Yan M, Zhou J, Wu Q, Tiong YW, Lam HT, Zhang J, Tong YW (2023) A closed loop case study of decentralized food waste management: system performance and life cycle carbon emission assessment. Sci Total Environ 899:165783. https://doi.org/10.1016/j.scitotenv.2023.165783/
27. Mersico L, Aureli S, Foschi E (2025) Exploring the role of digital platforms in promoting value co-creation: evidence from the Italian municipal solid waste management system. Sustain Account Manag Policy J. https://doi.org/10.1108/SAMPJ-03-2024-0317/
28. Paletta A, Aureli S, Mersico L (2023) Digital Platform for Knowledge sharing and value co-creation in waste recycling: the case of Alpha. https://www.researchgate.net/publication/371609276_Digital_Platform_for_Knowledge_sharing_and_value_co-creation_in_waste_recycling_the_case_of_Alpha/
29. Ferriz-Papi JA, Lee A, Alhawamdeh M (2024) Examining the challenges for circular economy implementation in construction and demolition waste management: a comprehensive review using systematic methods. Buildings 14(5):1237. https://doi.org/10.3390/buildings14051237/
30. Saka A, Taiwo R, Saka N, Oluleye B, Dauda J, Akanbi L (2024) Integrated BIM and machine learning system for circularity prediction of construction demolition waste. https://doi.org/10.48550/arXiv.2407.14847/

31. Wu Z, Pei T, Bao Z et al (2025) Utilizing intelligent technologies in construction and demolition waste management: from a systematic review to an implementation framework. Front Eng Manag 12:1–23. https://doi.org/10.1007/s42524-024-0144-4/
32. United-BIM (2024) BIM services. https://www.united-bim.com/bim-services/
33. New York State (2011). Business plan for centralized access to consistent cadastral data. https://www.fgdc.gov/grants/2010CAP/InterimFinalReports/171-10-4-NY-FinalReport.pdf/
34. Doe J (2023) Implementing transit-oriented-development in low density urban areas. The case study of Turin-Torre Pellice. Master's thesis, Politecnico di Torino. https://webthesis.biblio.polito.it/20995/1/tesi.pdf/
35. GTT (2022) Public transport operations in Turin: integration and efficiency. Turin Transport Company. https://www.gtt.torino.it/en/public-transport/
36. Regional Government of Piedmont (2021) Regional planning and funding for transit-oriented development. Piedmont Regional Government. https://www.regione.piemonte.it/web/sites/default/files/media/documenti/2021-12/oecd_piedmont.pdf/
37. Siemiatycki M (2011) Urban transportation public–private partnerships: drivers of uneven development? Environ Plan A 43:1707–1722. https://doi.org/10.1068/a43572/
38. Creß C, Bing Z, Knoll A (2023) Intelligent transportation systems using roadside infrastructure: a literature survey. IEEE Trans Intell Transp Syst:1. https://doi.org/10.1109/TITS.2023.3343434/
39. Dmitrieva E, Pathani A, Pushkarna G, Acharya P, Rana M, Surekha P (2024) Real-time traffic management in smart cities: insights from the traffic management simulation and impact analysis. BIO Web Conf 86. https://doi.org/10.1051/bioconf/20248601098/
40. Rivera R, Amorim M, Reis J (2021) Public transport systems and its impact on sustainable smart cities: a systematic review. https://doi.org/10.1007/978-3-030-78570-3_3/
41. European Commission (2013) Guidelines on smart ticketing (Urban ITS Expert Group). European Union. https://transport.ec.europa.eu/document/download/f545c05c-371b-4122-bfa4-9eb5ba8b1f6f_en?filename=2013-urban-its-expert_group-guidelines-on-smart-ticketing.pdf/
42. SUMPs-Up (2019) Turin's transformative Sustainable Urban Mobility Plan inspires a prestigious EU project to pay a visit. Turin press release. https://sumps-up.eu/fileadmin/user_upload/SUMPs_in_Action/SUMPs-Up_-_Turin_Press_Release_final.pdf/
43. National Bureau of Statistics of China (2010) Statistical yearbook of China 2010. China Statistics Press. https://www.stats.gov.cn/sj/ndsj/2010/indexeh.htm/
44. Jiang T-t, Pan JF, Pu X-M, Wang B, Pan J-J (2015) Current status of coastal wetlands in China: degradation, restoration, and future management. Estuar Coast Shelf Sci 164. https://doi.org/10.1016/j.ecss.2015.07.046/
45. United Nations Environment Programme (2011) Guidelines for integrated management of coastal and marine areas. https://www.unep.org/resources/report/guidelines-integrated-management-coastal-and-marine-areas/
46. SIDA (The Swedish International Development Cooperation Agency) (2011) Spatial planning in the coastal zone of the East Asian seas region: integrating emerging issues and modern management approaches. https://climatechange.issuelab.org/resources/17770/17770.pdf/
47. Ministry of Ecology and Environment of the People's Republic of China (2016) Twelfth Five-Year Plan for ecological and environmental protection. http://english.mee.gov.cn/Resources/Plans/National_Fiveyear_Plan/201606/P020160601356854927248.pdf/
48. National Development and Reform Commission of China (2021) Outline of the Fourteenth Five-Year Plan for national economic and social development of the People's Republic of China. https://en.ndrc.gov.cn/policies/202203/P020220315511326748336.pdf/
49. National Marine Environmental Forecasting Center (2022) Development of intelligent monitoring systems for coastal management. http://www.nmefc.cn/
50. Government of India (2016) Sagarmala: national perspective plan. Ministry of Ports, Shipping and Waterways. https://sagarmala.gov.in/about-sagarmala/national-perspective-plans/
51. Ministry of Earth Sciences, Government of India (2019) Draft national policy on the blue economy. https://incois.gov.in/documents/Blue_Economy_policy.pdf/

52. World Bank (2020) The blue economy: a sustainable approach to ocean and coastal development. https://documents1.worldbank.org/curated/en/965641473449861013/pdf/AUS16344-REVISED-v1-BlueEconomy-FullReport-Oct3.pdf/
53. National Institute of Ocean Technology (2021) Integrated coastal surveillance system: a technological approach for coastal security. https://www.niot.res.in/
54. International Solid Waste Association (2024) Global waste management outlook 2024: beyond an age of waste, turning rubbish into a resource. https://wedocs.unep.org/bitstream/handle/20.500.11822/44939/global_waste_management_outlook_2024.pdf?sequence=1&isAllowed=y/
55. Maalouf A, Mavropoulos A (2023) Re-assessing global municipal solid waste generation. Waste Manag Res 41(4):936–947. https://doi.org/10.1177/0734242X221074116/
56. Choudhury M, Sahoo S, Samanta P, Tiwari A, Tiwari A, Chadha U et al (2022) COVID-19: an accelerator for global plastic consumption and its implications. J Environ Public Health 2022(1):1066350. https://doi.org/10.1155/2022/1066350
57. Goswami S, Choudhury M, Agarwal S (eds) (2025a) Electronic waste: impact on health, animals, and the environment. CRC Press. https://doi.org/10.1201/9781003582311
58. Goswami S, Choudhury M, Samanta P (eds) (2025b) Micro-nano plastics exposure, environmental degradation and public health crisis. Springer Singapore. https://doi.org/10.1007/978-981-96-5047-7
59. Roy and Choudhury (2022) Bioeconomy and green plastic production. In: Baskar C, Ramakrishna S, Daniela La Rosa A (eds) Encyclopedia of green materials. Springer, Singapore. https://doi.org/10.1007/978-981-16-4921-9_189-
60. Bhardwaj LK, Rath P, Jain H, Choudhury M (2025) Exploring the effects of e-waste on soil, water quality and human health. Discov Civil Eng 2(1):12. https://doi.org/10.1007/s44290-025-00167-2
61. Torralba J, Iriarte D, Tourret D, Meza J (2024) Using multicomponent recycled electronic waste alloys to produce high entropy alloys. Intermetallics 164:108128. https://doi.org/10.1016/j.intermet.2023.108128/
62. Vuppaladadiyam SSV, Vuppaladadiyam AK, Sahoo A, Urgunde A, Murugavelh S, Šrámek V, Pohořelý M, Trakal L, Bhattacharya S, Sarmah AK, Shah K, Pant KK (2024) Waste to energy: trending key challenges and current technologies in waste plastic management. Sci Total Environ 913:169436. https://doi.org/10.1016/j.scitotenv.2023.169436/
63. Choudhury M, Sharma A, Pervez A, Upadhyay P, Dutta J (2021) Growing menace of microplastics in and around the coastal ecosystem. In: Coastal ecosystems: environmental importance, current challenges and conservation measures. Springer International Publishing, Cham, pp 117–137. https://doi.org/10.1007/978-3-030-84255-0_6
64. Choudhury M, Roy P (2025) Challenges with microplastic pollution in the regime of UN sustainable development goals. World Dev Sustain 6:100216. https://doi.org/10.1016/j.wds.2025.100216
65. Filho W, Barbir J, May J, May M, Swart J, Yang P, Dinis M, Aina Y, Bettencourt S, Charvet P, Azadi H (2025) Towards more sustainable oceans: a review of the pressing challenges posed by marine plastic litter. Waste Manag Res. https://doi.org/10.1177/0734242X251313927/
66. Pilapitiya PGC, Ratnayake A (2024) The world of plastic waste: a review. Clean Mater 11:100220. https://doi.org/10.1016/j.clema.2024.100220/
67. Maghmoumi A, Marashi F, Houshfar E (2020) Environmental and economic assessment of sustainable municipal solid waste management strategies in Iran. Sustain Cities Soc 59:102161. https://doi.org/10.1016/j.scs.2020.102161/
68. Tan TN, Samanta P, Choudhury M, Kalimuthu K, Dhar S, Kularajasingam J (2025) Environmental degradation of plastics: understanding the mechanisms. In: Goswami S, Choudhury M, Samanta P (eds) Micro-nano plastics exposure, environmental degradation and public health crisis, Sustainable development goals series. Springer, Singapore. https://doi.org/10.1007/978-981-96-5047-7_2
69. Duan Y (2025) Intelligent waste classification system integrating deep learning and robotic arm technology. In: Proceedings of the 2024 eighth international conference on electronic information technology and computer engineering, pp 579–585. https://doi.org/10.1145/3711129.3711229/

70. Luo H, Tyrrell H, Jingyang B, Muazu R, Long X (2024) Fundamental, technical and environmental overviews of plastic chemical recycling. Green Chem. https://doi.org/10.1039/D4GC03127J/
71. Choudhury M, Rajpal A, Goswami S, Chakravorty A, Raghavan V (eds) (2024) Analytical case studies on municipal and biomedical waste management: perspectives on sustainable development goals. CRC Press. https://doi.org/10.1201/9781003499695
72. Akram MA, Savitha R, Kinsella GK, Nolan K, Ryan BJ, Henehan GT (2024) Microbial and enzymatic biodegradation of plastic waste for a circular economy. Appl Sci 14(24):11942. https://doi.org/10.3390/app142411942/
73. Chakraborty D, Chatterjee S, Althuri A, Palani SG, Venkata Mohan S (2023) Sustainable enzymatic treatment of organic waste in a framework of circular economy. Bioresour Technol 370:128487. https://doi.org/10.1016/j.biortech.2022.128487/
74. Kasiński S, Dębowski M (2024) Municipal solid waste as a renewable energy source: advances in thermochemical conversion technologies and environmental impacts. Energies 17(18):4704. https://doi.org/10.3390/en17184704/

Resource Recovery in Urban Settings: Developing Circular Supply Chains for Sustainable Cities

Saurav Narayan, Puneet Pathak, Basona Khatun, Rakesh Choudhary, Michael Troilo, Chandrkant Raj, and Viola Vambol

Abstract Rapid urbanization is straining material, water, and energy systems; this chapter positions circular supply chains as the mechanism to reclaim these flows and build regenerative, climate-resilient cities. We synthesize principles—closing loops, cascading use, regenerative design, design for disassembly, and urban mining—into a scheme that links four recovery streams (organics, water, energy, materials/products) with enabling capacities: reverse logistics, digital traceability (IoT, AI, blockchain), distributed recovery hubs, and inclusive networks spanning municipalities, businesses, informal workers, and citizens. Governance levers (policy coherence, urban planning tools, PPPs, and blended/green finance) and performance dashboards (MFA, urban metabolism indicators) operationalize and scale interventions. Anticipated outcomes include reduced landfill dependence, lower emissions, and decreased virgin extraction, alongside the creation of green jobs and benefits to resilience. We also map persistent barriers—institutional gaps, technical and infrastructure limitations, cultural resistance, and market immaturity—and propose pathways using modular design, product passports, and data-driven decision-making to mainstream circularity across urban systems through adaptive governance and cross-sector partnerships worldwide.

S. Narayan (✉) · P. Pathak · B. Khatun
Department of Law, School of Legal Studies, Central University of Punjab, Bathinda, Punjab, India

R. Choudhary
Department of Civil Engineering, National Institute of Technology Delhi, New Delhi, India

M. Troilo
Busch School of Business, Catholic University of America, Washington, DC, USA

C. Raj
College of International Education, Jinzhou Medical University, Jinzhou City, Liaoning Province, People's Republic of China

V. Vambol
Department of Applied Ecology and Nature Management, National University Yuri Kondratyuk Poltava Polytechnic, Poltava, Ukraine

S. Majumdar et al. (eds.), *Sustainable Urban Future*, Urban Sustainability,
https://doi.org/10.1007/978-981-95-2976-6_10

Keywords Circular economy · Circular supply chains · Urban resource recovery · 5R strategies · Reverse logistics

1 Introduction

Resource recovery in the urban context refers to the process of extracting valuable materials, energy, and water from municipal waste, wastewater, and stormwater systems, aligning with principles of the circular economy (CE). It involves the application of advanced technologies and sustainable practices to reduce environmental impact while maximizing reuse potential [1–3]. This approach transforms urban waste streams into secondary resources, promoting environmental pliability and economic efficiency [4, 5].

Resource recovery not only mitigates pollution but also supports sustainable urban development by reintegrating recovered outputs into the production cycle. The role of CE principles in sustainable urban development is central to transforming cities into resilient, low-waste, and resource-efficient systems [6]. CE principles aim to keep resources in use for as long as possible, extract maximum value from them, and regenerate natural systems, thereby directly supporting the Sustainable Development Goals [7, 8].

CE principles play a vital role in sustainable urban development by promoting waste reduction and resource efficiency, designing pollution out, and adopting regenerative practices [9]. These principles support the creation of resilient urban systems by encouraging the reuse and recycling of materials, minimizing environmental impact, and returning valuable nutrients to the environment. They also contribute to supporting local economies and creating green jobs [10] by promoting circular business models, such as repair, remanufacturing, and sharing economies. Furthermore, circular approaches help cities enhance resilience to supply chain disruptions [11] and climate change while driving innovation and systems thinking through integrated urban planning and innovative technologies [12]. These principles contribute to improving the quality of life [13] by creating cleaner, more inclusive, and resource-conscious urban environments.

Circular supply chains play a crucial role in mitigating urban environmental challenges by transitioning from linear '*take-make-dispose*' models to regenerative systems that prioritize resource efficiency and waste minimization [14]. In densely populated urban settings, these supply chains facilitate the continuous circulation of materials through reuse, remanufacturing, and recycling [15], thereby significantly reducing dependence on landfills, carbon emissions, and raw material extraction [16]. Moreover, circular supply chains facilitate localized production and shorter logistics networks, resulting in reduced environmental footprints and enhanced urban resilience [17, 18]. By embedding sustainability and innovation into product life cycles, circular supply chains facilitate the development of low-carbon, resource-smart cities that align with long-term environmental and economic goals [19].

2 Principle of Circular Supply Chain in Cities

Closing loops means maximizing the use of resources by encouraging practices such as recycling, remanufacturing, and reuse. This ensures that materials and products are returned to the supply chain rather than being discarded [20]. In urban settings, this approach helps retain valuable materials within the city and decreases the reliance on new resource extraction [21]. *Cascading use* refers to the sequential use of materials across different applications, maximizing their value at each stage before they are discarded or used for energy recovery [22]. This approach helps minimize the need for raw materials and prolongs the usefulness of existing resources [23]. *Regenerative design* focuses on creating systems that actively restore and strengthen both ecological and social environments. It prioritizes design strategies and resource choices that renew natural systems and support long-term resilience, rather than simply minimizing damage [24]. In cities, this can include initiatives such as green rooftops and urban farming that reintroduce biodiversity, as well as the use of recycled or renewable materials in public infrastructure construction [25]. *The Designing for Disassembly and Remanufacturing* approach involves creating products and buildings in a manner that allows them to be easily dismantled, repaired, upgraded, or repurposed after their initial use [26]. In urban areas, it facilitates efficient reuse and remanufacturing, thereby reducing waste and the need for new materials [27]. *Urban mining* refers to the recovery of valuable materials from buildings, electronic waste, infrastructure, and other discarded urban resources, treating the city itself as a source of raw materials [28]. This practice decreases reliance on newly extracted resources, shortens supply chains, and helps reduce environmental impact [29].

3 The Urban Resource Challenges

3.1 Rapid Urbanization and Its Impact on Resource Consumption

Rapid urbanization significantly intensifies urban resource consumption by accelerating demand for materials, energy, water, and infrastructure. Over 75% of global natural resource use now occurs in cities, which also generate approximately 50% of worldwide waste and 60–80% of greenhouse gas emissions [30]. This surge in urban expansion leads to increased extraction of non-renewable resources, escalating energy consumption for housing, transportation, and industry, and enlarging waste streams that strain municipal waste management systems [31]. Consequently, rapid urbanization amplifies environmental pressures and underscores the urgent need for sustainable resource management strategies [32].

Fig. 1 Construction and demolition waste piled at a building site, including bricks, concrete, tiles, metal, and plastic debris at Bathinda, Punjab, India. (Photo by: Saurav Narayan)

3.2 Types of Urban Waste

Rapid urbanization has led to diverse types of urban waste, including **solid waste** (municipal refuse such as plastics, paper, and metals), **wastewater** (sewage and greywater), **construction and demolition debris** Fig. 1 (bricks, concrete, and other building materials), **organic waste** (food scraps and green residues), and **e-waste** (discarded electronics containing hazardous components) [33–39]. Proper categorization is essential for designing effective circular management strategies tailored to each waste stream, especially in regional contexts where community bin practices vary significantly. Without this differentiation, cities face overburdened landfills, environmental contamination, and health risks, highlighting the need for targeted recovery and recycling interventions [38].

3.3 Linear vs. Circular Resource Flows

Table 1 compares the linear and circular economy models across key aspects, including resource use, waste generation, economic focus, environmental impact, and urban practices. It highlights how the circular economy emphasizes sustainability through resource regeneration and waste minimization, in contrast to the traditional linear approach, which is centered on extraction, consumption, and disposal.

Table 1 Comparison of linear economy v. circular economy

Aspect	Linear economy (take-make-waste)	Circular economy (closed loops)
Resource Use	Extracts raw materials, uses, then discards [40]	Reuses, recycles, and regenerates resources [41]
Waste Generation	High, with most materials ending as waste [42]	Minimized, with waste seen as a resource [43]
Economic Model	Focus on short-term consumption [44]	Focus on long-term value retention [45]
Environmental Impact	Significant pollution, depletion of resources [46]	Reduced emissions, lower resource extraction [47]
Urban Example	Landfilling, incineration, linear infrastructure [48]	Recycling, composting, urban mining, sharing economy [49]

Table 2 Strategies of resource recovery

Strategy	Methods	Benefits	Reference
Organic Waste	Composting, anaerobic digestion, biochar, insect composting	Waste reduction, energy, soil, employment opportunities	[34, 50]
Water Recovery	Rainwater harvesting, greywater recycling, green infra	Water conservation, pollution control, ecosystem support	[51]
Energy Recovery	WtE tech, landfill gas, biogas, planning integration	Energy production, emissions cuts, economics	[52]
Materials and Product Recovery	Conventional recycling, urban mining, deconstruction	Resource conservation, circular economy, innovation	[53]

4 Resource Recovery Strategies in Urban Areas

The four fundamental areas of resource recovery in cities are: Organic Waste to Resource, Water Recovery, Energy Recovery, and Materials and Product Recovery as explained in Table 2.

5 Bibliometric Insights into Resource Recovery and Circular Supply Chains (2015–2024)

The bibliometric analysis, spanning 2015 to 2024 and based on Scopus data, reveals a significant upward trajectory in research output related to resource recovery and circular supply chains. Figure 2 highlights this growth, with the number of publications increasing from 33 in 2015 to an impressive 557 in 2024, totaling 2226 scholarly documents. This sharp increase reflects growing academic and industrial interest in sustainability, circular economy practices, and the efficient use of resources. Complementing this, Fig. 3, the thematic map, categorizes research into four quadrants based on relevance and development. Motor themes, such as the circular economy and sustainable development, are both central and mature, forming the intellectual core of the field. Niche themes, such as biomass and nonhuman factors, are well-developed but less integrated with mainstream topics. Emerging or declining themes, such as Industry 4.0 and closed-loop supply chains, are either in an early stage or require further academic integration. In contrast, basic themes like food waste and human-related topics are foundational but underdeveloped, indicating areas ripe for future research.

Figure 4 and Table 3 collectively portray the international collaboration landscape, illustrating how circular supply chain research has evolved into a global endeavor. The United Kingdom leads in collaborative efforts with 365 co-authored publications, followed by China (237) and the United States (207), all showing robust cross-continental partnerships. India, Italy, Germany, and Australia also emerge as significant contributors, helping bridge global knowledge exchange. The data highlights the pivotal role of European countries and the increasing contributions from the Global South, with growing involvement from nations such as Brazil, Indonesia, and South Africa. These figures collectively illustrate not only the academic momentum behind circular economy research but also the deepening international networks needed to tackle global sustainability challenges through cooperative innovation and policy-driven frameworks (Fig. 5).

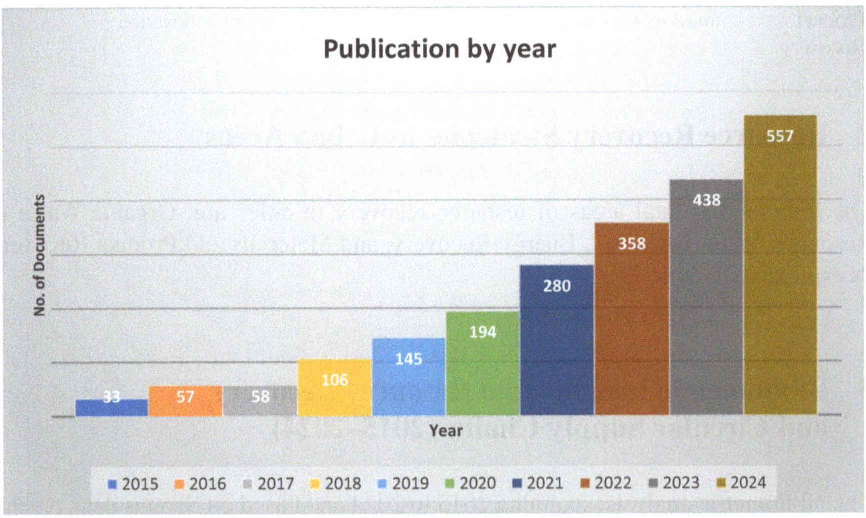

Fig. 2 Year wise Publication (2015–2024)

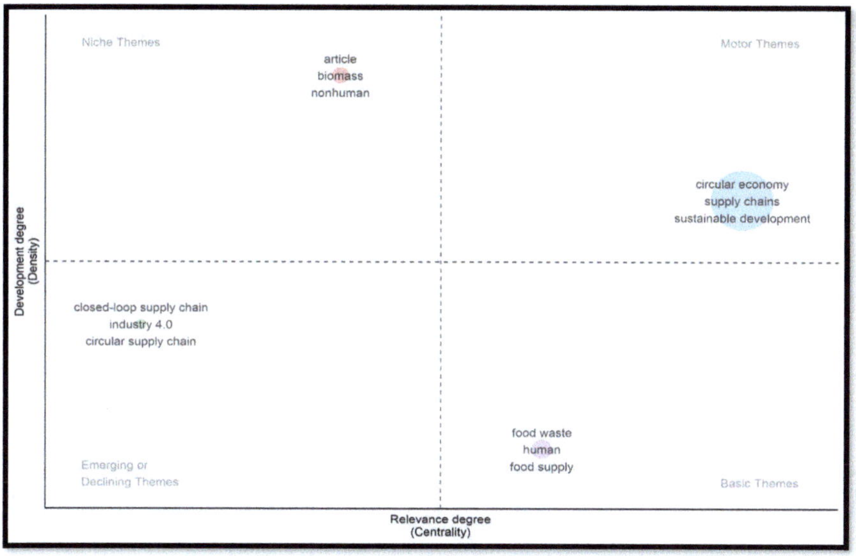

Fig. 3 Thematic map analysis of research on resource recovery and circular supply chains

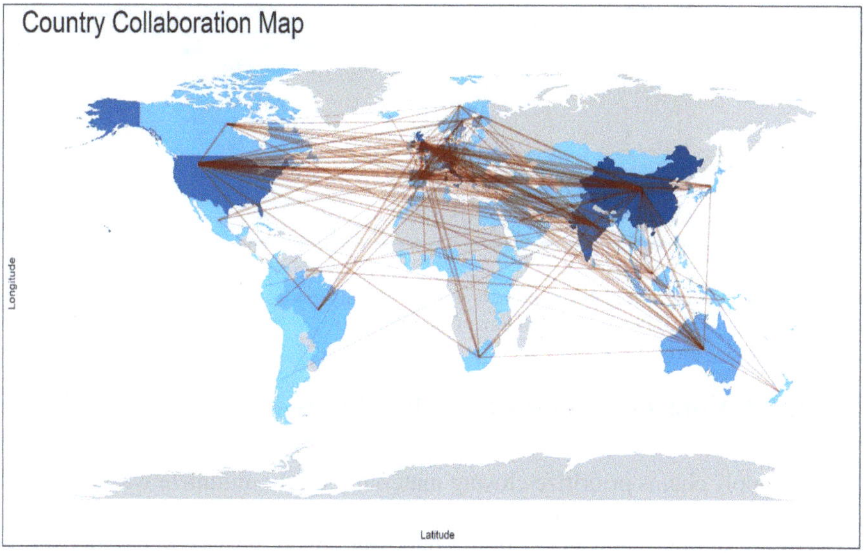

Fig. 4 Global research collaboration on resource recovery and circular supply chains

Table 3 Top collaborating countries

Rank	Country	Total collaborations
1	United Kingdom	365
2	China	237
3	United States	207
4	India	193
5	Italy	154
6	Germany	144
7	Australia	142
8	Netherlands	123
9	Spain	117
10	France	115

Fig. 5 An example of sustainable approach and ecofriendly environmental practices at Chandubi Wetland, Assam, India. (Photo by: Moharana Choudhury)

6 Foundations of Circular Supply Chains

Circular supply chains prioritize closing material loops to minimize waste, enhance resource efficiency, and promote the regeneration of natural systems [54]. In contrast to traditional supply chains that follow a one-way flow of resources, circular supply chains aim to maintain products, components, and materials in continuous use at their highest value.

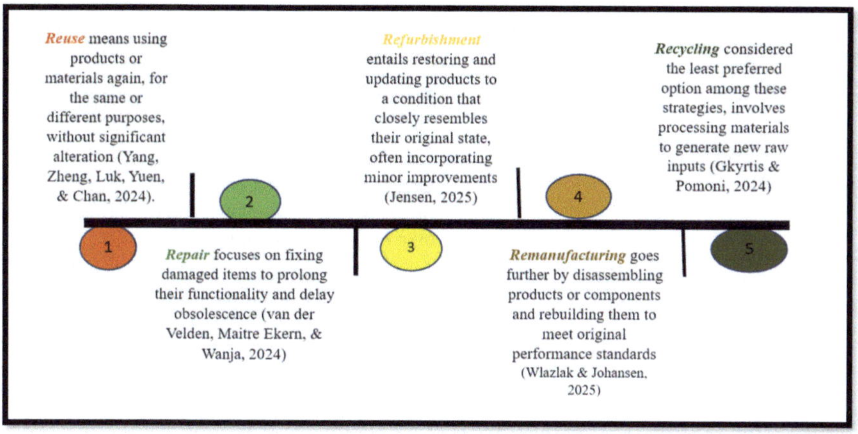

Fig. 6 Key concept of 5R. *Source* (Yang et al. [55], van der Velden et al. [56], Jensen [57], Wlazlak and Johansen [58], Gkyritsis and Pomoni [59])

6.1 Key Concepts: The 5Rs

In the context of circular supply chains, several strategies are employed to extend the lifecycle of products and materials while reducing environmental impact (Fig. 6).

6.2 Components of Circular Supply Chains in Urban Environments

Circular supply chains in urban contexts are characterized by dynamic, interconnected systems that support sustainable resource use and material recovery [60]. Distributed networks involving both local and global stakeholders facilitate the continuous circulation of products and materials, enhancing system resilience while minimizing environmental impact [61]. These networks are supported by multidirectional flows of information, goods, and capital, which enable efficient tracking, disassembly, and reintegration of products into the supply chain, especially within spatially intelligent wastewater and sewer systems that align with circular economy goals [34]. Circular inputs, such as recycled, renewable, or non-virgin materials, are increasingly incorporated into production processes to reduce dependency on finite resources [61]. Reverse logistics systems play a crucial role by enabling the collection, sorting, and redirection of used products for reuse, refurbishment, remanufacturing, or recycling [62]. Additionally, product life cycle management ensures that materials are strategically handled from the design phase through end-of-life to maximize recovery and reuse [63]. However, failure to recover and safely manage complex waste streams such as electronic waste can have severe consequences for

Table 4 Stakeholder contributions to circular supply chains

Who	What they do in a circular supply chain
Municipalities	Make rules, build things like recycling centers, and help manage how waste and resources move through the city [66]
Businesses	Create products that can be reused or recycled, invest in new technology, and run programs to take back or fix old items [67]
Informal Sectors (waste pickers, small-scale recyclers, and scrap dealers)	Collect and sort waste materials, often doing important jobs that official systems don't cover [68]
Citizens	Help by sorting waste, returning used products, reusing items, and choosing products that are better for the environment [69]

urban biodiversity and ecological health, as highlighted by recent findings on e-waste toxicity and its cascading effects on wildlife [64]. Lastly, waste valorization processes convert unavoidable by-products into valuable resources, supporting closed-loop systems and promoting overall sustainability within urban economies [65].

6.3 Stakeholders

Table 4 highlights the shared responsibility in creating a circular supply chain in cities. It demonstrates how various groups from government to individuals must collaborate. Each plays a unique part, whether through leadership, innovation, or everyday actions, making the system efficient, sustainable, and inclusive for long-term urban resilience.

6.4 Technologies Enabling Circular Supply Chains

Digital technologies play a pivotal role in enabling circular supply chains, especially within the complexity of urban environments. The Internet of Things (IoT) enables the real-time tracking of products and materials while monitoring key conditions such as temperature and location, to support optimized reuse, repair, and recycling [70]. Blockchain technology provides secure, transparent, and tamper-proof records of product origin, movement, and status, thereby enhancing traceability, ensuring regulatory compliance, and streamlining reverse logistics operations [71]. Artificial Intelligence (AI) contributes by analyzing vast datasets to forecast maintenance needs, optimize inventory management, and uncover opportunities for material recovery and reuse [72]. Collectively, these technologies provide the transparency, traceability, and operational efficiency necessary to scale circular supply chains effectively and sustainably.

7 Measuring Urban Circularity and Recovery Performance

Evaluating how well cities recover and manage resources relies on specific metrics and KPIs, such as rates of material reuse, recycling efficiency, and waste diversion. Tools such as urban metabolism dashboards, circular economy indicators, and material flow analysis (MFA) help track material flows—including inputs, usage, and outputs—providing valuable insights for identifying areas for improvement. Despite the usefulness of these tools, cities often face challenges related to data quality, consistency, and the lack of standardized methodologies, which can limit comparability and hinder informed decision-making [73]. However, geospatial analysis frameworks have shown promise in optimizing such evaluations for wastewater systems (Choudhary et al. 2025). Still, progress in data visualization and analytical platforms is making it increasingly feasible for urban areas to assess and improve their circularity in a more structured and effective way [74, 75].

8 Policy, Governance, and Financing

As noted earlier, creating a circular economy requires a coordinated effort and the allocation of resources. In this section, we discuss the essential elements of policy, governance, and financing that support the creation and maintenance of circular supply chains. These include policy frameworks, urban planning, and public-private partnerships (PPPs).

8.1 Matrix of Urban Circularity

The matrix of urban circularity offers a structured framework for cities to evaluate and improve their transition toward a circular economy by systematically analyzing material flows, stocks, and related indicators across urban systems. It combines material flow accounting with the monitoring of resource inputs, outputs, and waste across sectors, as well as material stock accounting, with a particular emphasis on building materials, which represent the largest share of urban stocks. By examining both direct and indirect circularity indicators, the matrix highlights key trends, supports comparisons with national and EU targets, and identifies opportunities to close resource loops through practices such as reuse and recycling [76]. This framework ultimately promotes more efficient use of resources, lowers environmental impacts, and aids urban decision-makers in crafting effective circular strategies [77].

8.2 Policy Frameworks for Circular Cities

The EU Circular Economy Action Plan (CEAP) serves as a comprehensive policy framework that guides the transition toward circular cities by promoting sustainable resource use and systemic change [78]. Central to the CEAP is a lifecycle approach, addressing every stage of a product's journey from design to end-of-life to support circular processes and responsible consumption [79]. The plan incorporates both legislative and non-legislative measures, including waste reduction targets, landfill diversion strategies, and requirements for the separate collection of textiles and biowaste [80, 81]. A sector-specific focus ensures targeted action in high-impact areas, including electronics, batteries, packaging, plastics, textiles, construction, and food systems [82, 83]. To drive innovation and support economic transformation, the CEAP has mobilized over EUR 10 billion in public funding, with a focus on research, development, and job creation. Its primary objectives are to establish sustainable products as the norm, reduce total waste—particularly non-recycled municipal waste by 50% by 2030, empower consumers and public institutions, promote circular practices in cities and regions, and position the EU as a global leader in circular economy implementation [84].

8.3 Urban Planning for Circularity

Table 5 is a table outlining key urban planning tools designed to support circular economy principles in city development.

Table 5 Urban planning tools for circularity

Tool	Description
Zoning Regulations [85]	Enable resource sharing, co-living, and protect local character
Economic Incentives [86]	Reform property taxes, introduce tradable land-use certificates, support brownfield renewal
Integrated Action Plans [86]	Package of measures for redevelopment, stakeholder roles, financing, and performance review

8.4 Public-Private Partnerships and Innovation Ecosystems

Public-Private Partnerships (PPPs) are essential for advancing circular economy initiatives by leveraging the strengths of both sectors [87]. These partnerships help distribute investment and operational risks, enhancing the financial attractiveness of circular projects. By combining public regulatory oversight and funding with private sector capital, expertise, and innovation, PPPs accelerate the adoption of new technologies and improve operational efficiency. Moreover, they foster long-term, stable commitments crucial for executing complex, multi-year projects. Beyond financial collaboration, PPPs play a key role in creating innovation ecosystems, where cooperation among public agencies, private enterprises, academia, and civil society drives systemic transformation and supports the adoption of circular business models and infrastructure [88].

8.5 Financing Models for Resource Recovery Infrastructure

Financing circular infrastructure, particularly for resource recovery, requires a diverse mix of financial instruments and strategies [89]. Blended finance models merge public, private, and philanthropic resources to reduce risk and mobilize substantial capital [90]. PPPs are commonly used to develop large-scale infrastructure such as recycling facilities and wastewater treatment plants. Additionally, green bonds and loans are increasingly issued by municipalities and private entities to fund sustainable waste management and energy recovery projects. Development banks also offer concessional and syndicated loans on favorable terms to attract private investment. Ensuring long-term project viability depends on revenue diversification, with income derived from user fees, the sale of recovered resources, energy generation, Green technology approaches and ecosystem services [91, 92]. Key to these financing approaches is a supportive policy environment and sustainable business models that address both capital and operational costs. Together, integrated policies, forward-thinking urban planning, collaborative partnerships, and innovative financing mechanisms are vital to enabling the development of circular, resource-efficient urban systems [93].

9 Circular Infrastructure and Technology Enablers

Circular infrastructure leverages technological enablers to drive resource efficiency and sustainability. Brilliant waste collection and sorting systems, powered by IoT sensors and AI, enable real-time monitoring of bin fill levels, optimize collection routes, and automate material separation, minimizing costs and environmental impact [94]. Integrating IoT with blockchain technology provides secure, transparent, and tamper-proof tracking of materials throughout their lifecycle, ensuring traceability and trust in circular supply chains [95, 96]. Urban design is increasingly supporting modular and flexible building systems, offering structures that can be easily adapted, dismantled, and reused thereby prolonging asset life and reducing construction waste [97, 98]. Digital platforms play a central role in facilitating sharing, repair, and reuse by democratizing access to product repair information, connecting users to local repair or sharing hubs, and supporting collaborative consumption, which keeps materials in use for longer and reduces landfill dependency [99].

10 Case Studies of Circular Supply

Table 6 presents examples from global cities and companies that have implemented practical models of circular supply chains tailored to their local contexts. These initiatives demonstrate how diverse urban settings, ranging from highly industrialized to rapidly developing, apply circular economy principles to manage waste, recover resources, and build more sustainable systems.

Table 6 Case studies

City	Circular strategy	Key focus area
Amsterdam	Circular construction and use of material banks [100]	Sustainable building materials and reuse infrastructure
Tokyo	Advanced recovery of electronic waste [101]	E-waste collection, refurbishment, and recycling
São Paulo	Decentralized composting of organic waste [102]	Community-based composting and waste reduction
Kigali	Involving informal workers in waste management [103]	Social inclusion and resource recovery
Pune (India)	Integration of informal waste pickers and decentralized composting [104]	Public–private partnerships and waste-to-resource systems
Cashify (India)	Digital platform for e-waste resale and reuse [105]	Consumer-driven electronics refurbishment and recycling
Material Recovery Facility (MRF)	Centralized sorting and recovery of recyclable waste [106]	Infrastructure for waste segregation, recycling, and formalization of waste streams
Hasiru Dala (India)	Integration of informal waste workers into formal waste management systems [107]	Social enterprise model that supports waste segregation, recycling, and dignified livelihoods for waste pickers with a strong emphasis on social justice [108]

11 Challenges

In Table 7, we summarize some of the major challenges of circular supply chains from the literature we have reviewed. We organize the challenges along themes, barriers, and a brief description.

12 Looking to the Future

A path to integrating circular economy principles into urban planning involves leveraging innovative technologies, decentralized infrastructure, and data-driven governance. Cities can embed circular metrics into innovative systems using tools such as product passports, IoT dashboards, and blockchain for enhanced traceability and transparency. Advanced urban metabolism modelling with digital twins enhances resource planning, while decentralized recovery hubs and modular supply chains improve flexibility and reduce emissions. Circular spending can also support climate adaptation through projects like greywater-fed wetlands. Strategies such as source-segregated organics for biogas, water reuse systems, on-site construction waste recycling, regional e-waste disassembly, and industrial symbiosis platforms further promote material recovery and efficiency. These approaches vary in complexity and timeline, requiring cities to tailor implementation based on their unique context and priorities.

13 Conclusion

In conclusion, embedding circular supply chains within urban resource systems represents a pivotal step toward creating regenerative, climate-resilient cities. By reclaiming materials, water, and energy through distributed recovery hubs, digital traceability, and inclusive stakeholder networks, municipalities can shift from linear consumption to restorative loops that safeguard ecosystems and stimulate green employment. Real-time data, policy coherence, and innovative finance will be central to scaling these models, while public awareness and engagement in the informal sector ensure social equity. Sustained research, cross-sector partnerships, and adaptive governance will enable cities to close resource loops, strengthen economic competitiveness, and advance the global sustainability agenda for future urban generations.

Table 7 Key barriers are to implementing circular economy in urban planning across institutional, technical, behavioural, and logistical themes

Theme	Barrier	Description
Institutional Barriers	Lack of supportive policies and regulations [109]	Cities lack comprehensive circular economy strategies, clear targets, and incentives. Existing regulations often favor linear systems, and fragmented governance across multiple agencies hinders coordinated action
	Insufficient governmental support [110]	Minimal funding, ambiguous or inconsistent regulations, and lack of standardization for reused products impede progress
	Weak policy coherence [111]	Conflicting policies (e.g., waste vs. bioenergy) and a lack of harmonization across jurisdictions create confusion and inefficiency
Technical Barriers	Inadequate infrastructure [112]	Most cities lack the necessary collection, sorting, processing, and reverse logistics systems to support circular flows. Existing infrastructure is often optimized for linear disposal, not reuse or recycling
	Technological limitations [113]	Circular design techniques are not automated or standardized, requiring significant resources for each project. There are also challenges in tracing materials and ensuring quality in remanufactured products
	Limited innovation and investment [114]	Insufficient investment in new recycling and recovery technologies, and the persistence of outdated infrastructure, slow the transition
Behavioral and Cultural Barriers	Low public awareness and engagement [115]	A lack of consumer information about product origins, durability, and recyclability limits demand for circular products
	Cultural resistance [116]	Planned obsolescence and ingrained linear consumption habits persist. Many citizens and businesses are not motivated or incentivized to adopt circular behaviors
	Fragmented collaboration [117]	Poor communication and limited willingness to collaborate across supply chains and value chains hinder systemic change
Infrastructure and Logistics Challenges	Reverse logistics gaps [118]	Efficient systems for returning used products for repair, refurbishment, or recycling are often missing or underdeveloped
	Market immaturity [119]	Markets for secondary materials are volatile, with inconsistent quality and limited demand, making circular business models less viable
	Logistical complexity [120]	Lack of harmonized transport and waste flows between municipalities leads to inefficiency and confusion

References

1. Capodaglio AG (2023) Urban wastewater mining for circular resource recovery: approaches and technology analysis. Water 15(22):3967. https://doi.org/10.3390/w15223967
2. Prasad MNV (2023) Resource recovery from urban flood, municipal and industrial wastewaters in the context of remediation technologies and circular economy. In: Smol M, Prasad MNV, Stefanakis AI (eds) Water in circular economy. Springer, pp 139–159. https://doi.org/10.1007/978-3-031-18165-8_8
3. Majumdar S, Goswami S, Sillanpää M, Choudhury M (2025) In: Choudhury M, Majumdar S, Goswami S, Sillanpää M (eds) Smart wastewater systems and climate change innovations through spatial intelligence, Chapter 1, vol 15. Royal Society of Chemistry, pp 1–13. https://doi.org/10.1039/9781837676880-00001
4. Lehmann S (2011) Resource recovery and materials flow in the city: zero waste and sustainable consumption as paradigm in urban development. J Green Build 6(3):88–105. https://doi.org/10.3992/jgb.6.3.88
5. Smol M, Kulczycka J, Kowalski Z (2021) Circular economy indicators in urban water systems: challenges and opportunities. Sustain Prod Consump 28:602–614. https://doi.org/10.1016/j.spc.2021.11.025
6. Munonye WC, Ajonye GO (2025) Energy-driven circular design in the built environment: rethinking architecture and infrastructure. Front Sustain Cities 7:1569362. https://doi.org/10.3389/frsc.2025.1569362
7. Bhattacharjee N, Narayan S (2024) From shelter to the streets: the feminine face of homelessness in contemporary democracies. Int J Human Rights 28(5):852–882. https://doi.org/10.1080/13642987.2024.2327374
8. Domenech T, Borrion A (2021) Embedding circular economy principles into urban regeneration and waste management: framework and metrics. Sustainability 14(3):1293. https://doi.org/10.3390/su14031293
9. Cervantes Puma GC, Salles A, Bragança L (2024) Nexus between urban circular economies and sustainable development goals: a systematic literature review. Sustainability 16(6):2500. https://doi.org/10.3390/su16062500
10. Sulich A, Sołoducho-Pelc L (2022) The circular economy and the green jobs creation. Environ Sci Pollut Res 29:14231–14247. https://doi.org/10.1007/s11356-021-16562-y
11. Fernando Y, Sroufe R, Tseng ML, Shaharudin MS, Wahyuni-TD IS, Jawo M (2025) The circular supply chains in smart cities: a catalyst for sustainable development. J Ind Prod Eng:1–25. https://doi.org/10.1080/21681015.2025.2497867
12. Iacovidou E, Hahladakis JN, Purnell P (2021) A systems thinking approach to understanding the challenges of achieving the circular economy. Environ Sci Pollut Res 28:24785–24806. https://doi.org/10.1007/s11356-020-11725-9
13. Sekulić M, Stojanović V, Pantelić M, Nađ I (2022) Impact of the circular economy on quality of life: a systematic literature review. Geograph Pannonica 26(1):79–92. https://doi.org/10.5937/gp26-36059
14. Tanveer U, Ishaq S, Huy TQ, Hoang TG (eds) (2024) Navigating the circular age of a sustainable digital revolution. IGI Global. https://doi.org/10.4018/979-8-3693-2827-9
15. Mohan N, Sinha S, Kumar D (2023) A review on remanufacturing, reuse, and recycling in supply chain—exploring the evolution of information technology over two decades. Int J Inf Manag Data Insights 3(1):100160. https://doi.org/10.1016/j.jjimei.2023.100160
16. Carissimi MC, Bin Hameed H, Creazza A (2024) Circular economy: the future nexus for sustainable and resilient supply chains? Sustain Fut 2024:100365. https://doi.org/10.1016/j.sftr.2024.100365
17. Alonso-Muñoz S, González-Sánchez R, Siligardi C, García-Muiña FE (2021) New circular networks in resilient supply chains: an external capital perspective. Sustainability 13(11):6130. https://doi.org/10.3390/su13116130

18. Meyer K, Görgens SJ, Persch C, Dröder K, Herrmann C (2023) Increasing resilience of material supply by decentral urban factories and secondary raw materials. Front Manuf Technol 3:1106965. https://doi.org/10.3389/fmtec.2023.1106965
19. Bergström B, Edstrand H (2021) Managing sustainability in strategic decision-making: Life cycle thinking for sustainable innovation—a case study at Volvo Cars. Master's thesis, Chalmers University of Technology, Chalmers ODR. https://hdl.handle.net/20.500.12380/302849
20. Triguero Á, Cuerva MC, Sáez-Martínez FJ (2022) Closing the loop through eco-innovation by European firms: circular economy for sustainable development. Bus Strateg Environ 31(2):733–749. https://doi.org/10.1002/bse.3024
21. Choy YK, Onuma A, Lee KE (2025) The nexus of industrial–urban sustainability, the circular economy, and climate–ecosystem resilience: a synthesis. Sustainability 17(6):2620. https://doi.org/10.3390/su17062620
22. Vural Gursel I, Elbersen B, Meesters KPH (2023) Monitoring circular biobased economy—systematic review of circularity indicators at the micro level. Resour Conserv Recycl 197:107104. https://doi.org/10.1016/j.resconrec.2023.107104
23. Mojica A, Serrano-Ruiz JC, Andres B, de la Torre R (2025) A conceptual framework for the upcycling supply chain in the wood sector. Sustainability 17(3):1006. https://doi.org/10.3390/su17031006
24. Plaves Y, Jacobs P, Uylaki T, Jonescu EE (2024) Regenerative design: research, practice, and implementation in the built environment. Proc Inst Civil Eng Urban Des Plann 177(3):96–108. https://doi.org/10.1680/jurdp.23.00036
25. Calheiros CSC, Stefanakis AI (2021) Green roofs towards circular and resilient cities. Circ Econ Sustain 1:395–411. https://doi.org/10.1007/s43615-021-00033-0
26. Munaro MR, Tavares SF (2025) Design for adaptability and disassembly: guidelines for building deconstruction. Construct Innov Inf Process Manag 25(2):665–687. https://doi.org/10.1108/CI-10-2022-0266
27. Shahhoseini A, Heydari S, Pedrammehr S (2023) Manufacturing and assembly for the ease of product recycling: a review. Designs 7(2):42. https://doi.org/10.3390/designs7020042
28. Pathak P, Rout PR (eds) (2021) Urban mining for waste management and resource recovery: sustainable approaches, 1st edn. CRC Press. https://doi.org/10.1201/9781003201076
29. Murthy V, Ramakrishna S (2022) A review on global e-waste management: urban mining towards a sustainable future and circular economy. Sustainability 14(2):647. https://doi.org/10.3390/su14020647
30. Ellen MacArthur Foundation, Arup (2019) Circular economy in cities: project guide. Ellen MacArthur Foundation. https://www.ellenmacarthurfoundation.org/assets/downloads/CE-in-Cities-Project-Guide_Mar19.pdf
31. Ministry of Housing and Urban Affairs (2022) Circular economy in waste management (Final report). Government of India. https://mohua.gov.in/pdf/627b8318adf18Circular-Economy-in-waste-management-FINAL.pdf
32. Yu P, Wei Y, Ma L, Wang B, Yung EHK, Chen Y (2024) Urbanization and the urban critical zone. Earth Crit Zone 1(1):100011. https://doi.org/10.1016/j.ecz.2024.100011
33. Bhardwaj LK, Rath P, Jain H, Choudhury M (2025) Exploring the effects of e-waste on soil, water quality and human health. Discov Civil Eng 2(1):1–16. https://doi.org/10.1007/s44290-025-00167-2
34. Choudhary R, Kumar A, Narayan S, Mishchenko I, Choudhury M, Bhardwaj LK, Ahmad A, Andrienko M, Gaman P, Boiko O (2025) Integrating fuzzy logic into life cycle assessment for sustainable municipal solid waste management in metro cities. Ecol Eng Environ Technol 26(7):316–331. https://doi.org/10.12912/27197050/205500
35. Estévez S, Arias A, Feijoo G, Moreira MT (2025) Methodological guide and roadmap to assess the compliance of wastewater treatment plants with sustainability and circularity criteria. Water Res 274:123125. https://doi.org/10.1016/j.watres.2025.123125
36. Lee J, Choi H, Kim J (2024) Environmental and economic impacts of e-waste recycling: a systematic review. Chem Eng J 494:152917. https://doi.org/10.1016/j.cej.2024.152917

37. Yang J, Zhang J, Du X, Gao T, Cheng Z, Fu W, Wang S (2025) Ammonia inhibition in anaerobic digestion of organic waste: a review. Int J Environ Sci Technol 22:3927–3942. https://doi.org/10.1007/s13762-024-06029-1
38. Ouro-Salim O (2024) Urban mining of e-waste management globally: literature review. Clean Waste Syst 9:100162. https://doi.org/10.1016/j.clwas.2024.100162
39. Bonifazi G, Grosso C, Palmieri R, Serranti S (2025) Current trends and challenges in construction and demolition waste recycling. Curr Opin Green Sustain Chem 53:101032. https://doi.org/10.1016/j.cogsc.2025.101032
40. Campbell G, Gedikli A, Koondhar MA (2025) Circular economy and mining. In: Shahbaz M, Sharma GD, Gedikli A, Erdoğan S (eds) Global pathways for efficient waste management and inclusive economic development. Springer, pp 155–175. https://doi.org/10.1007/978-981-96-5569-4_8
41. Pavloudakis F, Roumpos C, Spanidis P-M (2024) Planning the closure of surface coal mines based on circular economy principles. Circ Econ Sustain 4(1):75–96. https://doi.org/10.1007/s43615-023-00278-x
42. Ashesh A (2023) E-waste recycling: transition from linear to circular economy. In: Management of electronic waste: resource recovery, technology and regulation. Wiley, pp 191–201. https://doi.org/10.1002/9781119894360.ch9
43. Bandh SA, Malla FA, Wani SA, Hoang AT (2023) Waste management and circular economy. In: Bandh SA, Malla FA (eds) Waste management in the circular economy. Springer, Cham, pp 1–17. https://doi.org/10.1007/978-3-031-42426-7_1
44. Tiwari S, Si Mohammed K (2024) Unraveling the impacts of linear economy, circular economy, green energy and green patents on environmental sustainability: empirical evidence from OECD countries. Gondwana Res 135:100018. https://doi.org/10.1016/j.gr.2024.07.018
45. Srivastava S, Dwivedi A, Mangla SK, Dey BL, Koufopoulos DN (2024) Mindfully fashioned: sustaining style through product value retention. J Retail Consum Serv 81:103992. https://doi.org/10.1016/j.jretconser.2024.103992
46. Kandpal V, Jaswal A, Santibanez Gonzalez EDR, Agarwal N (2024) Challenges and opportunities for sustainable energy transition and circular economy. In: Sustainable energy transition. Springer, Cham. https://doi.org/10.1007/978-3-031-52943-6_11
47. Sardianou E, Nikou V, Kostakis I (2024) Harmonizing sustainability goals: empirical insights into climate change mitigation and circular economy strategies in selected European countries with SDG13 framework. Sustainability 16(1):296. https://doi.org/10.3390/su16010296
48. Chaudhari US, Watkins DW, Handler RM, Reck BK, Johnson AT, Hossain T, Hartley DS, Thompson VS, Peralta A, Shonnard DR (2025) Systems analysis and optimization of circular PET packaging supply chains in the United States: environmental and socioeconomic impacts. J Adv Manuf Process. https://doi.org/10.1002/amp2.70008
49. Al-Sari' MI, Haritash AK (2024) Municipal organic solid waste management in the concept of urban mining and circular economy: a model from Palestine. J Mater Cycles Waste Manag 26(5):2980–2995. https://doi.org/10.1007/s10163-024-02018-1
50. Babu R, Veramendi PMP, Rene ER (2021) Strategies for resource recovery from the organic fraction of municipal solid waste. Case Stud Chem Environ Eng 3:100098. https://doi.org/10.1016/j.cscee.2021.100098
51. Number Analytics (2025) Urban water conservation strategies. Number Analytics. https://www.numberanalytics.com/blog/urban-water-conservation-strategies. Accessed 23 Jul 2025
52. Padmanabhan S, Giridharan K, Stalin B, Kumaran S, Kavimani V, Nagaprasad N, Jule LT, Krishnaraj R (2022) Energy recovery of waste plastics into diesel fuel with ethanol and ethoxy ethyl acetate additives on circular economy strategy. Sci Rep 12:5330. https://doi.org/10.1038/s41598-022-09148-2
53. Discovery Alert (2025) Urban mining: recovering valuable materials from our cities. Discovery Alert. https://discoveryalert.com.au/news/urban-mining-anthropogenic-stock-2025/. Accessed 23 Jul 2025
54. Kennedy S, Linnenluecke MK (2022) Circular economy and resilience: a research agenda. Bus Strateg Environ 31(6):2754–2765. https://doi.org/10.1002/bse.3004

55. Yang Y, Zheng B, Luk C, Yuen K, Chan A (2024) Towards a sustainable circular economy: understanding the environmental credits and loads of reusing modular building components from a multi-use cycle perspective. Sustain Prod Consump 46:543–558. https://doi.org/10.1016/j.spc.2024.02.027
56. van der Velden M, Maitre-Ekern E, Wanja DK (2024) The role of independent repair in a circular and regenerative economy. Circ Econ Sustain 4(4):2981–3006. https://doi.org/10.1007/s43615-023-00304-y
57. Jensen HH (2025) 4th and 5th R—maximizing product lifecycles: integrating refurbishment and remanufacturing strategies. In: Circular economy opportunities and pathways for manufacturers (Future of business and finance, Chapter 7). Springer, Cham. https://doi.org/10.1007/978-3-031-75279-7_7
58. Wlazlak P, Johansen K (2025) Development of remanufacturing systems at original equipment remanufacturers. Int J Interact Des Manuf. https://doi.org/10.1007/s12008-025-02314-6
59. Gkyrtis K, Pomoni M (2024) An overview of the recyclability of alternative materials for building surface courses at pavement structures. Buildings 14(6):1571. https://doi.org/10.3390/buildings14061571
60. Brglez K, Perc M, Kovačič Lukman R (2023) The complexity and interconnectedness of circular cities and the circular economy for sustainability. Sustain Dev 32(3):2049–2065. https://doi.org/10.1002/sd.2766
61. Ellen MacArthur Foundation (2024) Building a circular supply chain [Web page]. Ellen MacArthur Foundation. https://www.ellenmacarthurfoundation.org/supply-chains/overview
62. Adesoga TO, Olaiya OP, Onuma EP, Ajayi OO, Olagunju OD (2024) Review of reverse logistics practices and their impact on supply chain sustainability. Int J Sci Res Arch 12(2):162–168. https://doi.org/10.30574/ijsra.2024.12.2.1216
63. Institute of Sustainability Studies (2024) What is a circular supply chain? Institute of Sustainability Studies. https://instituteofsustainabilitystudies.com/insights/lexicon/what-is-a-circular-supply-chain/
64. Chowdhury PR, Goswami S, Choudhary R, Choudhury M (2025) Electronic wastes, its impact on wildlife and biodiversity. In S Goswami, M Choudhury & S Agarwal (eds) Electronic waste: impact on health, animals, and the environment (1st ed.), pp 65–71. https://doi.org/10.1201/9781003582311-8
65. Gueboudji Z (2024) Waste valorization techniques. In: Anouzla A, Souabi S (eds) Generation of energy from municipal solid waste (Chapter 2). Springer, Cham. https://doi.org/10.1007/978-3-031-74334-4_2
66. Jensen H (2024) For manufacturers, the circular economy strengthens supply chains. World Economic Forum. https://www.weforum.org/stories/2024/02/how-manufacturers-could-lead-the-way-in-building-the-circular-economy/
67. Oracle (2023) What is a circular supply chain? Oracle. https://www.oracle.com/scm/circular-supply-chain/. Accessed Jul 2025
68. Ampong RO, Miezah K, Banunle A, Amuah EEY, Amoah EG (2024) Comparative analysis of the effectiveness and efficiency of formal and informal waste collectors in urban waste management system. Heliyon 10(20):e39017. https://doi.org/10.1016/j.heliyon.2024.e39017
69. Ortega Alvarado IA, Pettersen IN (2023) The role given to citizens in shaping a circular city. Urban Geogr 45(4):611–630. https://doi.org/10.1080/02723638.2023.2221097
70. Bin Abu Sofian ADA, Lim HR, Manickam S, Ang WL, Show PL (2023) Towards a sustainable circular economy: algae-based bioplastics and the role of Internet-of-Things and machine learning. ChemBioEng Rev. https://doi.org/10.1002/cben.202300028
71. Schmidt JL, Sehnem S, Spuldaro JD (2024) Blockchain and the transition to the circular economy: a literature review. Corp Soc Responsib Environ Manag 31(3):2010–2032. https://doi.org/10.1002/csr.2674
72. Tutore I, Parmentola A, Costagliola di Fiore M, Calza F (2024) A conceptual model of artificial intelligence effects on circular economy actions. Corp Soc Responsib Environ Manag 31(5):4772–4782. https://doi.org/10.1002/csr.2827

73. Singh S, Singh P, Sharma A, Choudhury M (eds) (2022) Agriculture Waste management and bioresource: the circular economy perspective. Wiley. https://onlinelibrary.wiley.com/doi/book/10.1002/9781119808428
74. Metabolism of Cities (2022) Urban circularity assessment method. CityLoops. https://cityloops.eu/fileadmin/user_upload/Materials/UCA/CityLoops_WP4_D4.4_Urban_Circularity_Assessment_Method_Metabolism_of_Cities.pdf
75. Minimum (2025) Material flow analysis (MFA). https://www.minimum.com/resources/material-flow-analysis-mfa
76. Bharathy N, Kowsalya P, David EM, Choudhury M, Khan PA, Aidoo MK et al (2025) Breaking barriers: innovative approaches and technologies for climate resilience: a case study from European countries. In: Choudhury M, Dixit G, Majumdar S (eds) Sustainable synergy: harnessing ecosystems for climate resilience. Climate change management. Springer, Cham. https://doi.org/10.1007/978-3-031-77957-2_12
77. Bellstedt C, Martín Carreño GE, Athanassiadis A, Chaudhry S (2022) Urban circularity assessment method. Metabolism of cities. https://cityloops.eu/fileadmin/user_upload/Materials/UCA/CityLoops_WP4_D4.4_Urban_Circularity_Assessment_Method_Metabolism_of_Cities.pdf
78. Ecer K, Güner O (2024) The European Union's policies and role in tackling climate change in the context of the European Green Deal. In: Açikalin ŞN, Erçetin ŞŞ (eds) The social consequences of climate change. Emerald Publishing, pp 163–185. https://doi.org/10.1108/978-1-83797-677-520241013
79. De Laurentiis V, Caldeira C, Sala S, Tonini D (2024) Life cycle thinking for the assessment of waste and circular economy policy: status and perspectives from the EU example. Waste Manag 179:205–215. https://doi.org/10.1016/j.wasman.2024.02.037
80. Michalovič M (2024) Legal regulation facilitating the transition to a circular economy in the legal system of Slovakia. J Agric Environ Law 19(36):57–80. https://doi.org/10.21029/JAEL.2024.36.57
81. Rajpal A, Choudhury M, Goswami S, Chakravorty A, Raghavan V (eds) (2024) Waste management and treatment: advances and innovations, 1st edn. CRC Press. https://doi.org/10.1201/9781003258377
82. Choudhury M, Roy P (2025) Challenges with microplastic pollution in the regime of UN sustainable development goals. World Dev Sustain 2025:100216. https://doi.org/10.1016/j.wds.2025.100216
83. Dey S, Santra M, Choudhury M, Ghosh AR, Samanta P (2024) Food waste generation and its industrial utilization: an overview. Environ Sci Pollut Res. https://doi.org/10.1007/s11356-024-34252-3
84. Ellen MacArthur Foundation (2022) The EU's circular economy action plan. In: Circular examples. https://www.ellenmacarthurfoundation.org/circular-examples/the-eus-circular-economy-action-plan
85. Circular City Centre (C3) & ICLEI—Local Governments for Sustainability (2025) Circular consumption in cities (European Circular Economy Stakeholder Platform). European Investment Bank & ICLEI. https://circulareconomy.europa.eu/platform/sites/default/files/2025-02/circular-consumption-in-cities.pdf
86. Ferber U, Preuß T (2008) Circular land use management in cities and urban regions: a policy mix utilizing existing and newly conceived instruments to implement an innovative strategic and policy approach (Difu-Paper). German Institute for Urban Affairs (Difu). http://www.udrzitelnavystavba.cz/WP1_papers/01_Ferber.pdf
87. Tanveer U, Hoang TG, Ishaq S, Khalid RU (2025) Public–private partnerships as catalysts for digital transformation and circular economy: insights from developing countries. Technol Forecast Soc Chang 219:124270. https://doi.org/10.1016/j.techfore.2025.124270
88. Sustainability Directory (2025) What role does public–private partnership play in promoting circular economy initiatives? Sustainability Directory. https://sustainability-directory.com/question/what-role-does-public-private-partnership-play-in-promoting-circular-economy-initiatives/

89. Taera EG, Lakner Z (2025) Sustainable finance: bridging circular economy goals and financial inclusion in developing economies. World 6(2):44. https://doi.org/10.3390/world6020044
90. Havemann T, Negra C, Werneck F (2022) Blended finance for agriculture: exploring the constraints and possibilities of combining financial instruments for sustainable transitions. In: Desa G, Jia X (eds) Social innovation and sustainability transition. Springer, pp 347–358. https://doi.org/10.1007/978-3-031-18560-1_23
91. Banerjee, Choudhury (2022) Green technology for wastewater treatment and remediation. In: Baskar C, Ramakrishna S, Daniela La Rosa A (eds) Encyclopedia of green materials. Springer, Singapore. https://doi.org/10.1007/978-981-16-4921-9_30-1
92. Gogoi, Choudhury (2022) Green materials synthesis for wastewater treatment. In: Baskar C, Ramakrishna S, Daniela La Rosa A (eds) Encyclopedia of green materials. Springer, Singapore. https://doi.org/10.1007/978-981-16-4921-9_31-1
93. Lazurko A, Drechsel P, Hanjra MA (2018) Financing resource recovery and reuse in developing and emerging economies: enabling environment, financing sources and cost recovery, Resource recovery & reuse series No. 11. International Water Management Institute. https://doi.org/10.5337/2018.220
94. Suvarnamma A, Pradeepkiran JA (2021) SmartBin system with waste tracking and sorting mechanism using IoT. Clean Eng Technol 5:100348. https://doi.org/10.1016/j.clet.2021.100348
95. Banerjee A, Venkatesh M (2018) Product tracking and tracing with IoT and blockchain (Insights Brief, Infosys-Oracle). Infosys Limited. https://www.infosys.com/oracle/Insights/documents/product-tracking-tracing.pdf. Accessed 23 Jul 2025
96. Wong EKS, Ting HY, Atanda AF (2023) Enhancing supply chain traceability through blockchain and IoT integration: a comprehensive review. Green Intell Syst Appl 4(1):11–28. https://doi.org/10.53623/gisa.v4i1.355
97. Hamida MB, Jylhä TE, Remøy HT, Gruis VH (2022) Circular building adaptability and its determinants: a literature review. Int J Build Pathol Adapt 41(6):47–69. https://doi.org/10.1108/IJBPA-11-2021-0150
98. Kyrö R, Jylhä T, Peltokorpi A (2019) Embodying circularity through usable relocatable modular buildings. Facilities 37(1–2):75–90. https://doi.org/10.1108/F-12-2017-0129
99. MET Bhujbal Knowledge City (2024) Understanding modern circular architecture: principles, benefits, and examples. MET School of Architecture and Interior Design. https://www.metbkcsoaid.ac.in/post/understanding-modern-circular-architecture-principles-benefits-and-examples. Accessed 23 Jul 2025
100. Marin J, Alaerts L, Van Acker K (2020) A materials bank for circular Leuven: how to monitor 'messy' circular city transition projects. Sustainability 12(24):10351. https://doi.org/10.3390/su122410351
101. Liu X, Yu J, Okubo K (2021) Analysis of the efficiency of various waste electrical and electronic equipment's collection routes: a case study focusing on collection route for waste mobile phones in the Tohoku area of Japan. Recycling 6(1):13. https://doi.org/10.3390/recycling6010013
102. Amato-Lourenço LF, Cunha França G, Seckler MM, Mauad T (2024) Enhancing urban waste sustainability through community-driven composting in São Paulo megacity. Environ Challenges 14(8):100864. https://doi.org/10.1016/j.envc.2024.100864
103. Squire JN, Nkurunziza J (2021) Urban waste management in post-genocide Rwanda: an empirical survey of the city of Kigali. J Asian Afr Stud 57(4):760–772. https://doi.org/10.1177/00219096211035430. (Original work published 2022)
104. Estrada M, Galvin M, Maassen A, Hörschelmann K (2022) Catalysing urban transformation through women's empowerment in cooperative waste management: the SWaCH initiative in Pune, India. Local Environ 28(7):852–866. https://doi.org/10.1080/13549839.2022.2090532
105. Garg N (2025) Finding the right platform to sell old electronic items. Cashify. https://www.cashify.in/finding-the-right-platform-to-sell-old-electronic-items
106. Ministry of Housing and Urban Affairs, Central Public Health and Environmental Engineering Organisation (2020) Advisory on material recovery facility (MRF) for municipal solid waste

[PDF]. Swachh Bharat Mission—Urban. https://www.sbmurban.org/storage/app/media/pdf/SBM%20Advisory%20on%20MRF%20for%20MSW.pdf
107. Chandran P (2021) Extended producer responsibility and informal waste workers: from integration to sustainable inclusion [Policy brief]. Hasiru Dala. https://hasirudala.in/wp-content/uploads/2021/06/Extended-Producer-Responsibility-and-Informal-Waste-Workers.pdf
108. Lal R, Baeraiya B, Thakur R, Gautam RK, Narayan S, Yadav D, Pathak P, Singha S (2024) Traditional knowledge in drug development and the rights of indigenous peoples: a legal and ethical perspective. J Drug Alcohol Res 13:6408. https://doi.org/10.4303/JDAR/236408. https://www.researchgate.net/profile/Saurav-Narayan-2/publication/392626701_Traditional_Knowledge_in_Drug_Development_and_the_Rights_of_Indigenous_Peoples_A_Legal_and_Ethical_Perspective/links/684b24ced0be921dfef74c62/Traditional-Knowledge-in-Drug-Development-and-the-Rights-of-Indigenous-Peoples-A-Legal-and-Ethical-Perspective.pdf
109. Kazancoglu İ, Sagnak M, Mangla SK, Kazancoglu Y (2021) Circular economy and the policy: a framework for improving the corporate environmental management in supply chains. Bus Strateg Environ 30(1):590–608. https://doi.org/10.1002/bse.2641
110. Trevisan AH, Lobo A, Guzzo D, Gomes LAV, Mascarenhas J (2023) Barriers to employing digital technologies for a circular economy: a multi-level perspective. J Environ Manag 332:117437. https://doi.org/10.1016/j.jenvman.2023.117437
111. Mubarik MS, Kontoleon A, Shahbaz M (2024) Beyond the hurdles: exploring policy obstacles in the path to circular economy adoption. J Environ Manag 370:122667. https://doi.org/10.1016/j.jenvman.2024.122667
112. Thirumal S, Udawatta N, Karunasena G, Al-Ameri R (2024) Barriers to adopting digital technologies to implement circular economy practices in the construction industry: a systematic literature review. Sustainability 16(8):3185. https://doi.org/10.3390/su16083185
113. Vimal KEK, Sivakumar K, Kandasamy J et al (2023) Barriers to the adoption of digital technologies in a functional circular economy network. Oper Manag Res 16:1541–1561. https://doi.org/10.1007/s12063-023-00375-y
114. Lehmann C, Cruz-Jesus F, Oliveira T, Damásio B (2022) Leveraging the circular economy: investment and innovation as drivers. J Clean Prod 360:132146. https://doi.org/10.1016/j.jclepro.2022.132146
115. Gonella JSL, Godinho Filho M, Campos LMS, Ganga GMD (2024) People's awareness and behaviours of circular economy around the world: literature review and research agenda. Sustain Account Manag Policy J 15(5):1118–1154. https://doi.org/10.1108/SAMPJ-08-2022-0413
116. Musova Z, Musa H, Rech F (2025) Circular economy challenges: a bibliometric exploration of cognitive and behavioral barriers. J Organiz Change Manag. https://doi.org/10.1108/JOCM-01-2025-0011
117. Munaro MR, Tavares SF (2023) A review on barriers, drivers, and stakeholders towards the circular economy: the construction sector perspective. Clean Respons Consump 8:100107. https://doi.org/10.1016/j.clrc.2023.100107
118. Beames A, Claassen GDH, Akkerman R (2021) Logistics in the circular economy: challenges and opportunities. In: Rezaei J (ed) Strategic decision making for sustainable management of industrial networks, vol 8. Springer, pp 1–14. https://doi.org/10.1007/978-3-030-55385-2_1
119. Uhrenholt JN, Kristensen JH, Rincón MC, Adamsen S, Jensen SF, Waehrens BV (2022) Maturity model as a driver for circular economy transformation. Sustainability 14(12):7483. https://doi.org/10.3390/su14127483
120. Jacobs C, Soulliere K, Sawyer-Beaulieu S, Sabzwari A, Tam E (2022) Challenges to the circular economy: recovering wastes from simple versus complex products. Sustainability 14(5):2576. https://doi.org/10.3390/su14052576
121. Chaiya C, Pinthong P (2024) Integrating environmental and socioeconomic factors for a sustainable circular economy in Thailand. Sustainability 16(23):10748. https://doi.org/10.3390/su162310748

Performance Metrics for Circular Economy: Techniques for Evaluating Effectiveness and Measuring Progress

Saurav Narayan, Basona Khatun, S. Hansi Munasinghe, Shobhit Raj Pandey, Rakesh Choudhary, Michael Troilo, and Tanay Sandeep Birje

Abstract Anchored in the accelerating shift from '*take–make–dispose*' systems, this chapter synthesizes how circular economy (CE) performance is currently measured and how it should evolve. We cluster metrics into five families—material flows, environmental impacts, economic value, social outcomes and systemic circularity—and map them to stakeholder decisions across policy, business, finance and community arenas. Core frameworks (ISO 59000 series, EU monitoring set, GRI disclosures) and leading tools (LCA, MFA, Circulytics, MCI, Circularity Gap indicators) are interrogated for comparability, completeness and practicality. Sector vignettes (electronics, construction, food, fashion, mobility) translate abstractions into deployable indicators. From this analysis, we expose recurring barriers—data gaps, fragmentation, narrow environmental focus—and outline four future directions: global harmonisation, ESG convergence, AI-enabled predictive modelling and justice-centred social metrics. The result is a sequenced roadmap that equips researchers, regulators and investors with a transparent, decision-ready measurement architecture to scale

S. Narayan (✉)
Department of Law, School of Legal Studies, Central University of Punjab, Bathinda, Punjab, India

B. Khatun
Department of Law, School of Legal Studies, Central University of Punjab, Bathinda, Punjab, India

S. H. Munasinghe
University of Kelaniya, Colombo, Sri Lanka

S. R. Pandey
Department of Law, School of Legal Studies, Babasaheb Bhimrao Ambedkar University, Lucknow, Uttar Pradesh, India

R. Choudhary
Department of Civil Engineering, National Institute of Technology Delhi, New Delhi, India

M. Troilo
Busch School of Business, Catholic University of America, Washington, DC, USA

T. S. Birje
Faculty of Science and Engineering, Macquarie University, Macquarie Park, NSW, Australia

S. Majumdar et al. (eds.), *Sustainable Urban Future*, Urban Sustainability,
https://doi.org/10.1007/978-981-95-2976-6_11

circular transitions with speed, accountability and equity across sectors, regions and communities.

Keywords Circular economy · Performance metrics · Sustainability assessment · Indicator frameworks · Systems evaluation

1 Introduction

The circular economy (CE) concept has become increasingly appealing in global policy discussions as an attractive alternative to the traditional linear economic model of production and consumption, characterized by the '*take, make, dispose*' approach. CE or circular approaches prioritize reduction, reuse, recycling, and regenerative design to manage resources so resources can be continuously used while mitigating waste and environmental impact [1, 2]. As a consequence of circular economy strategies adopted by increasingly substantial global organizations such as policymakers, businesses, and communities, there is increasing urgency to develop the measures and benchmarks necessary to assess and measure accountability, allocate investment expenditure, and develop policies for a circular economy [3, 4]. Despite this urgency, measuring circularity has numerous theoretical and methodological complications. CE metrics often lack standardization, in-sector adaptability, and multidimensional potential, which limit their utility in providing a basis for evidence-based decision-making [5]. Although some measures have demonstrated the importance of sustainability objectives that focus on CE, many performance indicators concern material flows or environmental outcomes without paying equal attention to the economic, social, and systemic dimensions of circularity [6]. Moreover, a gap remains in the availability and quality of data in the sustainability sector, especially in developing regions, which contributes to the challenge of tracking change [7].

In response to these challenges, researchers and organizations have begun to develop new tools, digital platforms, and composite indices to provide a more comprehensive and dynamic view of CE performance [8]. These metrics include the Material Circularity Indicator (MCI), circularity gap indicators, and lifecycle-based assessments. This chapter aims to explore the current landscape of circular economy metrics, considering their categories, tools, and applications, while also discussing their challenges and potential for future development [9, 10] (Fig. 1).

Fig. 1 E-waste from appliances in Colombo, Sri Lanka—a challenge for circular tracking. (Photo by: S. Hansi Munasinghe)

2 Bibliometric Analysis on Circular Economy and Performance Metrics (2015–2024)

A bibliometric analysis was conducted using the Scopus database with the search string TITLE-ABS-KEY (circular economy OR performance metrics) AND PUBYEAR > 2014 AND PUBYEAR < 2025, focusing on literature published between 2015 and 2024. This query was designed to capture scholarly contributions that discuss either the concept of the circular economy or the development and application of performance metrics over the past decade. The selected time frame encompasses a period of significant growth in sustainability-focused research, during which circular economy principles have gained prominence across various disciplines, and performance measurement has become increasingly crucial for evaluating progress. The analysis examines key indicators, including the annual volume of publications, the dominant thematic structure, and the leading journals. The results, as presented in Figs. 2, 3, and 4, reveal a growing convergence between circular economy practices and the need for robust performance evaluation tools. Together, these understandings provide a structured overview of the intellectual landscape, highlighting how academic attention has evolved toward integrating metrics-driven accountability within circular economic systems.

The results reveal a sharply increasing research interest in these topics over the past decade. Starting with just 21 documents in 2015, the annual publication volume grew steadily, reaching 65 by 2019. A notable acceleration occurred from 2020 onward—125 documents in 2020, rising to 142 in 2021, and continuing this upward trajectory to 173 in 2022 and 222 in 2023. The year 2024 saw a dramatic spike, culminating in 391 publications. The percentage change from the beginning of the period to the end is a staggering 1761%, with a 76% increase from 2023 to 2024. The compound annual growth rate (CAGR) is 34%. This consistent growth reflects

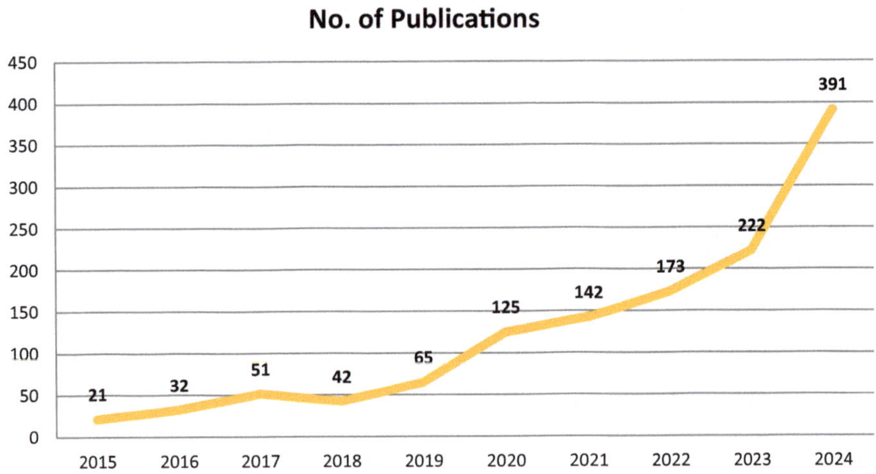

Fig. 2 Number of publications (2015–2024)

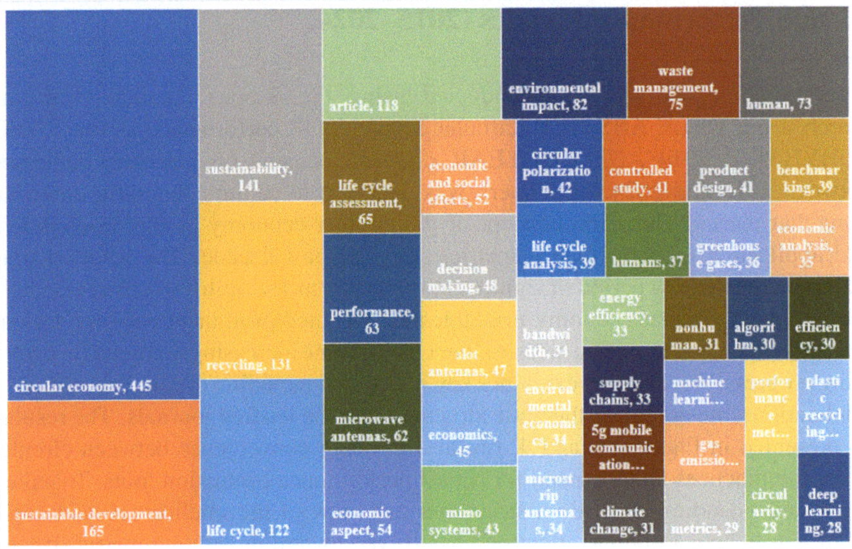

Fig. 3 Thematic structure of circular economy and performance metrics

the expanding academic and policy focus on embedding circular economy principles and performance assessment tools across various sectors. These trends, illustrated in Fig. 2, suggest that performance metrics are being increasingly integrated into circular economy research, driven by a global push for sustainability, accountability, and data-informed decision-making.

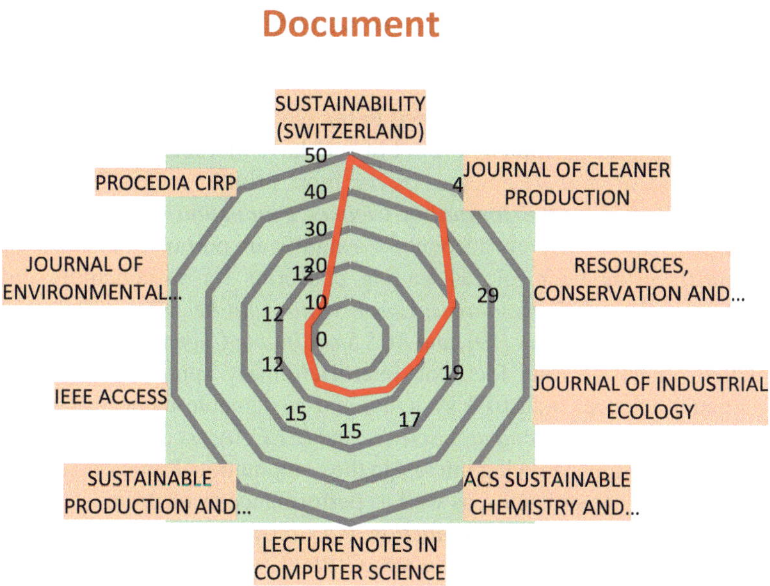

Fig. 4 Top 10 most prolific sources publishing on circular economy and performance metrics

The thematic structure of the literature retrieved was further explored through a keyword co-occurrence analysis, visually represented in a tree map (Fig. 3). The analysis reveals that '*circular economy*' dominates the research discourse, appearing in 445 documents (16%), signifying its central role in the investigated body of work. Other prominent keywords include '*sustainable development*' (165; 6%), '*sustainability*' (141; 5%), '*recycling*' (131; 5%), and '*life cycle*' (122; 4%), reflecting the close thematic relationship between circularity, environmental goals, and resource optimization. The presence of terms such as '*waste management*' (75), '*environmental impact*' (82), and '*life cycle assessment*' (65) indicates a strong methodological and operational focus within the literature. Similarly, keywords like '*performance*' (63), '*decision making*' (48), '*economic aspect*' (54), and '*benchmarking*' (39) highlight the growing emphasis on evaluating and quantifying circular practices through robust performance metrics and economic frameworks.

Notably, the emergence of '*supply chains*' (33), '*efficiency*' (30), '*gas emissions*' (29), and '*metrics*' (29), as well as the direct appearance of '*performance metrics*' (29), points to a growing awareness of maintaining environmental sustainability in operations and logistics in both research and practice. These terms suggest increasing efforts to integrate circular economy principles within industrial and supply chain contexts using measurable indicators. The diversity of other lower-frequency but still significant keywords—such as '*climate change*,' '*carbon dioxide*,' '*optimization*,' '*machine learning*,' and '*deep learning*'—signals the interdisciplinary and evolving nature of the field, especially the incorporation of emerging technologies and data analytics into circular strategies. The thematic structure reflects a clear progression from foundational concepts like recycling and sustainability toward more sophisticated, data-driven approaches to performance assessment within the circular economy framework.

The dissemination of research related to the circular economy and performance metrics is concentrated within a core set of scholarly journals. As shown in Fig. 4, *Sustainability (Switzerland)* emerges as the most prolific source, contributing 49 documents, followed by the *Journal of Cleaner Production* with 42 publications. These two journals collectively account for a significant portion of the total research output, underscoring their role as leading platforms for sustainability research. Other key journals include *Resources, Conservation and Recycling* (29 documents), *Journal of Industrial Ecology* (19), and *ACS Sustainable Chemistry and Engineering* (17), which reflect a strong focus on interdisciplinary approaches that integrate environmental science, industrial systems, and sustainable engineering practices. Notably, *Lecture Notes in Computer Science* and *IEEE Access* also appear among the top sources, indicating a growing interest in the computational sciences for modeling and assessing circular strategies, as well as performance evaluation.

The presence of *Sustainable Production and Consumption* (15), *Journal of Environmental Management* (12), and *Procedia CIRP* (12) further supports the relevance of production efficiency, environmental governance, and industrial processes in the ongoing discourse on the circular economy. These findings underscore the multifaceted nature of the research domain, which encompasses environmental, technical, and managerial perspectives to promote sustainable transitions, supported by measurable performance indicators.

3 Understanding the Need for Circular Economy Metrics

3.1 Distinction Between Linear and Circular Performance Indicators

Linear economy indicators center on the traditional 'take-make-dispose' model, quantifying aspects such as resource extraction, production, consumption, and disposal. Standard metrics include total waste generated, the use of virgin materials, and landfill rates. These indicators primarily measure efficiency and throughput, yet they lack incentives for resource recirculation or waste reduction [11, 12].

In contrast, circular economy indicators aim to extend the lifecycle of resources by emphasizing reuse, recycling, durability, and the potential for repair. Metrics such as the Material Circularity Indicator (MCI), Circularity Gap Metric, and Linear Flow Index (LFI) assess the effectiveness of reintroducing materials into productive use. In recent times, various types are creating challenges; municipal, biomedical and hazardous waste and emerging contaminants are the among the priority concern for proper handling disposal and management [13–16]. The fundamental distinction lies in their focus: linear indicators track depletion and disposal, whereas circular indicators prioritize resource retention, waste minimization, and overall system sustainability [17–19].

4 Roles of Metrics in Policy-Making, Business Strategy, and Community Engagement

Metrics play a pivotal role across multiple domains, including policy-making, business strategy, and community engagement. In policy-making, they enable governments to establish baselines, monitor progress toward sustainability goals, and develop effective regulatory frameworks, such as extended producer responsibility and recycling targets. They also support compliance with international agreements and guide public investment in infrastructure and innovation.

In business strategy, metrics help companies identify inefficiencies, optimize resource utilization, and reduce operational costs. They are crucial for Environmental, Social, and Governance (ESG) reporting, attracting sustainability-driven investors, and achieving certifications like Cradle to Cradle. Moreover, they facilitate benchmarking against industry standards and competitors, fostering innovation in product design and supply chains.

Within community engagement, metrics enhance transparency and accountability, thereby strengthening trust with consumers and local stakeholders. They also promote social benefits, including job creation in recycling and remanufacturing sectors, and encourage collaborative efforts among businesses, NGOs, and citizens to advance local sustainability initiatives, supporting greater social equity and inclusion [20–22] (Fig. 5).

4.1 Types of Stakeholders

Metrics that measure materials flow are important assessment tools for the efficiency and sustainability of resource use in economic systems. The Material Circularity Indicator (MCI) evaluates the restorative materials flows by determining the amount of material will be cycled back for use instead of disposed of [23]. Resource productivity represents the economic worth generated for each unit of resource procurer thereby promoting more efficient use and reuse of materials [24]. Input/output ratios can provide insight into sustainability by comparing the proportion of recycled or renewable input to the total inputs made; and/or evaluating how much of the output is reused, recycled or disposed of [25]. More recently, advances of geospatial analysis of waste-water treatment operations have depicted how tracking spatial flows are beneficial for assessing circularity in these infrastructure systems [26]. Finally, the percentage of product recyclability evaluates the percentage of a product's material

Fig. 5 Segregate waste, support Swachh Bharat. Clean Airport, Clean India, at Indira Gandhi International Airport, New Delhi, India. (Photo by: Saurav Narayan)

that could be typically recycled at its end-of-life and may provide a good indicator of circularity potential [27] (Table 1).

Table 1 Stakeholders and their Information

Stakeholder	Information	Example
Government	Foundational statistics, development tracking, compliance data, societal impact figures [28]	Crafting policies, enforcing regulations, submitting reports to global organizations
Businesses	Analysis of resource flows, data on waste minimization, circular economy indicators, certification preparedness [29]	Business strategy, ESG disclosures, fostering innovation, managing risks
NGOs	Environmental/social impact indicators, openness benchmarks, community involvement data [30]	Promoting advocacy, raising public awareness, assessing project outcomes
Financial Sector	ESG metrics, risk evaluations, insights on investment consequences [31]	Making responsible investment choices, minimizing financial risks
Communities	Employment statistics, public health and environmental wellbeing data [32]	Designing community projects, driving local progress, planning with citizen input
Academia	Extensive databases, clarity in methods [33]	Academic research, shaping policies, innovating technologies

5 Categories of Circular Economy Metrics

5.1 Material Flow Metrics

Material flow metrics are crucial tools for assessing the efficiency and sustainability of resource utilization within economic systems. The Material Circularity Indicator (MCI) assesses how restorative material flows are by measuring the extent to which materials are cycled back into use rather than being discarded [23]. Resource productivity captures the economic value generated per unit of resource input, promoting the efficient use and reuse of materials [24]. Input/output ratios offer insight into sustainability by comparing the proportion of recycled or renewable inputs to total inputs, or evaluating how much of the output is reused, recycled, or disposed of [25]. Additionally, the percentage of product recyclability measures the share of a product's material that can be effectively recycled at the end of its life, providing a clear indicator of its circular potential [27].

5.2 Environmental Impact Metrics

Environmental impact metrics are crucial for evaluating the sustainability of products and processes throughout their entire lifecycle. The carbon footprint quantifies total greenhouse gas emissions across a product's or organization's lifecycle, offering insight into the climate implications of circular strategies [34]. Water and energy usage metrics measure both total and circular (reused or recycled) consumption, underscoring the efficiency and sustainability of resource use [35]. Lifecycle Assessment (LCA) provides a comprehensive evaluation of environmental impacts from raw material extraction through to end-of-life, guiding decisions that aim to reduce adverse ecological effects and enhance overall sustainability performance [36].

5.3 Economic Metrics

Economic metrics are essential for evaluating the financial performance and viability of circular economy practices. Circular Value Added reflects the additional economic Value created through activities such as remanufacturing, recycling, and product-as-a-service models, highlighting the contribution of circular strategies to economic growth [37]. Cost savings from resource efficiency quantify the financial benefits gained by minimizing waste, utilizing recycled materials, or extending product lifespans, emphasizing operational advantages [38]. Return on Circular Investment (ROCI) evaluates the profitability of investments in circular initiatives, providing crucial support for developing robust business cases and promoting the broader adoption of circular strategies [39].

5.4 Social Metrics

Social metrics provide insight into the human and societal impacts of circular economy initiatives. Job creation in circular sectors refers to the employment generated through activities such as recycling, repair, and remanufacturing, highlighting the potential for economic inclusion and workforce development [40]. Health and community well-being assess the positive effects of circular practices on public health, safety, and quality of life, such as reductions in pollution and the promotion of safer, more sustainable products [41, 42]. Equity in access to circular infrastructure assesses the fairness and inclusivity of community access to services such as recycling, repair, and reuse, ensuring that the benefits of circular transitions are distributed equally across all segments of society [43].

5.5 Systemic Metrics

System-level metrics provide a comprehensive view of circular economy performance across broader geographic or institutional scales. Urban Metabolism Analysis investigates the flow of materials and energy within cities or regions, revealing patterns and inefficiencies that inform targeted circular interventions at the systemic level [13, 15, 44]. Circularity Gap Indicators quantify the disparity between existing circular practices and their potential, whether at global, national, or local scales, thus identifying the progress made and the gaps that remain [45]. These metrics are essential for guiding policy, planning, and investment toward more sustainable and circular urban and regional systems (Fig. 6).

6 Tools and Methodologies for Assessment

Table 2 describes key circular economy assessment tools and methodologies, highlighting their purpose, scope, outputs, and practical applications across product, organizational, and national levels.

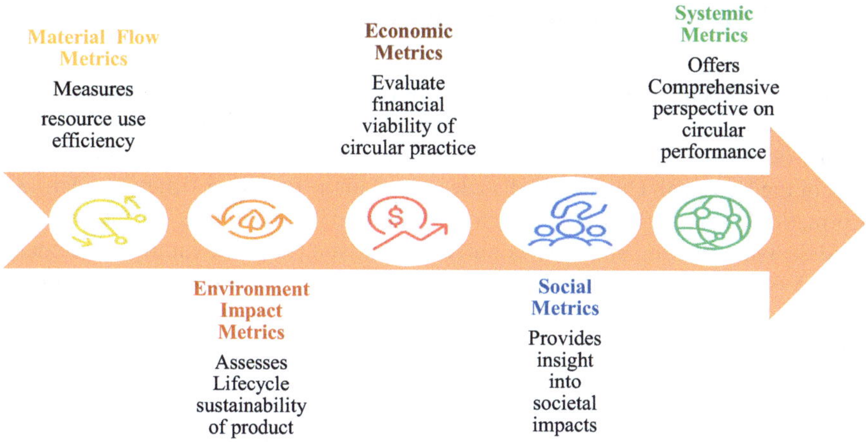

Fig. 6 Categories of circular economy metrics

Table 2 Key circular economy tools by scope, purpose, and outputs

Tool/methodology	Purpose	Scope/scale	Key outputs	Use cases
Life Cycle Assessment (LCA) [46]	Assess environmental impacts from cradle-to-grave	Product/Process level	CO_2 water use, resource depletion data	Eco-design, sustainability reporting, compliance
Material Flow Analysis (MFA) [47, 48]	Track physical material flows in a system	Organization, city, country	Input/output flows, stock changes, waste flows	Resource efficiency, waste management
Input-Output Analysis [49]	Map sectoral economic and material interactions	Macroeconomic/sectoral level	Input-output matrix, environmental impact modeling	National analysis, emissions accounting
Product-Level Metrics [50]	Measure product circularity characteristic	Product level	Recyclability index, EoL recovery rate, repair score	Product design, certification, marketing
Corporate Tools (Circulytics, CTI) [51]	Assess organizational circularity performance	Company/corporate level	Circularity scores, gap analysis	Strategy, monitoring, investor communications

7 Benchmarking and Reporting Frameworks

Reporting and benchmarking strategies: The two practices are significant in testing the efficiency and the feasibility of the circular economy (CE) efforts of different regions or sectors. The recently proposed Circular Economy Monitoring Framework of the European Union, which is described in the literature, offers standardized information on production, consumption, waste, and innovations as well as evidence-based decision-making and inter-country comparison [52]. The family of standards of ISO 59000 include the ISO 59004 (guiding principles) and ISO 59010 (implementation guidance) documents and ISO 59020 (measurement framework) that may be applied to the implementation and evaluation of the CE systems across the globe [53, 54]. These rules complement the lifecycle assessment process of ISO 14040 so as to add harmony in quantifying the CE. Moreover, the concept of circularity is included in plans of world sustainability of the United Nations Sustainable Development Goals (SDGs, to be more exact, SDG 12) through responsible production and consumption [55]. The Circularity Gap Report prepared by the Ellen MacArthur Foundation has been added to the global and national benchmarking reports by providing a clue of the reuse material progress and reporting the main issues of circularity bottlenecks [45]. Some countries like Japan and the Netherlands have localized their scorecards to reflect the local situation to support more strategic positioning and inclusion of the local environment. These frameworks are anchored in transparency, comparability, and investment-readiness of circular transitions in their aggregate.

8 Frameworks and Standards for CE Evaluation

8.1 ISO Standards

ISO standards play a crucial role in establishing a unified framework for implementing and measuring circular economy practices globally. Developed by the ISO/TC 323 technical committee on circular economy, these standards provide consistent guidance across sectors and regions [56]. ISO 59004 outlines key vocabulary, principles, and transition guidelines, emphasizing sustainable resource management, systemic resilience, stakeholder engagement, and continuous improvement [53]. ISO 59010 provides practical guidance on integrating circular principles into business models and value networks, fostering cross-sector collaboration and ongoing performance evaluation [57]. ISO 59020 presents a structured approach to measuring circularity at regional, organizational, and product levels, standardizing data collection, indicator selection, and reporting [54]. Aligned with the UN Sustainable Development Goals (SDGs), ISO 59020 promotes transparency, supports informed decision-making, and enables organizations to minimize resource use while enhancing accountability.

8.2 European Commission's CE Monitoring Framework

The European Commission's Circular Economy Monitoring Framework serves as a comprehensive tool for tracking the EU's transition towards a circular economy. It encompasses key areas such as production, consumption, waste management, secondary raw materials, and competitiveness and innovation. By offering harmonized indicators, the framework enables policymakers and stakeholders to assess system-wide circularity identify progress and gaps, and design targeted policy interventions to address these areas. It plays a vital role in aligning EU strategies with circular economy objectives, fostering evidence-based decision-making and cross-sectoral collaboration [58].

8.3 Global Reporting Initiative (GRI) Standards for Circularity

The GRI Standards offer specific disclosures highly relevant to circular economy practices, including those related to waste generation, material use, and resource efficiency. Organizations adopt these standards to report transparently on their circular performance, enhancing accountability and enabling consistent, comparable assessments across sectors. By aligning with GRI, companies support informed stakeholder engagement, build trust, and demonstrate their commitment to sustainable and circular resource management [59].

8.4 Doughnut Economics and Sustainable Development Goals (SDG) Integration

Doughnut Economics, developed by Kate Raworth, presents a comprehensive framework that integrates planetary boundaries with social foundations to guide sustainable and equitable economic development. It urges cities and organizations to evaluate their circularity within both ecological ceilings and social minimums, ensuring that economic activity remains within safe and just limits [60]. Complementing this approach, the integration of the Sustainable Development Goals [61] within circular economy standards, particularly ISO 59020 and ISO 59004, reinforces alignment with global objectives. These standards support targets related to responsible consumption and production, climate action, and social equity, embedding circularity within broader sustainability agendas [55].

9 Case Examples of Circular Metrics

Alongside technological benchmarks, ecological consequences of waste streams, particularly e-waste, highlight an urgent need for biodiversity-integrated circular indicators [62]. Sector-specific circular economy metrics offer focused assessments of how efficiently materials are used, waste is minimized, and resources are recovered within various industries. In *construction*, these metrics emphasize the reuse of materials, adaptable designs, and the reduction of embodied carbon [63]. For *electronics*, the focus lies on ease of repair, product collection rates, and the incorporation of secondary raw materials [64]. In *agriculture and food systems*, indicators evaluate food waste reduction, nutrient recovery, and the use of recycled resources such as water [65]. The *fashion and textile* sector monitors factors like the percentage of recycled materials used, the durability of garments, and the effectiveness of take-back programs [66]. Meanwhile, in *automotive and manufacturing*, key metrics include rates of remanufacturing, use of recycled inputs, and the integration of renewable energy [27]. These customized metrics enable industries to track progress, meet regulatory requirements, and foster circular innovation.

Leading organizations are setting benchmarks in circular economy measurement through targeted, practical case studies described in Table 3. Dell integrates recycled content tracking into its electronics manufacturing, while the Ellen MacArthur Foundation's Circularity Gap Index assesses global resource cycles. Unilever focuses on packaging recyclability in consumer goods. Metabolic applies urban metabolism analysis for strategic urban planning, and Toyota measures vehicle recovery rates to improve automotive circularity. These best practices demonstrate how tailored metrics can drive sector-specific sustainability and support broader circular transitions.

Table 3 Examples of best practices in circular economy metrics by leading organizations

Organization/project	Circular metric used	Application area
Dell Technologies [67]	Recycled content percentage	Electronics manufacturing
Ellen MacArthur Foundation—Circularity Gap Report [52]	Circularity Gap Index	Global resource use
Unilever [68]	Packaging recyclability rate	Consumer goods packaging
Metabolic (Netherlands) [69]	Urban metabolism analysis	City-wide sustainability strategy
Toyota [70]	End-of-life vehicle recovery rate	Automotive recycling

10 Challenges in Measuring Circular Economy Performance

Measuring CE progress faces numerous challenges as described in Table 4. The lack of standardized metrics hampers comparability, while tools often overlook social and systemic dimensions. Data limitations, particularly in lower-income regions, further restrict the effectiveness of measurement. Metrics typically fail to adapt across sectors or regions, emphasize quantitative over qualitative insights, and inadequately capture long-term outcomes. Their poor alignment with strategic policy frameworks reduces their practical relevance, and competing frameworks create fragmentation, complicating the coherent application and reducing the overall impact in advancing circularity goals.

Table 4 Barriers to effective circular economy metrics

Aspect	Description
Lack of Standardized Metrics [71]	The absence of established indicators for a CE has created disparate assessments and minimal comparability across regions and sectors. Advanced techniques such as fuzzy logic integrated with life cycle assessment frameworks have been proposed to enhance uncertainty management and scenario comparability in circular waste management systems [72]. The absence of established indicators for a CE has created disparate assessments and minimal comparability across regions and sectors
Narrow Focus of Existing Tools [73]	CE tools and approaches emphasize material efficiency or environmental outputs: they often ignore issues of social equity, behavioral aspects, and the systemic dimension
Data Gaps and Quality Issues [74]	The restricted access to high-quality, non-aggregated, or real-time data, in particular for low- and middle-income regions, ultimately limits our ability to develop reliable metrics
Sectoral and Regional Variability [75]	Most metrics do not allow enough flexibility for differences seen across industrial processes, infrastructure, or cultural contexts
Overemphasis on Quantitative Metrics [76]	Technical metrics (such as material flows) often focus heavily on quantitative assessments and neglect qualitative insights, important behavioral factors affecting change, and user engagement essential to support systemic responses on the circularity level
Difficulty Capturing Long-term Impacts [77]	Most measures look at short-term outputs and indicators based on resilience and regeneration are still underdeveloped
Poor Integration with Policy and Strategy [78]	The limited relationship of CE metrics to strategic planning tools and processes (e.g., ESG frameworks, national reporting) limits their influence and impact
Fragmentation Across Frameworks [76]	The presence of multiple competing frameworks (ISO, EU, GRI, Circulytics) limits the integration and potential impact of CE metrics and creates confusion and duplication

11 Future Directions

Build a true '*nervous system*' for the circular economy through four sequenced moves. First, harmonize metrics globally: co-create a modular Core Circular Metrics List ($\leq$25 indicators) that aligns ISO 59020, the EU framework, GRI and industry dashboards, pilot it across G20 nations by 2028, and embed it in national statistics by 2030. Second, hardwire these metrics into mainstream ESG and impact-investing regimes (ISSB, SASB, SFDR) so return-on-circular-investment, material circularity and climate co-benefits shape capital flows and ratings. Third, pivot from rear-view reporting to foresight with AI-enabled modelling: federated data, generative BOM twins, graph neural networks and reinforcement learning to surface loops, simulate policies and optimize interventions in real time. Fourth, lock in equity by integrating social and cultural indicators—via a Just Circularity Index, participatory data methods and mandatory Social Circularity Impact Statements. Harmonization supplies the data substrate, ESG integration mobilizes finance, AI delivers predictive intelligence, and social metrics guarantee justice—together accelerating a circular transition at the speed and scale required. AI-Enabled Predictive Circularity Modelling.

Shift from ex-post reporting to ex-ante scenario analytics by coupling federated data platforms with AI and machine-learning models:

- Generative-BOM twins reconstruct bill-of-materials data where suppliers are opaque, improving lifecycle assessment inputs.
- Graph neural networks detect hidden material loops or industrial symbiosis matches in real-time. Case studies in industrial cities like Kryvyi Rih demonstrate how AI-driven environmental predictions can enhance CE modeling accuracy, particularly in emissions and air quality forecasting [79].
- Reinforcement-learning agents optimise policy portfolios (e.g., EPR fees + deposit-return rates) for maximum circular value added under budget constraints. Explainable AI models, such as SHAP-based ensembles, are also advancing predictive capacities in water quality forecasting offering transferable insights for environmental CE metric systems [80].

12 Policy Implications and Decision-Making

Metrics play a vital role in shaping and assessing policy, providing both quantitative and qualitative insights that support goal-setting, performance tracking, and continuous improvement (Insight7 2025). In public procurement, they go beyond cost and speed to encompass compliance, efficiency, and effectiveness—ensuring alignment with policy goals and enabling strategic performance enhancement [81]. For developing countries, strong metric and monitoring systems are essential for managing equitable transitions, particularly in climate adaptation and resilience, by linking funding to real-world needs, promoting accountability, and supporting inclusive growth through the tracking of social, economic, and equity outcomes [82].

13 Conclusion

Within the frame of this chapter, the objective—to evaluate how circular-economy (CE) performance is measured and to outline next-generation solutions—is achieved by (1) a 2015–2024 bibliometric scan evidencing rapid growth, (2) a stakeholder–decision matrix clarifying data needs, (3) classification of metrics into five families with representative tools, (4) critical reading of core frameworks (ISO 59000 series, EU set, GRI, Doughnut–SDG links), (5) sector vignettes translating theory to practice, and (6) synthesis of eight barriers plus four future directions (harmonisation, ESG convergence, AI modelling, equity metrics). This breadth defines the scope—spanning product to system levels and material to social dimensions—and directly mirrors the title *Performance Metrics for Circular Economy: Techniques for Evaluating Effectiveness and Measuring Progress.* Consequently, the chapter culminates in a unified, data-driven roadmap that equips researchers, regulators, investors, and communities to scale CE transitions with speed, comparability, and justice.

References

1. Abu-Bakar H, Charnley F (2024) Developing a strategic methodology for circular economy roadmapping: a theoretical framework. Sustainability 16(15):6682. https://doi.org/10.3390/su16156682
2. Behal A, Islam S, Tan CSL (2025) Exploring circular economy practices in school uniforms: a study on parental perspectives from Australia. Sustainability 17(11):5158. https://doi.org/10.3390/su17115158
3. Herrador M, Van ML (2024) Circular economy strategies in the ASEAN region: a comparative study. Sci Total Environ 908:168280. https://doi.org/10.1016/j.scitotenv.2023.168280
4. Osei A, Agyemang AO, Cobbinah J (2025) Strategic sustainability initiatives and the circular economy: insights from firm-level targets, board dynamics, stakeholder pressure, and digital transformation. Bus Ethics Eur Rev. https://doi.org/10.1111/beer.12818
5. Bukowski L, Werbinska-Wojciechowska S (2025) Towards Maintenance 5.0: resilience-based maintenance in AI-driven sustainable and human-centric industrial systems. Preprints. https://doi.org/10.20944/preprints202507.0345.v1
6. Tognato de Oliveira C, Groff Andrade Oliveira G (2023) What circular economy indicators really measure? An overview of circular economy principles and sustainable development goals. Resour Conserv Recycl 190:106850. https://doi.org/10.1016/j.resconrec.2022.106850
7. Tripathi S, Bachmann N, Brunner M et al (2024) Assessing the current landscape of AI and sustainability literature: identifying key trends, addressing gaps and challenges. J Big Data 11:65. https://doi.org/10.1186/s40537-024-00912-x
8. European Commission (2020) Circular economy action plan: for a cleaner and more competitive Europe. https://eur-lex.europa.eu/legal-content/EN/TXT/?uri=CELEX%3A52020DC0098

9. Dey S, Santra M, Choudhury M, Ghosh AR, Samanta P (2024) Food waste generation and its industrial utilization: an overview. Environ Sci Pollut Res.1–20. https://doi.org/10.1007/s11356-024-34252-3
10. Singh S, Singh P, Sharma A, Choudhury M (eds) (2022) Agriculture waste management and bioresource: the circular economy perspective. Wiley. https://doi.org/10.1002/9781119808428
11. Mondal T, Choudhury M, Kundu D, Dutta D, Samanta P (2023) Landfill: an eclectic review on structure, reactions and remediation approach. Waste Manag 164:127–142. https://doi.org/10.1016/j.wasman.2023.03.034
12. Paul S, Choudhury M, Deb U, Pegu R, Das S, Bhattacharya SS (2019) Assessing the ecological impacts of ageing on hazard potential of solid waste landfills: a green approach through vermitechnology. J Clean Prod 236:117643. https://doi.org/10.1016/j.jclepro.2019.117643
13. Choudhury M, Adhikari MD, Agarwal S, Samanta P, Sharma A, Kundu D, Kumar S (2025) Pharmaceuticals and personal care products in soil: sources, impacts and myco-remediation strategies. Emerg Contam 2025:100488. https://doi.org/10.1016/j.emcon.2025.100488
14. Choudhury M, Rajpal A, Goswami S, Chakravorty A, Raghavan V (eds) (2024) Analytical case studies on municipal and biomedical waste management: perspectives on sustainable development goals. CRC Press. https://doi.org/10.1201/9781003499695
15. Choudhury M, Choudhary R, Sharma D, Bhattacharjee B, Majumdar S, Chegini S et al (2025) Spatial intelligence integration in smart wastewater systems: advancing efficiency and sustainability in urban sewer networks, Chapter 11. In: Choudhury M, Majumdar S, Goswami S, Sillanpää M (eds) Smart wastewater systems and climate change: innovations through spatial intelligence, vol 15. Royal Society of Chemistry, pp 155–169. https://doi.org/10.1039/9781837676880-00155
16. Rajpal A, Choudhury M, Goswami S, Chakravorty A, Raghavan V (eds) (2024) Waste management and treatment: advances and innovations, 1st edn. CRC Press. https://doi.org/10.1201/9781003258377
17. Hoover Solutions (2024) Circular vs. linear economy: the key differences. Hoover CS. https://hooversolutions.com/circular-vs-linear-economy-the-key-differences/. Accessed 14 Jul 2025
18. PACE Circular (Platform for Accelerating the Circular Economy) (2020) Circular metrics for business: the landscape. PACE Circular. https://pacecircular.org/sites/default/files/2021-03/5faa4d272e1a82a1d9126772_20201029%20-%20BCG%20Metrics%20-%20White%20Papers%20-%20The%20Landscape%20-%20210_x_297_mm%20-%20bleed_3_mm.pdf
19. Rocchi L, Paolotti L, Cortina C, Boggia A, Tenuta A (2021) Measuring circularity: an application of modified material circularity indicator to agricultural systems. Agric Food Econ 9(9). https://doi.org/10.1186/s40100-021-00182-8
20. Meegle (2025) Circular economy metrics. https://www.meegle.com/en_us/topics/circular-economy/circular-economy-metrics. Accessed 14 Jul 2025
21. Nirmani IAP (2024) Understanding the role of community participation in enhancing circular economy outcomes in developing regions. World J Adv Res Rev 23(3):3034–3043. https://doi.org/10.30574/wjarr.2024.23.3.2935
22. Voukkali I, Papamichael I, Loizia P, Lekkas DF, Rodríguez-Espinosa T, Navarro-Pedreño J, Zorpas AA (2023) Waste metrics in the framework of circular economy. Waste Manag Res 41(12):1741–1753. https://doi.org/10.1177/0734242X231190794
23. Poolsawad N, Chom-in T, Samneangngam J, Suksatit P, Songma K, Thamnawat S, Kanoksirirath S, Mungcharoen T (2023) Material circularity indicator for accelerating low-carbon circular economy in Thailand's building and construction sector. Environ Prog Sustain Energy 42(5):e14105. https://doi.org/10.1002/ep.14105
24. Blomsma F, Brennan G (2017) The emergence of circular economy: a new framing around prolonging resource productivity. J Ind Ecol 21(3):603–614. https://doi.org/10.1111/jiec.12603
25. Meglin R, Kytzia S, Habert G (2021) Regional circular economy of building materials: environmental and economic assessment combining material flow analysis, input-output analyses, and life cycle assessment. J Ind Ecol 26(2):586–602. https://doi.org/10.1111/jiec.13205
26. Choudhary R, Kumari S, Kumar A, Kumar P, Choudhury M, Sharma D et al (2025) Optimizing wastewater management through geospatial analysis, Chapter 6. In: Choudhury M, Majumdar

S, Goswami S, Sillanpää M (eds) Smart wastewater systems and climate change: innovations through spatial intelligence, vol 15. Royal Society of Chemistry, pp 72–87. https://doi.org/10.1039/9781837676880-00072
27. AIMultiple (2025) Top 12 circular economy metrics to assess businesses. https://research.aimultiple.com/circular-economy-metrics/. Accessed 15 Jul 2025
28. Daramola OM, Apeh CE, Basiru JO, Onukwulu EC, Paul PO (2024) Environmental law and corporate social responsibility: assessing the impact of legal frameworks on circular economy practices. Int J Soc Sci Excep Res 3(1):63–79. https://doi.org/10.54660/IJSSER.2024.3.1.63-79
29. Parchomenko A, Nelen D, Gillabel J, Rechberger H (2019) Measuring the circular economy: a multiple correspondence analysis of 63 metrics. J Clean Prod 210:200–216. https://doi.org/10.1016/j.jclepro.2018.10.357
30. Organisation for Economic Co-operation and Development (2021) Social impact measurement for the social and solidarity economy (LEED Paper No. 2021/05). OECD Publishing. https://doi.org/10.1787/d20a57ac-en
31. El Wassal KA (2023) Circular economy and sustainable finance: a bibliometric review. Res Int Bus Financ 65:102045. https://doi.org/10.1016/j.ribaf.2023.102045
32. Lamprom W (2025) Circular economy and community participation in Chao Phraya River conservation: evidence from Thai urban riverside communities. Clean Respons Consump 2025:100299. https://doi.org/10.1016/j.clrc.2025.100299
33. Merli R, Preziosi M, Acampora A (2018) How do scholars approach the circular economy? A systematic literature review. J Clean Prod 178:703–722. https://doi.org/10.1016/j.jclepro.2017.12.112
34. Bherwani H, Nair M, Niwalkar A, Balachandran D, Kumar R (2022) Application of circular economy framework for reducing the impacts of climate change: a case study from India on the evaluation of carbon and materials footprint nexus. Energy Nexus 5:100047. https://doi.org/10.1016/j.nexus.2022.100047
35. Smol M, Mejia A, Gastaldi M, D'Adamo I (2024) Environmental indicators for assessment of circular economy (CE) implementation in the water and wastewater sector. Bus Strateg Environ. https://doi.org/10.1002/bse.4090
36. Gallo F, Manzardo A, Camana D, Mazzi A, Scipioni A (2024) Integration of a circular economy metric with life cycle assessment: methodological proposal of compared agri-food products. Int J Life Cycle Assess 29:1359–1379. https://doi.org/10.1007/s11367-022-02130-0
37. Sada S, Ikeda Y (2021) Regional economic integration via detection of circular flow in international value-added network. PLoS One 16(8):e0255698. https://doi.org/10.1371/journal.pone.0255698
38. Onyeaka H, Tamasiga P, Nwauzoma UM, Miri T, Juliet UC, Nwaiwu O, Akinsemolu AA (2023) Using artificial intelligence to tackle food waste and enhance the circular economy: maximising resource efficiency and minimising environmental impact: a review. Sustainability 15(13):10482. https://doi.org/10.3390/su151310482
39. Murali M, Kayal P, Maiti M (2025) Should you invest in the companies that promote the circular economy idea? Manag Environ Qual 36(4):877–894. https://doi.org/10.1108/MEQ-07-2022-0193
40. Nademi Y, Sedaghat Kalmarzi H (2025) Breaking the unemployment cycle using the circular economy: sustainable jobs for sustainable futures. J Clean Prod 488:144655. https://doi.org/10.1016/j.jclepro.2025.144655
41. Sharma S, Bangur P (2025) Circling to wellness: health implications of transitioning to a circular economy. Circ Econ Sustain 5:1835–1864. https://doi.org/10.1007/s43615-025-00507-5
42. Tripathi S, Agarwal S, Choudhary R, Choudhury M, Padoin N, van Hullebusch ED et al (2025) Strengthening disaster resilience: emergency response strategies in wastewater management, Chapter 10. In: Choudhury M, Majumdar S, Goswami S, Sillanpää M (eds) Smart wastewater systems and climate change: innovations through spatial intelligence, vol 15. Royal Society of Chemistry, pp 135–154. https://doi.org/10.1039/9781837676880-00135

43. Ashton WS, Fratini CF, Isenhour C, Krueger R (2022) Justice, equity, and the circular economy: introduction to the special double issue. Local Environ 27(10–11):1173–1181. https://doi.org/10.1080/13549839.2022.2118247
44. Ragazou K, Zournatzidou G, Sklavos G, Sariannidis N (2024) Integration of circular economy and urban metabolism for a resilient waste-based sustainable urban environment. Urban Sci 8(4):175. https://doi.org/10.3390/urbansci8040175
45. Bahramimianrood B, Xie S, Malaibari M, Abdoli S (2024) Reviewing circularity indicators for a sustainable transition to a circular economy. Procedia CIRP 131:829–834. https://doi.org/10.1016/j.procir.2024.01.144
46. International Organization for Standardization. (2006). ISO 14040:2006 Environmental management—Life cycle assessment— Principles and framework. https://www.iso.org/standard/37456.html
47. Binder, C. R. (2007). From material flow analysis to material flow management Part II: The role of structural agent analysis. *Journal of Cleaner Production, 15*(17), 1605–1617. https://doi.org/10.1016/j.jclepro.2006.08.017
48. Sendra, C., Gabarrell, X., & Vicent, T. (2007). Material flow analysis adapted to an industrial area. *Journal of Cleaner Production, 15*(17), 1706–1715. https://doi.org/10.1016/j.jclepro.2006.08.019
49. Leontief, W. (1986). Input-output economics (2nd ed.). Oxford University Press. https://liremarx.noblogs.org/files/2020/02/Wassily-Leontief-Input-Output-Economics-Oxford-University-Press-USA-1986.pdf
50. International Organization for Standardization. (2013). ISO 14021:2013 Environmental labels and declarations—Self-declared environmental claims (Type II environmental labelling). https://www.iso.org/obp/ui#iso:std:iso:14020:ed-2:v1:en
51. Ellen MacArthur Foundation. (2021). Circulytics: Measuring circular economy performance. Retrieved July 23, 2025, from https://www.ellenmacarthurfoundation.org/resources/circulytics/resources
52. Rigamonti L, Mancini E (2021) Life cycle assessment and circularity indicators. Int J Life Cycle Assess 26:1937–1942. https://doi.org/10.1007/s11367-021-01966-2
53. Heras-Saizarbitoria I, Boiral O, Testa F (2024) ISO 59000 standards for the circular economy: a call for accuracy. Circ Econ Sustain 4:1669–1675. https://doi.org/10.1007/s43615-024-00370-w
54. Schmidt M (2025) The actual goals and limits of circular economy—a critical perspective (Chapter 16). In: Araujo JB, Hinderer H, Viere T, Woidasky J (eds) Circular economy: from review to assessment. Wiley-VCH. https://doi.org/10.1002/9783527847358.ch16
55. Fonseca MJL (2024) Localizing the sustainable development goals: the case of Porto. Doctoral dissertation, Universidade do Porto. ProQuest Dissertations & Theses Global. https://www.proquest.com/openview/6ae21a17d459b2886df6795600a20666/1?pq-origsite=gscholar&cbl=2026366&diss=y
56. Patti S (2023) Eco-efficiency, circularity measurement and assessment (Chapter 7). In: Circular economy and policy. Springer. https://doi.org/10.1007/978-3-031-43324-5_7
57. Morris KC, Ferrero V, Hapuwatte B, Last N, Mathur N (2024) Standards as enablers for a circular economy. In: Nasr N (ed) Handbook of circular economy (Chapter 1). Wiley. https://doi.org/10.1002/9781394214297.ch1
58. Llorente-González LJ, Vence X (2019) Decoupling or 'decaffing'? The underlying conceptualization of circular economy in the European Union monitoring framework. Sustainability 11(18):4898. https://doi.org/10.3390/su11184898
59. Opferkuch K, Caeiro S, Salomone R, Ramos TB (2021) Circular economy in corporate sustainability reporting: a review of organisational approaches. Bus Strateg Environ 30(8):4011–4032. https://doi.org/10.1002/bse.2854
60. Siriwardhane D (2025) Doughnut economics: seven ways to think like a 21st-century economist: Raworth, Kate. Chelsia Green Publishing, Vermont, 2017 (ISBN 9781603586740, Price US $35.24). World Fut 81(4):318–327. https://doi.org/10.1080/02604027.2025.2480876

61. Bhattacharjee N, Narayan S (2024) From shelter to the streets: the feminine face of homelessness in contemporary democracies. Int J Human Rights 28(5):852–882. https://doi.org/10.1080/13642987.2024.2327374
62. Chowdhury PR, Goswami S, Choudhary R, Choudhury M (2025) Electronic wastes, its impact on wildlife and biodiversity. In: Electronic waste (Chapter 8). CRC Press. https://doi.org/10.1201/9781003582311-8
63. Moraga, G., Huysveld, S., Mathieux, F., Blengini, G. A., Alaerts, L., Van Acker, K., de Meester, S., & Dewulf, J. (2019). Circular economy indicators: What do they measure? *Resources, Conservation and Recycling, 146*, 452–461. https://doi.org/10.1016/j.resconrec.2019.03.045
64. Shevchenko, T., Esfandabadi, Z. S., Ranjbari, M., Saidani, M., Mesa, J., Shevchenko, S., Yannou, B., & Cluzel, F. (2024a). Metrics in the circular economy: An inclusive research landscape of the thematic trends and future research agenda. *Ecological Indicators, 165,* 112182. https://doi.org/10.1016/j.ecolind.2024.112182
65. Shevchenko, T., Esfandabadi, Z. S., Ranjbari, M., Saidani, M., Mesa, J., Shevchenko, S., Yannou, B., & Cluzel, F. (2024b). Metrics in the circular economy: An inclusive research landscape of the thematic trends and future research agenda. *Ecological Indicators, 165*, 112182. https://doi.org/10.1016/j.ecolind.2024.112182
66. Systemiq & acatech – National Academy of Science and Engineering. (2021). An analysis of available circular economy metrics to support policy- and decision-making (CE Metrics Report). https://www.systemiq.earth/wp-content/uploads/2021/08/CE-Metrics-Report_Final.pdf
67. Aggarwal R, Singh H, Garg P (2023) Dell Technologies Inc.: innovating toward circularity. J Inf Technol Teach Cases 15(1):133–141. https://doi.org/10.1177/20438869231215089. (Original work published 2025)
68. Gray M (2022) Packaging design in the circular economy: reuse models in fast-moving consumer goods. Master's thesis, University of Texas at Austin. Texas ScholarWorks. https://doi.org/10.26153/tsw/45987
69. Tsui T, Derumigny A, Peck D, van Timmeren A, Wandl A (2022) Spatial clustering of waste reuse in a circular economy: a spatial autocorrelation analysis on locations of waste reuse in The Netherlands using global and local Moran's I. Front Built Environ 8:954642. https://doi.org/10.3389/fbuil.2022.954642
70. Bhari B, Yano J, Sakai S (2021) Comparison of end-of-life vehicle material flows for reuse, material recycling, and energy recovery between Japan and the European Union. J Mater Cycles Waste Manag 23:644–663. https://doi.org/10.1007/s10163-020-01154-8
71. Avdiushchenko A, Zając P (2019) Circular economy indicators as a supporting tool for European regional development policies. Sustainability 11(11):3025. https://doi.org/10.3390/su11113025
72. Choudhary R, Kumar A, Narayan S, Mishchenko I, Choudhury M, Bharadwaj LK et al (2025) Integrating fuzzy logic into life cycle assessment for sustainable municipal solid waste management in metro cities. Ecol Eng Environ Technol 26(7):316–331. https://doi.org/10.12912/27197050/205500
73. Chen L-H, Hung P, Ma H-W (2020) Integrating circular business models and development tools in the circular economy transition process: a firm-level framework. Bus Strateg Environ 29(5):1887–1898. https://doi.org/10.1002/bse.2477
74. Serna-Guerrero R, Ikonen S, Kallela O, Hakanen E (2022) Overcoming data gaps for an efficient circular economy: a case study on the battery materials ecosystem. J Clean Prod 374:133984. https://doi.org/10.1016/j.jclepro.2022.133984
75. Corona B, Shen L, Reike D, Carreón JR, Worrell E (2019) Towards sustainable development through the circular economy—a review and critical assessment on current CE practices. J Clean Prod 239:117828. https://doi.org/10.1016/j.jclepro.2019.117828
76. Munonye WC (2025) Towards circular economy metrics: a systematic review. Circ Econ Sustain. https://doi.org/10.1007/s43615-025-00604-5
77. Kulakovskaya A, Knoeri C, Radke F, Blum NU (2023) Measuring the economic impacts of a circular economy: an evaluation of indicators. Circ Econ Sustain 3(2):657–692. https://doi.org/10.1007/s43615-022-00190-w

78. Rajput S, Singh SP (2021) Industry 4.0—challenges to implement circular economy. Benchmarking 28(5):1717–1739. https://doi.org/10.1108/BIJ-12-2018-0430
79. Halaktionov M, Bredun V, Choudhary R, Goroneskul M, Kumar A, Ouiya F, Sydorenko V, Markina L (2025) AI-enhanced air quality assessment and prediction in industrial cities: a case study of Kryvyi Rih, Ukraine. Ecol Eng Environ Technol 26(6):45–56. https://doi.org/10.12912/27197050/203725
80. Choudhary R, Kumar A, Priyadharsini C, Mude M, Naik M, Moharana C, Khan NA (2025) Predicting water quality index using stacked ensemble regression and SHAP based explainable artificial intelligence. Sci Rep 15(1):9463. https://doi.org/10.1038/s41598-025-09463-4
81. Organisation for Economic Co-operation and Development. (2023). Metrics in public procurement: Aligning incentives and performance. OECD Publishing. https://www.oecd.org/publications/metrics-public-procurement
82. Macquarie, R., & Tyson, M. (2025). Monitoring frameworks for just transitions in developing economies. *Climate Policy Journal, 22*(1), 45–60. https://www.climatepolicyjournal.org/just-transition-framework

Shaping Tomorrow: Emerging Trends and Innovations in Urban Waste Management and Circular Economy Integration

Monu Bhardwaj, Prernaa Sharma, Rekha Goswami, Utkarsh Verma, Aditya Verma, and Jonti Deuri

Abstract Urban waste management has become an increasingly critical challenge in the face of rapid urbanization, population growth, and resource scarcity. This chapter explores the latest trends and technological innovations transforming the urban waste management landscape. It explores how the integration of circular economy principles offers sustainable solutions for minimizing waste, conserving resources, and reducing environmental impacts. These technologies are not only operationally efficient but also provide policymakers and stakeholders with the information needed to make informed decisions. It highlights the crucial role of urban local bodies, public-private partnerships, and community engagement in establishing sustainable waste management practices. It further discusses the potential of recycling, upcycling, and waste-to-energy technologies in driving the transition to a circular economy. Successful implementations from case studies of cities around the world are illustrated, providing insights into scaling and replication. The chapter emphasizes the systemic approach through policy reforms, innovative financing mechanisms, and capacity-building initiatives to remove barriers to the adoption of a circular economy. It aims to create a future where urban waste is no longer perceived as a problem, but as an opportunity to recapture value through resource recovery and regeneration. This could lead to making cities more sustainable, reducing their carbon footprint, and contributing significantly to the goals set within global environmental and socio-economic objectives.

M. Bhardwaj
School of Management, Graphic Era Hill University, Dehradun, Uttarakhand, India

P. Sharma (✉) · R. Goswami
Department of Environmental science, Graphic Era Hill University, Dehradun, Uttarakhand, India

U. Verma · A. Verma
School of Computer Science and Engineering, Graphic Era Hill University, Dehradun, Uttarakhand, India

J. Deuri
Faculty of Engineering and Technology, Sharda University, Andijan, Uzbekistan

S. Majumdar et al. (eds.), *Sustainable Urban Future*, Urban Sustainability,
https://doi.org/10.1007/978-981-95-2976-6_12

Keywords Circular economy · Waste management · Smart technologies · Waste to energy · Sustainability

1 Introduction

Over the last few decades, our understanding of waste and its environmental impact has evolved. Waste generation continues to rise, and it is scientifically proven that our world cannot sustain the uncontrolled disposal of waste anymore [1]. Additionally, laws and regulations are focusing on waste and circular economy practices and strategies that need to be integrated more fully within local systems. New and improved technologies are emerging and maturing that can help manage waste more efficiently and beneficially for individuals, societies, and the planet [2, 3]. As the scientific community has explored the subject of waste management, new approaches, methods, regulations, and technologies have emerged [4]. The literature on waste management and environmental waste is vast, making it nearly impossible to review thoroughly and comprehensively. The first policies and strategies related to waste management were already in place in Europe in the early 1970s. Since then, it felt like the empirical knowledge and technical expertise had all but grown up. Laws and regulations (both widely in Europe and, at least on paper, across the globe) focus on waste and circular economy practices and strategies. Since waste is an integral part of local systems, it is expected that such practices and strategies will be increasingly integrated into local systems and governance [5]. However, it is at the local level that the theoretical knowledge and practical findings about waste are less well understood within the scientific community (Fig. 1 and Table 1).

On the other hand, new tools are becoming available for analyzing local systems in an integrated view, such as industrial ecology and input-output analysis. Furthermore,

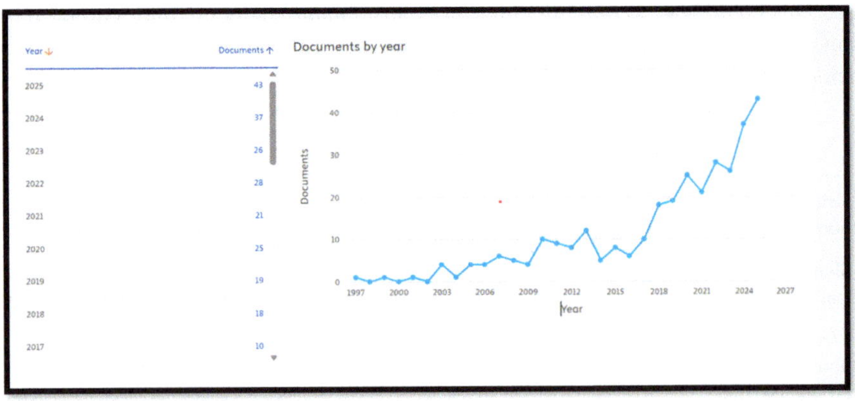

Fig. 1 Document per year

Table 1 Article inclusion and exclusion criteria

Selection criteria	Exclude	Include
Database: "Scopus" Date of Search: "21 June 2025" Period of Publications: 2017–2025		
Waste Management "OR" Circular Economy "OR" Smart Technologies "OR" Waste to Energy "AND" Sustainability	–	182
Subject area: "Air Quality and waste management, and sustainability and Technology"	10	162
Publication type: "Articles, Conference Paper, Book Chapter, and Review"	14	154
Language screening: "Include documents published in English only"	1	145

new and disruptive technologies have emerged and matured [6]. This work focuses on the local waste and material flow system of a medium-sized industrial region, where many systems are activated through technologies. However, only a few are conducive to waste management. This methodology will enable policymakers and analysts to compare the current system with alternative hypothetical scenarios or those of other regions. A more effective local system could become a viable target to aim toward in order to avoid the lock-in effect [7]. On another note, the place-based notion of waste and its environmental, economic, and social implications are underscored here, focusing on two particular aspects. Waste generation is modelled along with socio-economic aspects. Furthermore, the composition and fate of common waste types are analyzed in detail and from multiple perspectives. By doing so, the aim is to provide an empirical base that could be used elsewhere as representative of similar areas while acknowledging that these systems are vastly different and that local conditions intimately regulate the system [8].

2 The Concept of Circular Economy

Cities across the globe have begun to shift away from the traditional linear approach to urban management, based on a "take, make, use, dispose" model, towards more sustainable, circular approaches. This transition is driven in part by new opportunities for cities and in part by a pressing need, particularly as cities confront identified environmental capacity limits and rising urban demand [9]. Complex cities and urban areas are particularly susceptible to the compounding effects of population growth and urbanization, especially when faced with increased waste and more complex and costly urban operations and waste management systems, which may warrant a rethink in terms of both how cities are designed and how they are managed. Current urban design systems encourage and perpetuate wasteful habits and lifestyles that not only waste resources and greenhouse gas emissions but also waste much of the original investment into constructing these urban systems [4, 10]. If urban areas and urban systems were designed and managed differently, these areas could be transformed into significant assets of flourishing value and well-being for their communities and the planet. Urbanization and industrialization have intensified environmental problems, as urban areas house nearly half of the global population, consume 60–80% of industrial materials and energy, and generate most of the world's waste. To address this, the circular economy (CE) model is gaining momentum as demonstrated by the growth of published academic literature and public attention. CE promotes the idea of more efficient resource use, recycling, and regenerative systems. Water and wastewater cleaning is a crucial aspect of constructing circular systems. This is exemplified by 20 countries that have committed to CE on their national agendas, including CE from organic waste, during a time of growing interest in waste recycling and sustainable development.

Moreover, sustainable urban transformation must consider not only individual buildings but also entire stocks of buildings, neighborhoods, and metropolitan areas. The complexity of measuring success on this broader scale necessitates multi-criteria evaluation methods to examine buildings' environmental performance within the context of future urban development. To achieve urban sustainability and resilience over time, holistic solutions are necessary (Fig. 2).

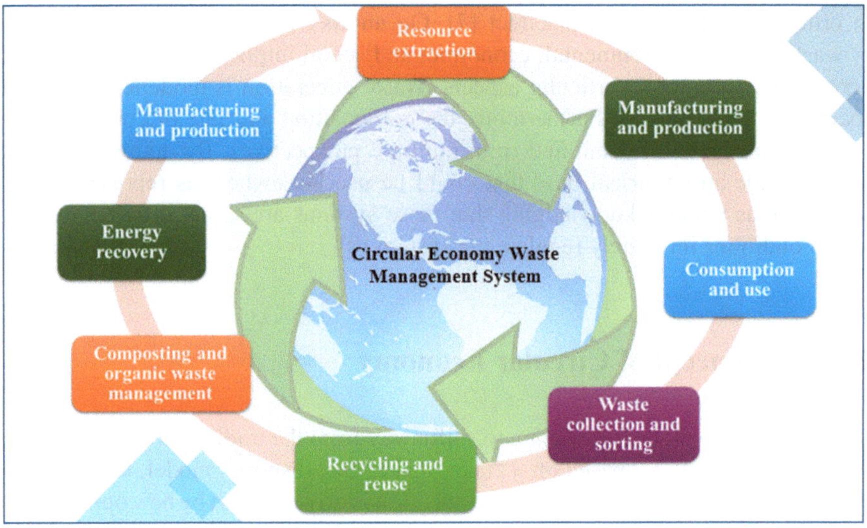

Fig. 2 Circular economy waste management system steps

3 Technological Innovations in Waste Management

Many organizations in the waste management space are expanding their role to provide utilities, equipped with first-hand data and vertically integrated systems throughout the value chain. They have adopted a "capacities and shortages" mindset to enhance performance indicators and have designed small footprint decentralized recycling systems. These systems typically incorporate local transportation with global logistics. They utilize revenue optimization and decision support tools to address the variability in the physical properties of waste and market fluctuations [11]. The stakeholders typically hold roles such as supplier, producer, brand holder, utility provider, and converter, encompassing both the material and energy spaces. In the plastic value chain, a range of potential emergent technologies was highlighted during exploratory analyses of the pipeline. The emphasis in this paper is broadly on chemical processes, but there are also innovations surrounding mechanical and general-purpose technologies. Positive public sentiment is benefiting current developments related to sustainable plastic in the context of various plastic strategies, and technological advancements are occurring within four global regions, with a focus on recycling and maritime initiatives [12, 13].

Geographically, low- and middle-income countries are still heavily dependent on exporting waste and often lack the most basic waste management functions. The economies experiencing the most development are increasing their use of mechanical processing and utilizing local feedstock; however, they are still far behind richer countries. Developed countries are leaders in attractive emerging technologies and have a sophisticated technology profile with overall waste and waste exports [14] (Fig. 3).

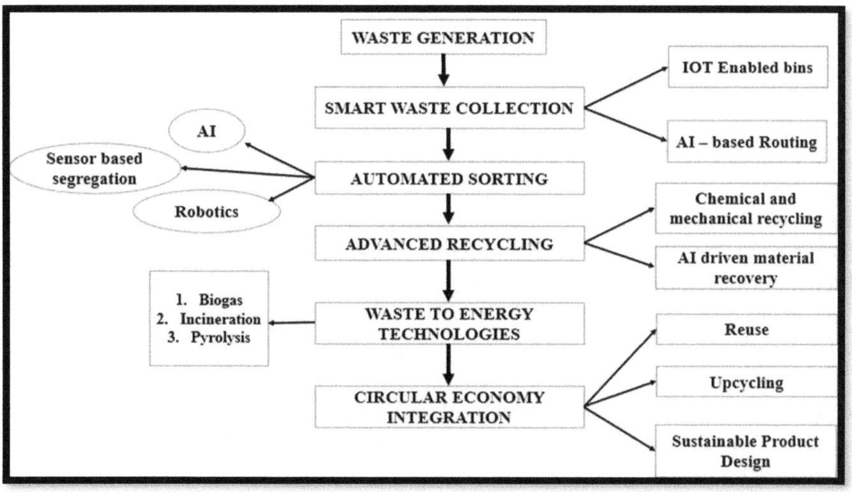

Fig. 3 Flowchart for technological innovations in waste management

3.1 Smart Technologies for Waste Collection and Sorting

Human society generates massive amounts of municipal solid waste (MSW) every day, and the amounts created are increasing at an upward trend. Our current waste disposal methods, such as landfilling and incineration are generally easier, but they pose direct environmental and health hazards due to pollution of the soil, water, and air [15]. In the context of waste disposal, there has been a growing emphasis on waste classification. Waste classification involves sorting waste by type, such as organic waste, paper, plastic, metal, and glass, to ensure safer and more efficient disposal. Organic waste can be composted, paper waste can be recycled into new paper, plastic waste can be processed into chemical fiber materials, and metal waste can be melted down and reused [16]. Nations around the globe have invested colossal amounts in waste-to-energy, composting, and sorting plants. That said, each method has its limitations: incineration can still lead to environmental pollution, composting can result in further unpleasant odors, and sorting waste typically requires human input and is laborious [17]. However, the growth and development of sensing and information technology have enabled the adoption of more sophisticated and automated waste management systems. These specific systems feature sensors and intelligent characteristics for waste classification, enabling personalized waste collection at the household level. Automatic systems reduce human input and labor, improving collection efficiency [18]. These technologies are being increasingly implemented in urban planning. In some cities, their buildings are being equipped with innovative waste storage systems that separate and categorize waste into five different types. For residential waste disposal, sensors must identify the resident to access the specific waste storage area and confirm that the waste has been authorized using RFID tags. In Cardston, the system was equipped with LED indicators to indicate whether disposal was successful, with green indicating acceptance and red indicating an error [19]. All the innovations increase waste sorting and promote a better, more secure environment.

3.2 Waste-to-Energy Technologies

The rapid urbanization and systematic shift in rural-urban lifestyles in the twenty-first century have contributed to a substantial increase in municipal solid waste (MSW), which is projected to rise to 3.4 billion tonnes (a 67% increase from 2015) primarily in urban areas by 2050 [20]. In 2015, only 13.5% of the waste generated was recycled, and 5.5% was recovered as energy [20]. Urban waste, however, can be viewed as an urban resource, and waste-to-energy (WTE) technology, whether thermal or biological, can convert that waste into electricity, heat, or biogas. Waste generation and collection dynamics have shifted due to various issues related to demographics, climate, and lifestyle. Several models have been suggested for forecasting future waste generation (Estimates of future waste generation can be conducted utilizing per capita, regression, and waste models), which grossly rely upon historical data and can also be computationally inefficient. Per capita models are poor at distinguishing the different trends in urban zones. Given all the issues regarding forecasting models, cities desperately need novel forecasting methods. Local and linear optimization models should be developed to enhance efficiency in waste collection, transportation, and resource recovery, especially for small and medium cities [8].

3.3 Biological Treatment of Waste

Waste disposal encompasses all private and public waste disposal activities, which include much of recycling and the valorization of bioresources. One potential opportunity is the urban bio-refinery, where biological processes, such as anaerobic digestion conversion, are applied to items of waste, like fat, oil, and grease in sewer discharge, to produce bio-products, including hydrogen [21]. Business models in this area are based on creating relationships with companies that have an interest in recovering the waste stream of various waste products. The business models would provide direction in terms of recovering or composting material (a waste material at which value can be applied), and what technologies would need to be developed to recover and generate income from the sale of recovered commodities and bio-products [22]. There is an increasing recognition of biological treatment as the preferred strategy for converting organic waste into biofuels and valuable by-products. Due to urbanization and the need for sustainable waste practices, biological treatment is being rapidly adopted [23]. These strategies need to be clearly defined, and there are still significant challenges in developing viable business models and scalable technology that can work effectively and profitably. One example of a business model is to develop bio-refineries in urban contexts that process waste from neighbouring cities. This requires work with regional waste processors, as well as planning considerations for capacity. Advanced modeling is used to evaluate how cities are designed in terms of system performance and optimal partners and costs to accommodate various waste streams [24] (Fig. 4).

3.4 Chemical Recycling Innovations

Georgia, Lugar—Innovative chemical recycling schemes are being investigated to improve urban waste processing and assist a circular economy. Federal and regional programs are focused on producing hydrogen fuel from organic waste, starting with a program to build an urban waste management plant. The urban waste management plant will also assist in sorting and recycling plastic, paper, glass, and metal, thereby decreasing the amount of waste sent to landfills in Tbilisi [25].

Chemical recycling processes such as glycolysis, hydrolysis, and methanolysis are evaluated for post-consumer municipal solid waste. Dispersive Raman and the paper spray ambient mass spectrometry are examples of advanced analytical processes for

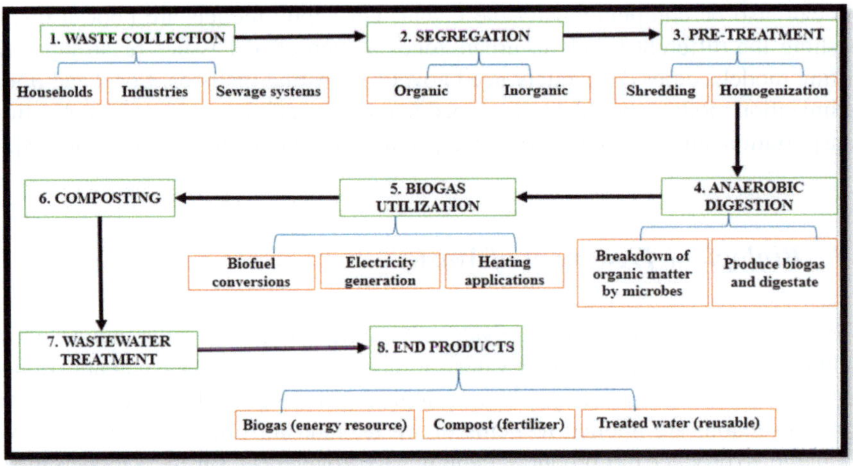

Fig. 4 Steps for biological treatment of waste in urban areas

identifying additives in polyethylene terephthalate (PET) waste. Additionally, the catalytic degradation process of plastics, such as polyethylene, polypropylene, and polystyrene, using metal-imidazolium ionic liquids under different conditions has been reported. Some general trends related to waste recycling and greenhouse gas emissions are also being examined primarily in Russian Federation federal subjects [26], highlighting the impact of waste management practices on the environment and informing future models for waste management.

4 Challenges in Urban Waste Management

Recycling has been acknowledged for more than 2000 years, when even the Romans recycled, until they stopped during military losses—recycling recently regained importance, particularly in some parts of the project, during the economic crisis of 1975–1976 and the first oil crisis, which pushed natural resource and energy prices higher [27]. As of 1986, economic protocols required EU member states to recycle one ton of packaging waste, to be reused to the specific capacity of each country. The Waste Framework Directive was published in 2000 to support the effective use of waste. In 2008, a trailblazing regulation was introduced regarding the charge of an electronic fee for all new products and packaging (domestic or imported) to transfer the cost of recycling waste to the manufacturer [28]. Even with these new strategies, dealing with rubbish is made even more challenging by the growing amount of packaging waste. These challenges to waste management are complicated by urbanization, as waste is often closer to people's homes, and urban waste systems are less effective. The high maintenance or cleaning costs of waste services related to urban

Fig. 5 Unsegregated urban waste open dumping is concern for environment. (Photo By: Moharana Choudhury)

waste, and the confused division of responsibility between commercial waste agents and public waste authorities. Waste management systems are often poorly designed, with a limited understanding of environmental behavior or culture. Many households mix recyclable waste and hazardous waste with regular rubbish. However, local councils often fail to take the time or effort to educate the public or enforce awareness of better waste management practices. In many systems, avoidable waste is combined with recyclable waste, and the simplicity of mixed waste collection services does not effectively convey a clear message or practice with easy-to-understand sorting guidelines; therefore, rubbish and waste continue to be displayed poorly without conviction [29]. There is a practical applicability gap between the practitioners, where situations have policy structures, but gaps in behaviour adoption in public continue to undermine effective urban recycling systems (Fig. 5).

5 Circular Economy Models and Strategies

The Circular Economy (CE) agenda encourages maximizing value from resources by promoting the reuse of materials across technical and natural systems. CE seeks to minimize the extraction of primary resources, landfill use, and environmental contamination, while also generating economic benefits such as innovation, job creation, and new business opportunities [30]. Increasing global awareness of CE has led many countries to develop CE strategies with the increased goal of the industry "doing more with less." In addition to the expected growth from a 71.6% urban population in 2012 to an 84.1% urban population by 2050 in Europe, the European Commission has also been developing policies to achieve sustainable urban development and resource efficiency. However, the approach still faces several key limitations. Firstly, spatial planning remains under the control of individual EU Member States, and the Commission has limited control over these matters—commissioners and government officials can only provide advice. Secondly, there are still no realizable circular design approaches that extend beyond the urban periphery, and large-scale approaches are yet limited, given the past and present advocacy of the CE model [31]. The successful implementation of CE also involves addressing the absence of clear, performance-based frameworks at the metropolitan/regional levels. Policies must be formulated to balance trade-offs and promote interactions in a science-based decision-making framework. The built environment especially complicates the metropolitan shift; urban settings amplify the interdependencies of socio-economic and environmental systems. Understanding these interdependencies is crucial for developing practical CE approaches that are applicable across diverse cities and regions [32, 33]. Moving forward, we need foresight, interdisciplinary collaboration, and adaptive governance to navigate complexity and effect fundamental transformation (Fig. 6).

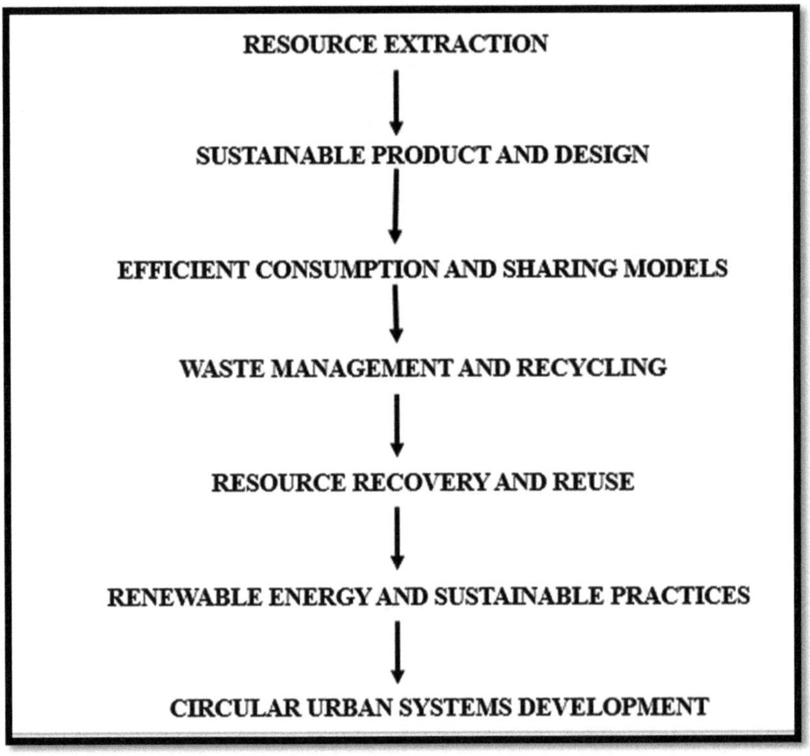

Fig. 6 Circular economy approach in urban development

6 Waste Reduction and Prevention

The new priorities in urban waste management and the integration of circular economy approaches are increasingly focusing on the collection of recyclables and the recovery of resources, sometimes overlooking the broader context of waste management actions. Eco-industrial networks can emerge as a valuable tool for minimizing waste because actors are involved in connected systems where waste from one actor becomes an input for the next. Unfortunately, waste is still frequently treated as a monolithic category, which undermines effective strategies. Zero-waste approaches—especially source reduction and recycling—challenge the linear economy and promote different ways of thinking about waste [34]. Efforts to reclaim valuable materials from waste streams considered of lesser value are also being made. It has been reported that technologies like plasma gasification are effective in handling various types of waste, including agricultural waste and biomedical waste, and are efficient due to their high process throughput and energy generation capabilities [35–37]. The zero-waste village project is a holistic urban design that integrates the processes of organic waste treatment (composting), construction waste management, recyclable waste recycling, and biomedical waste disposal into a unified space. Workers focus on separating and recycling waste, utilizing three plasma technologies as an effective waste treatment process to enhance operational efficiency [38] and secure a social license to operate.

The rapid pace of urbanization and industrialization, along with growing population-led waste generation and environmental limitations, are further pressing issues faced by the Asia-Pacific region. Cities and countries are seeking to address these issues through sustainable waste management strategies, including zero-waste plans and the phasing out of incineration facilities [39]. By-product synergy parks limit their industry's accessibility to external resources and facilitate material exchanges between firms that might not otherwise have the potential to receive materials. The use of integrated systems demonstrates a more sustainable approach to waste management and the increasing importance of innovative and decentralized systems.

7 Resource Recovery and Upcycling

Resource recovery and upcycling can be pursued by creating new platforms and co-designing process chains. It is essential to address distinctions in supply chains, process chains, and upcycling processes. Measurable material properties can inform us about specific treatment steps and lead to a substantial improvement in the properties of intermediate and final products. Upcycling in construction benefits from the pan-European market, with no hazardousness requirements and no regulations for recycling processes. Asia and Sub-Saharan Africa are emerging markets. Here, locally available waste materials are already used to produce non-standard building elements.

Super use and resource recovery can be tangibly pursued through standardized online platforms and by co-designing corresponding process chains, where a group of operators successively treat a supply chain of post-consumer waste materials to create an upcycled product. The treated material loses wear resistance but not its original mechanical properties. It is difficult to predict the performance of the new sample. The potential of the upcycled material is demonstrated by comparing the measurable properties of intermediate products, both before and after treatment, and the final products [40]. A proper experimental approach is presented, mainly featuring basic material tests and addressing intermediate material prerequisites. The benefit is shown in the choice of the up-cycling process. A longer-term example is the result of collaboration with and support from an industry partner.

8 Regulatory Frameworks and Policies

Urban waste management is becoming increasingly important, as urban areas are the source of a wide variety of waste fractions. These fractions are the result of the functioning of urban systems, which are intended as highly complex artificial ecosystems where interactions occur among a large number of social, physical, and biological elements and processes. Global trends are causing urban areas in the developing world to grow rapidly, and they account for approximately 90% of the world's waste. Moreover, as climate change mechanisms have already been set in motion, they will exacerbate existing global inequalities and resource distribution within countries, which will likely have a more significant impact on poorer countries. Urban waste management has a direct impact on street pollution, greenhouse gas (GHG) emissions, and ground pollution, with all of their dramatic consequences for human health.

The European Union has introduced specific policies, such as the Landfill Directive and the Packaging and Packaging Waste Directive, to address the issue of waste management in urban areas. The formulation of the so-called waste hierarchy is a core element of these policies. Other policies, such as the Urban Waste Water Treatment Directive, the Bathing Waters Directive, and the Birds Directive, are part of a broader policy framework aimed at preventing and controlling environmental pollution. On the other hand, the association of pollution with a decrease in economic productivity and the reinforcement of eco-innovations in the realm of the existing waste regulations framework, supported by further environmental and resource scarcity events, as the ones envisioned by some of the experts regarding the future of water or phosphorus consumption [41].

9 International Case Studies in Urban Waste Management

The rapid growth of urban populations in recent decades has significantly increased the volume of municipal solid waste in urban areas, thereby exacerbating environmental degradation. We must make more urgent changes to production and consumption to adopt more environmentally responsible forms. This includes government policies that facilitate stronger regulations for both the circular economy and urban waste management. In addition to that, the Negrete-Cardoso et al. [2, 3] dirtier urban metabolisms paper identified four key challenges for future studies and practice [2, 3] related to these matters—(1) developing waste management practices that reduce the environmental impact of urbanization whilst achieving on the energy and material recovery; (2) exploring and innovating business models that commence novel recycling chain management networks; (3) enabling consumers to normalize social responsibility while moderating excess consumption and (4) devising innovative interdisciplinary, long-term, public-private partnerships, that involve citizen engagement's for citizens to engage address the issues of urban waste at the individual.

Long-term strategies for the future are also concerned with trends like developing circular economy policies out of... contractual agreements with the private sector, developing micro-finance options that promote urban industrialization in informal economies, supporting municipal waste and waste management enterprises that strengthen and promote viable clean energy options while providing options such that reflect urban waste through pricing [42]. Urban solid waste is linked to global environmental concerns. From a bottom-up perspective, urban solid waste management is increasing due to high urban population densities, high economic consumption rates, and the economy's emphasis on production over consumption, thereby exacerbating environmental pressures throughout production and consumption pathways [41]. From a top-down perspective, metropolitan solid waste is a negative externality related to large-scale urban production in an economy with external demand. Indicators of sustainable development, such as GDP growth, carrying capacity, and urban population density, suggest that the environment remains under threat. Around the world, urban waste management is increasingly adopting circular economy principles, technologies, and public-private cooperation. The leaders are Germany and Sweden, through their sophisticated waste-to-energy systems and commitment to recycling as much as possible in the largest volume possible. China's 'Zero Waste Cities' program seeks to include informal waste pickers in a formal system. India has adopted a decentralized model for managing solid waste and producing biogas, which is still supported by the Swachh Bharat Mission campaigns. Brazil is formalizing informal waste collectors, while the United States is investing in technology, such as smart waste bins, to reduce reliance on landfills [41, 43]. Such trends demonstrate an ongoing global response to urban waste management, with a focus on adopting waste management solutions that prioritize sustainability, technology, and social inclusion (Table 2).

Table 2 Recent trends in Urban waste management

Region	Country	Recent trends in urban waste management
Europe	Germany	Advanced waste-to-energy technologies, high recycling rates, and strict EPR policies under the EU Circular Economy Action Plan
	Sweden	Pioneering waste-to-energy systems, achieving near-zero landfill dependency, and promoting circular economy models
Asia	China	"Zero waste cities" initiative focusing on waste reduction, recycling, and integrating informal waste pickers into formal systems
	India	Decentralized waste management, biogas production from organic waste, and initiatives like Swachh Bharat Mission
Africa	South Africa	Community-based waste management models and microfinance-supported small-scale recycling enterprises
	Kenya	Leveraging microfinance and informal sector integration for sustainable waste management
Americas	Brazil	Integrating informal waste collectors into formal recycling chains and promoting circular economy practices
	United States	Investing in smart waste management technologies and circular supply chains to reduce landfill dependency

10 Barriers to Circular Economy Integration

Sustainability has evolved from being primarily an environmental issue focused on waste collection to a comprehensive management concept encompassing three key pillars: environmental, economic, and social. This extended commitment has led to a greater awareness of waste worldwide. Europe is leading the way in the development of new urban waste management practices that can be replicated by other parts of the world, including developing nations. This paper examines the current urban waste 'crisis' in the EU, particularly in light of the circular economy principles that are now reshaping and directing all EU policies. It is also important to note that while being 'circular' is both a complex and a process that lies at the intersection of theory and practice, it also presents an opportunity to develop a more sustainable roadmap. Data based on reported EU municipal waste from 2014 to 2017 show a slight increase to 224 million tonnes, which also includes the multidimensional aspect of waste, amounting to a total of approximately 224 million tonnes. Finding the opportunity in the circular economy, there is an apparent commitment to reusing products before resources are wasted. Given the specific structure of each local and regional context, transitioning to a circular economy through waste management largely relies on the willingness of local and regional actors to implement good practices that deliver real sustainability and reduce environmental impacts. Through either policy or practice, urban waste issues will not be easily solved, simply because of the complexity of the issues surrounding waste management in all polities. To take a holistic process through effective futures, action is stated! Since action must be an active consideration among the two parties to the promise, the actors also need to be supported by a way of utilizing the multiple governance levels to be engaged on various fronts, and a similar adaptation of policies and tools to the particular urban realities being examined.

11 Economic and Environmental Benefits of Circular Economy

- Cities contribute to over 70% of greenhouse gas emissions globally, and circular economy (CE) models can be adopted to drive more sustainable waste management while attracting innovation and new business models.
- Moving from a linear city to a circular city helps address the rapid pace of waste growth in cities and the negative environmental and health impacts from conventional waste management.
- Developing better policy, making advances with the allocation of public resources, and developing partnerships to address complex urban waste challenges will be needed with the coming leaders of cities.
- Developing continuous monitoring, better recycling, longer lifecycles of materials, emerging technologies that can disassemble building materials and recover them safely from waste streams will be crucial in addressing urban environmental and health challenges.

11.1 Social Impacts and Community Engagement

- **Limited Land and Urban Land Use**: A crucial aim of circular economy (CE) in cities is to unlock brownfields and wastelands to halt sprawl. Often these areas are platforms of social and spatial inequity, enabling mechanisms of isolation, exclusion and economic marginalization.
- **Unused Resources**: Wastelands, though stigmatized as waste, had value in embedded resources that can re-enter urban flows of materials, from a CE perspective.
- **Monitoring and Governance**: A successful transitional model to a CE requires adjusting monitoring systems to report or track value in land resource systems, at both regional and local levels. The existing urban planning systems more often inhibit sustainable form of livable development that follow existing laws favouring incumbent land users to work creatively, to consider new forms of land use innovation and circular development.
- **Vision for Transition**: A successful transitional model to a urban CE requires new forms of land use policies, enabled regional governance structures for land, and collaborative planning systems that draw-in diverse urban areas into a coherent circular system of the city.

11.2 Education and Awareness Campaigns

- **Timeliness of Waste Management**: There still remain challenges to waste disposal with an overreliance of landfill and informal litter picking undermining circular economy
- **Challenges for the Informal Sector**: Informal waste management relies on insecurity, upholding poverty and opposing social justice and sustainability.
- **Collective Responsibility**: Most waste managers, policy makers, businesses, institutions and people must work together as a collective to progressively, innovatively treat waste differently to prevent waste from being sent to rural landfills.
- **Expanding Global Organizations**: Organizations like the United Nations Environment Programme (UNEP), World Wildlife Fund (WWF), The Ocean Cleanup, WasteAid, and Let's Do It! World, have roles to further expand attention, education, better waste management systems in communities, especially in developing states.
- **Systematic Change**: Transitioning to CE requires organized, interconnected urban waste systems focused on maximizing benefits for all people and dismantling the existing open linear waste model.

11.3 Collaboration and Partnerships in Circular Economy Initiatives

- **Urban environments as innovation hubs**: Cities are essential testing locations for circular economy innovations, through partnerships between government, business, academia, and civil society.
- **Resource efficiency**: Cities consume 75% of world resources and are responsible for 60–80% of greenhouse gas emissions, and yet the majority of the products used in a city are wasted within 6 months and now considered waste or downcycled materials.
- **Global perspectives**: Solutions should also consider challenges created by a rapidly growing global urban population; this work cannot stop at the borders of city limits.
- **Regulatory support**: Development of effective regulations, supportive policies and coordinated research and industry activity is needed to ensure that circular strategies can be applied at scale.

11.4 Circular Economy in Developing Countries

- **Urbanization Trends**: By the year 2030, two-thirds of the world's population will be living in urban areas and most will be in low- and middle-income countries (LAMICs).
- **Urban Advantage and Challenges**: Urbanization has been likened to economic growth, health, and education improvements. However, the realization of the 'urban advantage' translates into physical infrastructure and social protection. Quick urbanization in LAMICs has resulted in environmental pollution, congestion, squatter settlement development, contaminated substances, poor waste management, and more.
- **Waste Generation**: Most cities can account for around 70% of global waste, while LAMIC cities disproportionately add to this share of waste.
- **Unsustainable Current Models**: Current waste management systems in the Global South are unsustainable, as this waste system revels in rapid consumerism with dynamic resources developed around economic stimuli.
- **CE as an Alternative**: The circular economy is a new paradigm that seeks to keep materials in the economy, create more jobs, and support sustainable cities. However, CE research does not engage with local contexts for cities in LAMICs.
- **Case Study**—Delhi: A case study in Delhi highlights informal workers' contributions, by documenting industrial zones, engagement with key stakeholders, to understand the role of informal economy waste recovery on urban waste management and CE implementation.
- **Conceptually Integrating**: The study draws together the pace of urban growth, CE discussion promoting cradle to cradle ideas, and observable evidence of informal sector contributions to inform policy aimed at encouraging waste life reduction, resource productivity, and equitable development.

12 Future Research Directions

- The Circular Economy (CE) model has arisen as an alternative to the traditional linear economy, which works to design out waste and pollution, regenerate ecosystems, and maintain product life [44].
- CE focuses on how resources are used efficiently; minimizing the negative environmental effects of waste and pollution; and sustainably increasing resource use efficiency through practices such as reuse, remanufacturing, recycling, and renewable energy.
- Transitioning to the Circular Economy will require substantial shifts in society and be foundational in supporting global sustainable development goals.
- The amount of plastic produced worldwide is set to almost double from 2015 to 2040, mostly from food packaging but also in transport and construction.
- Temperature exerts influence on polymer production rates, affecting downstream applications and polymer breakdown pathways.

- Additional plastic production contributes to greenhouse gas (GHG) emissions from feedstock production, energy use in production, methane (CH4) release from landfill sites, emissions from incineration, and emissions from environmental photooxidation.
- Policymakers are looking for new technological pathways to sustainability in the value chain of plastics, as well as in waste management.
- There is an urgent need to achieve carbon neutrality to decrease GHG emissions from the sector, aligned with the commitments set forth in the Paris Agreement.
- These promising emerging technologies can provide the opportunity for a circular economy transition in plastics by enabling lower polymer emissions and reducing impacts along polymer degradation pathways.

13 Conclusion

In conclusion, the journey towards a circular economy (CE) in urban waste management, as evidenced by the case of Naples, Italy, underscores the challenges and opportunities inherent in this transition. Despite the growing awareness of circular economy principles, Naples continues to heavily rely on landfills, highlighting the complexity of shifting to a circular waste management system. The research emphasizes the crucial role of collaborative efforts involving local governments, private entrepreneurs, NGOs, and policymakers in enforcing CE regulations and overcoming governance hurdles. Engaging stakeholders from diverse sectors is essential, yet conflicting interests and resource disparities present obstacles to effective waste management and recycling. Additionally, the prevalence of illegal waste disposal disrupts the circular flow of materials, posing a significant barrier to implementing CE in urban settings. Furthermore, the situation in Naples serves as a reminder that successful implementation of circular economy strategies necessitates robust regulatory support, infrastructure enhancements, and cross-sector collaboration. Addressing illegal waste disposal, governance gaps, and systemic inefficiencies through policy-driven incentives, public awareness campaigns, and innovative waste processing technologies is imperative for a sustainable future. It is evident that the transition to a circular waste management system requires comprehensive efforts to tackle these challenges. Future initiatives should prioritize policy-driven incentives, public awareness campaigns, and innovative technology solutions for waste processing to pave the way for a more sustainable future.

References

1. Halkos G, Petrou KN (2016) Efficient waste management practices: A. A review. Department of Economics, University of Thessaly, Munich Personal Repec Archive
2. Negrete-Cardoso M, Rosano-Ortega G, Álvarez-Aros EL, Tavera-Cortés ME, Vega-Lebrún CA, Sánchez-Ruíz FJ (2022) Circular economy strategy and waste management: a bibliometric

analysis in its contribution to sustainable development, toward a post-COVID-19 era. Environ Sci Pollut Res 29(41):61729–61746
3. Negrete-Cardoso M, Rosano-Ortega G, Leobardo Álvarez-Aros E, Elena Tavera-Cortés M, Arturo Vega-Lebrún C, Javier Sánchez-Ruíz F (2022) Circular economy strategy and waste management: a bibliometric analysis in its contribution to sustainable development, toward a post-COVID-19 era. Environ Sci Pollut Res Int 30(25):67565–67581
4. Chowdhury R, Barman S, Choudhury M, Kim KH, Bhattacharya SS (2024) Earthworm modifies microbial community and functional genes for lignocellulosic waste valorization: isolating plant-growth-promoting bacteria via next generation sequencing. Int Biodeterior Biodegradation 193:105854. https://doi.org/10.1016/j.ibiod.2024.105854
5. Wilson DC (2023) Learning from the past to plan for the future: an historical review of the evolution of waste and resource management 1970–2020 and reflections on priorities 2020–2030—the perspective of an involved witness. Waste Manag Res 41(12):1754–1813
6. Rajpal A, Choudhury M, Goswami S, Chakravorty A, Raghavan V (eds) (2024) Waste management and treatment: advances and innovations. CRC Press. https://doi.org/10.1201/9781003258377
7. Li J, Chen J, Sheng B, Li P, Yang P, Feng DD, Qi J (2021) Automatic detection and classification system of domestic waste via multimodel cascaded convolutional neural network. IEEE Trans Industr Inform 18(1):163–173
8. Izquierdo-Horna L, Kahhat R, Vázquez-Rowe I (2022) Reviewing the influence of sociocultural, environmental and economic variables to forecast municipal solid waste (MSW) generation. Sustain Prod Consump 33:809–819
9. Singh S, Singh P, Sharma A, Choudhury M (eds) (2022) Agriculture waste management and bioresource: the circular economy perspective. Wiley. https://doi.org/10.1002/9781119808428
10. Dey S, Santra M, Choudhury M, Ghosh AR, Samanta P (2024) Food waste generation and its industrial utilization: an overview. Environ Sci Pollut Res. https://doi.org/10.1007/s11356-024-34252-3
11. Mayer C (2021) The future of the corporation and the economics of purpose. J Manag Stud. https://doi.org/10.1111/joms.12660
12. Choudhury M, Roy P (2025) Challenges with microplastic pollution in the regime of UN sustainable development goals. World Dev Sustain 6:100216. https://doi.org/10.1016/j.wds.2025.100216
13. Choudhury M, Sharma A, Pervez A, Upadhyay P, Dutta J (2021) Growing menace of microplastics in and around the coastal ecosystem. In: Coastal ecosystems: environmental importance, current challenges and conservation measures. Springer International Publishing, Cham, pp 117–137. https://doi.org/10.1007/978-3-030-84255-0_6
14. Sigov A, Ratkin L, Ivanov LA, Xu LD (2022) Emerging enabling technologies for industry 4.0 and beyond. Inf Syst Front 26:1585–1595
15. Marín-Beltrán I, Demaria F, Ofelio C, Serra LM, Turiel A, Ripple WJ et al (2022) Scientists' warning against the society of waste. Sci Total Environ 811:151359
16. Paul S, Choudhury M, Deb U, Pegu R, Das S, Bhattacharya SS (2019) Assessing the ecological impacts of ageing on hazard potential of solid waste landfills: a green approach through vermitechnology. J Clean Prod 236:117643. https://doi.org/10.1016/j.jclepro.2019.117643
17. Melakessou F, Kugener P, Alnaffakh N, Faye S, Khadraoui D (2020) Heterogeneous sensing data analysis for commercial waste collection. Sensors (Basel) 20(4):978
18. Choudhary R, Kumar A, Narayan S, Mishchenko I, Choudhury M, Bharadwaj LK et al (2025) Integrating fuzzy logic into life cycle assessment for sustainable municipal solid waste management in metro cities. Ecol Eng Environ Technol 26(7):316–331. https://doi.org/10.12912/27197050/205500
19. Ao Y, Zhu H, Wang Y, Zhang J, Chang Y (2022) Identifying the driving factors of rural residents' household waste classification behavior: evidence from Sichuan, China. Resour Conserv Recycl 180:106159
20. Maalouf A, Mavropoulos A (2023) Re-assessing global municipal solid waste generation. Waste Manag Res 41(4):936–947

21. Burström T, Parida V, Lahti T, Wincent J (2021) AI-enabled business-model innovation and transformation in industrial ecosystems: a framework, model and outline for further research. J Bus Res 127:85–95
22. Moretto G, Valentino F, Pavan P, Majone M, Bolzonella D (2019) Optimization of urban waste fermentation for volatile fatty acids production. Waste Manag 92:21–29
23. Rathour RK, Behl M, Dhashmana K, Sakhuja D, Ghai H, Sharma N et al (2023) Non-food crops derived lignocellulose biorefinery for sustainable production of biomaterials, biochemicals and bioenergy: a review on trends and techniques. Ind Crop Prod 204:117220
24. Hina M, Chauhan C, Kaur P, Kraus S, Dhir A (2022) Drivers and barriers of circular economy business models: where we are now, and where we are heading. J Clean Prod 333:130049
25. Bade SO, Tomomewo OS, Meenakshisundaram A, Ferron P, Oni BA (2024) Economic, social, and regulatory challenges of green hydrogen production and utilization in the US: a review. Int J Hydrog Energy 49:314–335
26. Liu Y, Chen S, Chen AY, Lou Z (2021) Variations of GHG emission patterns from waste disposal processes in megacity Shanghai from 2005 to 2015. J Clean Prod 295:126338
27. Cosma IG (2018) Waste recycling, the new-old challenge for sustainable urban development. J Urban Landsc Plan 3:99–113
28. Maitre-Ekern E (2021) Re-thinking producer responsibility for a sustainable circular economy from extended producer responsibility to pre-market producer responsibility. J Clean Prod 286:125454
29. Bayram B, Ozkan G, Kostka T, Capanoglu E, Esatbeyoglu T (2021) Valorization and application of fruit and vegetable wastes and by-products for food packaging materials. Molecules 26(13):4031
30. Alhawari O, Awan U, Bhutta MKS, Ülkü MA (2021) Insights from circular economy literature: a review of extant definitions and unravelling paths to future research. Sustainability 13(2):859
31. Marques AC, Teixeira NM (2022) Assessment of municipal waste in a circular economy: do European Union countries share identical performance? Clean Waste Syst 3:100034
32. Amenta L, Attademo A, Remøy H, Berruti G, Cerreta M, Formato E et al (2019) Managing the transition towards circular metabolism: living labs as a co-creation approach. Urban Plan 4(3):5–18
33. Remøy H, Wandl A, Ceric D, Timmeren AV (2019) Facilitating circular economy in urban planning. Urban Plan 4(3):1–4
34. De Oliveira DB, Becker RW, Sirtori C, Passos CG (2021) Development of environmental education concepts concerning chemical waste management and treatment: the training experience of undergraduate students. Chem Educ Res Pract 22(3):653–661
35. Choudhury M, Adhikari MD, Agarwal S, Samanta P, Sharma A, Kundu D, Kumar S (2025) Pharmaceuticals and personal care products in soil: sources, impacts and myco-remediation strategies. Emerg Contam 2025:100488. https://doi.org/10.1016/j.emcon.2025.100488
36. Choudhury M, Majumdar S, Goswami S, Sillanpää M (eds) (2025) Smart wastewater systems and climate change: innovations through spatial intelligence, vol 15. Royal Society of Chemistry. https://doi.org/10.1039/9781837676880
37. Choudhury M, Rajpal A, Goswami S, Chakravorty A, Raghavan V (eds) (2024) Analytical case studies on municipal and biomedical waste management: perspectives on sustainable development goals. CRC Press. https://doi.org/10.1201/9781003499695
38. Mondal T, Choudhury M, Kundu D, Dutta D, Samanta P (2023) Landfill: an eclectic review on structure, reactions and remediation approach. Waste Manag 164:127–142. https://doi.org/10.1016/j.wasman.2023.03.034
39. Rehman E, Rehman S (2022) Modeling the nexus between carbon emissions, urbanization, population growth, energy consumption, and economic development in Asia: evidence from grey relational analysis. Energy Rep 8:5430–5442
40. Crippa D, Villa MC, Di Prete B, Ratti L, Rebaglio A, Zanini M, Zanotto F (2022) Verso un Progetto Circolare, Tra Architettura E Allestimento. Piattaformedigitali per Il Riuso. Towards a circular project, between architecture and exhibition design. Digital platforms for reuse practices. Agathón 12:234–245

41. Obersteg A, Arlati A, Acke A, Berruti G, Czapiewski K, Dąbrowski M, Knieling J (2019) Urban regions shifting to circular economy: understanding challenges for new ways of governance. Urban Plan 4(3):19–31
42. Voukkali I, Papamichael I, Loizia P, Zorpas AA (2024) Urbanization and solid waste production: prospects and challenges. Environ Sci Pollut Res 31(12):17678–17689
43. Awino FB, Apitz SE (2024) Solid waste management in the context of the waste hierarchy and circular economy frameworks: an international critical review. Integr Environ Assess Manag 20(1):9–35
44. Aristi Capetillo A, Bauer F, Chaminade C (2023) Emerging technologies supporting the transition to a circular economy in the plastic materials value chain. Circ Econ Sustain 3(2):953–982
45. Anttiroiko AV (2023) Smart circular cities: governing the relationality, spatiality, and digitality in the promotion of circular economy in an urban region. Sustainability 15(17):12680
46. Goel K, Andharia J (2023) Unpacking the circular economy: foregrounding the contributions of Workers in the Informal Economy in Delhi
47. Halog A, Anieke S (2021) A review of circular economy studies in developed countries and its potential adoption in developing countries. Circ Econ Sustain 1(1):209–230
48. Let's Do It! World (2023) Global cleanup and waste awareness campaigns. https://www.letsdoitworld.org
49. Mestre A, Cooper T (2017) Circular product design. A multiple loops life cycle design approach for the circular economy. Des J 20(Suppl 1):S1620–S1635
50. The Ocean Cleanup (2023) Removing plastic waste from oceans and rivers. https://www.theoceancleanup.com
51. United Nations Environment Programme (2023) Sustainable waste management initiatives. https://www.unep.org
52. Vogiazides L, Kawalerowicz J (2023) Internal migration in the time of Covid: who moves out of the inner city of Stockholm and where do they go? Popul Space Place 29(4):e41
53. WasteAid (2023) Community-driven waste management solutions. https://www.wasteaid.org
54. World Wildlife Fund (2023) Reducing plastic pollution and promoting recycling. https://www.worldwildlife.org

MIX
Papier aus verantwortungsvollen Quellen
Paper from responsible sources
FSC® C105338

If you have any concerns about our products,
you can contact us on
ProductSafety@springernature.com

In case Publisher is established outside the EU,
the EU authorized representative is:
Springer Nature Customer Service Center GmbH
Europaplatz 3, 69115 Heidelberg, Germany

Printed by Libri Plureos GmbH
in Hamburg, Germany